THEORY OF HIGH TEMPERATURE SUPERCONDUCTIVITY

A Conventional Approach

THEORY OF HIGH TEMPERATURE SUPERCONDUCTIVITY

A Conventional Approach

Todor M Mishonov
St Clement of Ohrid University at Sofia, Bulgaria

Evgeni S Penev
University of California, Santa Barbara, USA

NEW JERSEY • LONDON • SINGAPORE • BEIJING • SHANGHAI • HONG KONG • TAIPEI • CHENNAI

Published by

World Scientific Publishing Co. Pte. Ltd.
5 Toh Tuck Link, Singapore 596224
USA office: 27 Warren Street, Suite 401-402, Hackensack, NJ 07601
UK office: 57 Shelton Street, Covent Garden, London WC2H 9HE

British Library Cataloguing-in-Publication Data
A catalogue record for this book is available from the British Library.

THEORY OF HIGH TEMPERATURE SUPERCONDUCTIVITY
A Conventional Approach

ISBN-13 978-981-4343-14-5
ISBN-10 981-4343-14-5

Printed in Singapore by B & Jo Enterprise Pte Ltd

Todor M. Mishonov and Evgeni S. Penev

Preface

The discovery of high-temperature superconductivity (HTS) by Bednorz and Müller [1] in 1986 brought a flurry of investigations of the mystery of high-T_c cuprates. About 20 years later we have the first handbook [2] giving a global view of this phenomenon, with more than 2,500 references. Large number of books have been published on this subject and superconductivity became an integral part of university physics education [3]. The present status of our knowledge on high-T_c materials is based on more than 100,000 scientific articles, Fig. 0.1. Given this enormous activity, the question poses itself as to which principles of high-T_c physics can be explained in the volume of an introductory monograph.

This book starts at the level of standard courses on quantum mechanics and statistical physics and finishes with discussion of the perspectives for

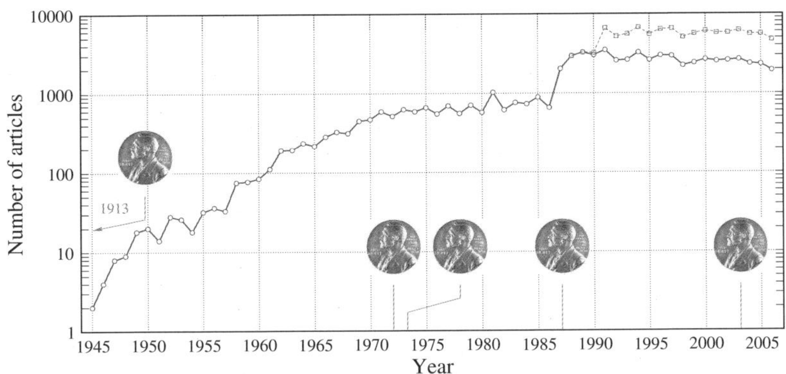

Fig. 0.1 Number of articles published in the period 1945–2006 containing 'supercon-duct*' in the title (○), or any of the title, abstract or the list of keywords (□) according to Science Citation Index Expanded (www.isiknowledge.com).

further understanding of HTS theory. The selection of topics to be covered was made on the basis of the aesthetic criterion for simplicity. We advocate the idea that the pairing mechanism is common for all cuprates and can be unraveled for levels of doping where the material properties are least perturbed by accessories such as stripes, pseudogap, space inhomogeneities, etc. This is the case of the overdoped cuprates that are the subject of the present book. For doping slightly above the optimal, the electron spectrum is well described by self-consistent band theory, outlined in the opening Chapter 1. For perovskites, the linear combination of atomic orbitals (LCAO) method is an adequate fundament to account for interaction effects.

More than 20 years of development of HTS physics have produced overwhelming hints that the electron exchange mediates the pairing. Among all exchange processes in condensed matter, the intraatomic *s*-*d* exchange is the most intensive one. In Chapter 2 we demonstrate how the *s*-*d* exchange can be incorporated into the standard Bardeen–Cooper–Schrieffer (BCS) scheme and derive the *d*-wave superconducting gap $\Delta_{\mathbf{p}}(T)$. The rationale of the correlation between the critical temperature T_c and the Cu4s energy level is the crucial test for the theory of HTS. The specific heat $C(T)$ and penetration depth $\lambda(T)$ functionals are expressed through the superconducting gap $\Delta_{\mathbf{p}}(T)$ in Chapter 3. It is shown that the analytical expressions for $C(T)$ and $\lambda(T)$ can be employed to fit and analyze experimental data for practically all superconductors with anisotropic gap. These formulas represent a generalization of the original isotropic-gap BCS results. Chapter 3 further offers a methodological derivation of parameters of the Ginzburg–Landau theory: penetration depth, coherence length, effective mass, and the heat capacity jump. We should emphasize that all results are derived from a microscopic Hamiltonian within the standard statistical physics methods. Special attention is also paid to the methodological derivation of the order parameter equation for anisotropic-gap superconductors. Chapter 4 is dedicated to some electrodynamic properties of high-temperature superconductors, such as the plasmons, predicted by one of the authors of this book [4, 5], and the electric-field effects giving access to the Cooper pair effective mass [6].

Due to the small coherence length, the fluctuation phenomena in high-temperature superconductors are much more pronounced compared to conventional superconductors. The investigation of these phenomena takes significant portion of HTS physics. Thus, Chapter 5 has a review character, focusing exclusively on the self-consistent Gaussian fluctuations that have broadest application. Using the ζ-function method for ultra-violet regularization we derive general expressions for the fluctuation heat capacity and magnetization. Discussion is also provided on how systematic investigations of the fluctuation phenomena can lead to reliable determination of fundamental material parameters of high-temperature superconductors. In Chapter 6, the kinetic equation for fluctuation Cooper pairs is derived within the time-dependent Ginzburg–Landau theory and applied to obtain expressions for the fluctuation conductivity. This standard kinetic approach is then employed in Chapter 7 to investigate fluctuation superconductivity in strong electric fields.

Another classical statistical problem — the electric field fluctuations between the CuO_2 layers — is considered in Chapter 8 and shown to explain the linear temperature dependence of the resistivity. Until recently, this dependence was interpreted as evidence for anomalous and unconventional behavior in the normal phase of high-temperature superconductors.

The final Chapter 9 treats the important technical problem of whether high-temperature superconductors can be used as active media for generation of electromagnetic waves in the THz range. So far this electromagnetic range is weakly used and known as a "THz gap" — a gap between the worlds of the transistor and the laser. Besides the plasmons that have already produced promising results [7,8], in this Chapter it is shown that the negative differential conductivity presents a new, yet unexplored opportunity for generation of THz oscillations.

If one searches Google for "mystery" and "high temperature superconductivity", the webpage hits are tens of thousands. The problem of HTS has been a true intellectual challenge for contemporary physics. What we believe has been achieved in the offered book could be phrased as "completing 'the Bardeen program'" — the mystery is revealed: we have the microscopic theory of HTS, we understand the observable properties and can predict technical applications.

Essential parts of each book chapter have been taught to undergraduate and graduate students in courses on superconductivity, statistical physics, solid state physics, and condensed-matter field theory. Starting from the level of such courses, students can reach the complexity of contemporary unsolved problems in HTS, e.g., the anisotropy of the scattering rate in the normal phase, etc. Bridging the gap between the textbook and monograph levels was the driving force for writing of this book.

Last but not least, the book is a testimonial in itself to the persistence and fruitfulness of a professor (T.M.)–student (E.P.) collaboration—one that has been almost entirely informal and dating back to the early 1990's.

Todor M. Mishonov *Sofia*
Evgeni S. Penev *Santa Barbara*

Contents

Chapter 1

Tight-binding modeling of layered perovskites

1.1 Introduction

After the discovery of the high-T_c superconductors the layered cuprates became one of the most studied materials in solid state physics. A vast range of compounds were synthesised and their properties comprehensively investigated. The electron band structure is of particular importance for understanding the nature of superconductivity in this type of perovskites [9]. Along this line one can single out the significant success achieved in the attempts to reconcile the photoelectron spectroscopy data [10, 11] and the band structure calculations of the Fermi surface (FS) especially for compounds with simple structure such as $Nd_{2-x}Ce_xCuO_{4-\delta}$ [12, 13]. A qualitative understanding, at least for the self-consistent electron picture, has been achieved and for the most electron processes in the layered perovskites one can employ adequate lattice models.

At the time when we actually started working on the subject of this book, there was not much analysis of the electronic band structures of the high-T_c materials in the terms of single analytical expressions available. This is something for which there was a clear need, in particular to help in the construction of more realistic many-body Hamiltonians. The aim of this chapter is to analyse the common features in the electron band structure of the layered perovskites within the tight-binding (TB) method (for a nice review see references [14–16]). In the following we shall focus on the metallic (being eventually superconducting) phase only, with the provision that the antiferromagnetic correlations, especially in the dielectric phase, could substantially alter the electron dispersion. It is shown that the linear combination of atomic orbitals (LCAO) approximation can be considered as an adequate tool for analysis of energy bands. Within the latter exact analytic results are obtained for the constant energy contours (CEC). These expressions are used to fit the FS of $Nd_{2-x}Ce_xCuO_{4-\delta}$ [12], $Pb_{0.42}Bi_{1.73}Sr_{1.94}Ca_{1.3}Cu_{1.92}O_{8+x}$ [17–21], and Sr_2RuO_4 [22] mapped in angle-resolved photoemission/angle-resolved ultraviolet spectroscopy (ARPES/ARUPS) experiments.

To address the conduction bands in the layered perovskites we start from a common Hamiltonian including the basis of valence states O2p, and Ru4$d\varepsilon$, or Cu3$d_{x^2-y^2}$, Cu4s, for ruthenates and cuprates, respectively. Despite the equivalent crystal structure of Sr_2RuO_4 [23] and $La_{2-x}Ba_xCuO_4$ [1, 24], the states in their conduction band(s) are, in some sense, complementary. In other words, for the CuO_2 plane the conduction band is of σ-character while for the RuO_2 plane the conduction bands are determined by π valence bonds. This is due to the separation into σ- and π-part of the Hamiltonian $H = H^{(\sigma)} + H^{(\pi)}$ in first approximation. The latter two Hamiltonians are studied separately.

This chapter is organized as follows. After a few remarks on the applicability of the TB model and a brief appology to the band theory in general, in Sec. 1.3 we consider the generic $H^{(4\sigma)}$ Hamiltonian of the CuO_2 plane [25–27] and $H^{(\pi)} = H^{(xy)} + H^{(z)}$ is then studied in Sec. 1.4. The results of the comparison with the experimental data are summarised in Sec. 1.5 and the last section discusses practical aspects in connecting the present theory with experiment.

1.2 Apology to the band theory

It is well-known that the electron band theory is a self-consistent treatment of the electron motion in the crystal lattice. Even the classical 3-body problem demonstrates strongly correlated solutions, so it is *a priori* unknown whether the self-consistent approximation is applicable when describing the electronic structure of every new crystal. However, the one-particle band picture is an indispensable stage in the complex study of materials. It is the analysis of experimental data using a conceptually clear band theory that reveals nontrivial effects: how strong the strongly correlated electronic effects are, whether it is possible to take into account the influence of some interaction-induced order parameter back into the electronic structure etc. Therefore the comparison of the experiment with the band calculations is not an attempt, as sometimes thought, to hide the relevant issues—it is a tool to reveal interesting and nontrivial properties of the electronic structure.

Many electron band calculations have been performed for the layered perovskites and results were compared to data due to ARPES experiments. The shape of the Fermi surface is probably the simplest test to check whether we are on the right track or some conceptually new theory should be used from the very beginning.

The tight-binding interpolation of the electronic structure is often used for fitting the experimental data. This is because the accuracy of that approximation is often higher than the uncertainties in the experiment. Moreover, the tight-binding method gives simple formulae which could be of use for experimentalists to see how far they can get with such a simple minded approach. The tight-binding parameters, however, have in a sense "their own life" independent of the *ab initio* calculations. These parameters can be fitted directly to the experiment even when, by some reasons, the electron band calculations could give wrong predictions. In this sense the tight-binding parameters are the appropriate intermediary between the theory and experiment. As for the theory, establishing of reliable one-particle tight-binding parameters is the preliminary step in constructing more realistic many-body Hamiltonians. The role of the band theory is, thus, quite ambivalent: on one hand, it is the final "language" used in efforts towards understanding a broad variety of phenomena; on the other hand, it is the starting point in developing realistic interaction Hamiltonians for sophisticated phenomena such as magnetism and superconductivity.

The tight-binding method is the simplest one employed in the electron band calculations and it is described in every textbook in solid state physics; the layered perovskites are now probably the best investigated materials [2, 28, 29] and the Fermi surface is a fundamental notion in the physics of metals. There is a consensus that the superconductivity of layered perovskites is related to electron processes in the CuO_2 and RuO_2 planes of these materials. It is not, however, fair to criticise a given study, employing the tight-binding method as an interpolation scheme to the first principles calculations, for not thoroughly discussing the many-body effects. The criticism should rather be redirected to the *ab initio* electronic structure calculations. An interpolation scheme cannot contain more information than the underlying theory. It is not erroneous if such a scheme works with an accuracy high enough to adequately describe both the theory and experiment.

In view of the above, we found it strange that there were no simple interpolation formulae for the Fermi surfaces available in the literature and the experimental data were being published without an attempt towards simple interpretation. One of the aims of the present chapter was initially to help interpret the experimental data by the tight-binding method as well as setting up notions for the analysis of the *ab initio* calculations.

1.3 Layered cuprates

The CuO_2 plane appears as a common structural detail for all layered cuprates. Therefore, in order to retain the generality of the considerations, the electronic properties of the bare CuO_2 plane will be addressed without taking into account structural details such as dimpling, orthorhombic distortion, double planes, surrounding chains etc. For the square unit cell with lattice constant a_0 three-atomic basis is assumed $\{\mathbf{R}_{\mathrm{Cu}}, \mathbf{R}_{\mathrm{O_a}}, \mathbf{R}_{\mathrm{O_b}}\} = \{\mathbf{0}, (a_0/2, 0), (0, a_0/2)\}$. The unit cell is indexed by vector $\mathbf{n} = (n_x, n_y)$, where n_x, n_y = integer. Within such an idealized model the LCAO wave function spanned over the $|\mathrm{Cu}3d_{x^2-y^2}\rangle$, $|\mathrm{Cu}4s\rangle$, $|\mathrm{O_a}2p_x\rangle$, $|\mathrm{O_b}2p_y\rangle$ states reads

$$\begin{aligned}\psi_{\mathrm{LCAO}}(\mathbf{r}) = \sum_{\mathbf{n}} \Big[& X_{\mathbf{n}} \psi_{\mathrm{O_a}2p_x}(\mathbf{r} - \mathbf{R}_{\mathrm{O_a}} - a_0\mathbf{n}) \\ & + Y_{\mathbf{n}} \psi_{\mathrm{O_b}2p_y}(\mathbf{r} - \mathbf{R}_{\mathrm{O_b}} - a_0\mathbf{n}) + S_{\mathbf{n}} \psi_{\mathrm{Cu}4s}(\mathbf{r} - \mathbf{R}_{\mathrm{Cu}} - a_0\mathbf{n}) \\ & + D_{\mathbf{n}} \psi_{\mathrm{Cu}3d_{x^2-y^2}}(\mathbf{r} - \mathbf{R}_{\mathrm{Cu}} - a_0\mathbf{n}) \Big],\end{aligned} \tag{1.1}$$

where $\Psi_{\mathbf{n}} = (D_{\mathbf{n}}, S_{\mathbf{n}}, X_{\mathbf{n}}, Y_{\mathbf{n}})$ is the tight-binding wave function in lattice representation.

The neglect of the differential overlap leads to an LCAO Hamiltonian of the CuO_2 plane

$$\begin{aligned}H = \sum_{\mathbf{n}} \Big\{ & D^{\dagger}_{\mathbf{n}}[-t_{pd}(-X_{\mathbf{n}} + X_{x-1,y} + Y_{\mathbf{n}} - Y_{x,y-1}) + \epsilon_d D_{\mathbf{n}}] \\ & + S^{\dagger}_{\mathbf{n}}[-t_{sp}(-X_{\mathbf{n}} + X_{x-1,y} - Y_{\mathbf{n}} + Y_{x,y-1}) + \epsilon_s S_{\mathbf{n}}] \\ & + X^{\dagger}_{\mathbf{n}}[-t_{pp}(Y_{\mathbf{n}} - Y_{x+1,y} - Y_{x,y-1} + Y_{x+1,y-1}) \\ & \quad - t_{sp}(-S_{\mathbf{n}} + S_{x+1,y}) - t_{pd}(-D_{\mathbf{n}} + D_{x+1,y}) + \epsilon_p X_{\mathbf{n}}] \\ & + Y^{\dagger}_{\mathbf{n}}[-t_{pp}(X_{\mathbf{n}} - X_{x-1,y} - X_{x,y+1} + X_{x-1,y+1}) \\ & \quad - t_{sp}(-S_{\mathbf{n}} + S_{x,y+1}) - t_{pd}(D_{\mathbf{n}} + D_{x,y+1}) + \epsilon_p Y_{\mathbf{n}}] \Big\},\end{aligned} \tag{1.2}$$

where the components of $\Psi_{\mathbf{n}}$ should be considered as being Fermi operators. The notations ϵ_d, ϵ_s, and ϵ_p stand respectively for the $\mathrm{Cu}3d_{x^2-y^2}$, Cu4s and O2$p\sigma$ single-site energies. The direct $\mathrm{O_a}2p_x \to \mathrm{O_b}2p_y$ exchange is denoted by t_{pp} and similarly t_{sp} and t_{pd} denote the Cu4s $\to$ O2p and O2p $\to$

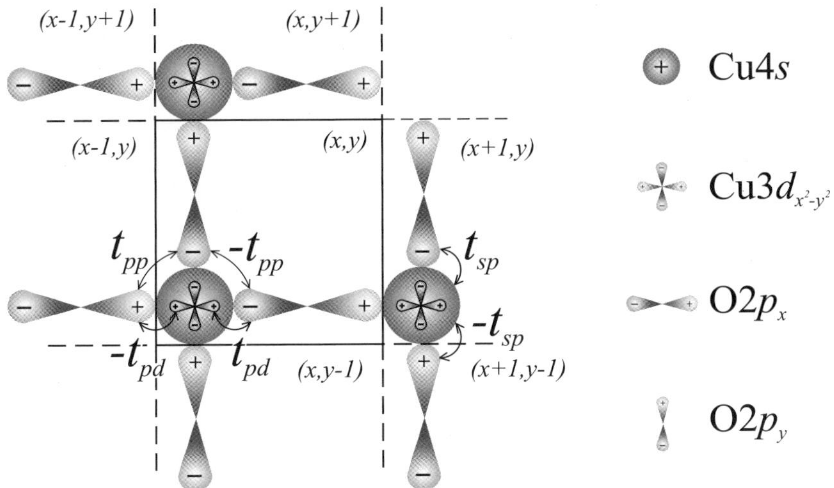

Fig. 1.1 Schematic of a CuO_2 plane (only orbitals relevant to the discussion are depicted). The solid square represents the unit cell with respect to which the positions of the other cells are determined. The indices of the wave function amplitudes involved in the LCAO Hamiltonian (1.2) are given in brackets. The rules for determining the signs of hopping integrals t_{pd}, t_{sp}, and t_{pp} are shown as well.

$Cu3d_{x^2-y^2}$ hoppings respectively. The sign rules for the hopping amplitudes are sketched in Fig. 1.1 — the bonding orbitals enter the Hamiltonian with a negative sign.

For the Bloch states diagonalizing the Hamiltonian (1.2)

$$\Psi_{\mathbf{n}} \equiv \begin{pmatrix} D_{\mathbf{n}} \\ S_{\mathbf{n}} \\ X_{\mathbf{n}} \\ Y_{\mathbf{n}} \end{pmatrix} = \frac{1}{\sqrt{N}} \sum_{\mathbf{p}} \begin{pmatrix} D_{\mathbf{p}} \\ S_{\mathbf{p}} \\ \mathrm{e}^{\mathrm{i}\varphi_a} X_{\mathbf{p}} \\ \mathrm{e}^{\mathrm{i}\varphi_b} Y_{\mathbf{p}} \end{pmatrix} \mathrm{e}^{\mathrm{i}\mathbf{p}\cdot\mathbf{n}}, \tag{1.3}$$

where N is the number of unit cells, we use the same phases as in references [25–27]:

$$\varphi_a = \frac{1}{2}(p_x - \pi), \quad \varphi_b = \frac{1}{2}(p_y - \pi). \tag{1.4}$$

Equation (1.3) describes the Fourier transform between the coordinate representation $\Psi_{\mathbf{n}} = (D_{\mathbf{n}}, S_{\mathbf{n}}, X_{\mathbf{n}}, Y_{\mathbf{n}})$, with $\mathbf{n}$ being the cell index, and the momentum representation $\psi_p = (D_p, S_p, X_p, Y_p)$ of the TB wave function (when used as an index, the electron quasi-momentum vector is denoted by p). Hence, the Schrödinger equation $\mathrm{i}\hbar d_t \hat{\psi}_{p,\alpha} = [\hat{\psi}_{p,\alpha}, \hat{H}]$ for

$\psi_{p,\alpha}(t) = \mathrm{e}^{-\mathrm{i}\epsilon t/\hbar}\psi_{p,\alpha}$, with α being the spin index $(\uparrow, \downarrow)$ (suppressed hereafter) and ϵ the band energy, takes the form

$$\left(H_p^{(4\sigma)} - \epsilon\mathbf{1}\right)\psi_p = \begin{pmatrix} -\varepsilon_d & 0 & t_{pd}s_x & -t_{pd}s_y \\ 0 & -\varepsilon_s & t_{sp}s_x & t_{sp}s_y \\ t_{pd}s_x & t_{sp}s_x & -\varepsilon_p & -t_{pp}s_xs_y \\ -t_{pd}s_y & t_{sp}s_y & -t_{pp}s_xs_y & -\varepsilon_p \end{pmatrix} \begin{pmatrix} D_p \\ S_p \\ X_p \\ Y_p \end{pmatrix} = 0, \quad (1.5)$$

where

$$\varepsilon_d = \epsilon - \epsilon_d, \quad \varepsilon_s = \epsilon - \epsilon_s, \quad \varepsilon_p = \epsilon - \epsilon_p, \quad (1.6)$$

and

$$\begin{aligned} s_x &= 2\sin(\tfrac{1}{2}p_x) & x &= \sin^2(\tfrac{1}{2}p_x) \\ s_y &= 2\sin(\tfrac{1}{2}p_y) & y &= \sin^2(\tfrac{1}{2}p_y) \\ 0 &\leqslant p_x,\ p_y \leqslant 2\pi. \end{aligned} \quad (1.7)$$

This 4σ-band Hamiltonian is generic for the layered cuprates, cf. Ref. [27]. We have also included the direct oxygen-oxygen exchange t_{pp} dominated by the σ amplitude. The secular equation

$$\det\left(H_p^{(4\sigma)} - \epsilon\mathbf{1}\right) = \mathcal{A}xy + \mathcal{B}(x+y) + \mathcal{C} = 0 \quad (1.8)$$

gives the spectrum and the canonical form of the CEC with energy-dependent coefficients

$$\begin{aligned} \mathcal{A}(\epsilon) &= 16(4t_{pd}^2t_{sp}^2 + 2t_{sp}^2t_{pp}\varepsilon_d - 2t_{pd}^2t_{pp}\varepsilon_s - t_{pp}^2\varepsilon_d\varepsilon_s) \\ \mathcal{B}(\epsilon) &= -4\varepsilon_p(t_{sp}^2\varepsilon_d + t_{pd}^2\varepsilon_s) \\ \mathcal{C}(\epsilon) &= \varepsilon_d\varepsilon_s\varepsilon_p^2. \end{aligned} \quad (1.9)$$

Hence, the explicit CEC equation reads as

$$p_y = \pm 2\arcsin\sqrt{y}, \quad \text{if} \quad 0 \leqslant y = -\frac{\mathcal{B}x + \mathcal{C}}{\mathcal{A}x + \mathcal{B}} \leqslant 1. \quad (1.10)$$

This equation reproduces the rounded square-shaped Fermi surface, centered at the (π, π) point, inherent for all layered cuprates. The best fit is achieved when $\mathcal{A}$, $\mathcal{B}$ and $\mathcal{C}$ are considered as fitting parameters. Thus, for a CEC passing through the $\mathsf{D} = (p_\mathrm{d}, p_\mathrm{d})$ and $\mathsf{C} = (p_\mathrm{c}, \pi)$ reference points, as indicated in Fig. 1.2, the fitting coefficients (distinguished by the subscript f) in the canonical equation

$$\mathcal{A}_f xy + \mathcal{B}_f(x+y) + \mathcal{C}_f = 0 \quad (1.11)$$

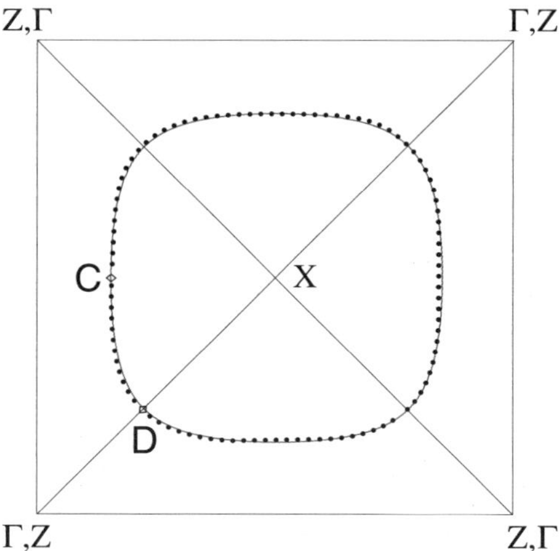

Fig. 1.2 LDA Fermi contour of $Nd_{2-x}Ce_xCuO_{4-\delta}$ (dotted line) calculated by Yu and Freeman [13] (courtesy of the authors), and the LCAO fit (solid line) according to Eq. (1.8). The fitting procedure uses C and D as reference points.

have the form

$$\begin{aligned} \mathcal{A}_f &= 2x_{\rm d} - x_{\rm c} - 1, & x_{\rm d} &= \sin^2(p_{\rm d}/2) \\ \mathcal{B}_f &= x_{\rm c} - x_{\rm d}^2, & x_{\rm c} &= \sin^2(p_{\rm c}/2) \\ \mathcal{C}_f &= x_{\rm d}^2(x_{\rm c}+1) - 2x_{\rm c}x_{\rm d}, \end{aligned} \tag{1.12}$$

and the resulting LCAO Fermi contour is quite compatible with the local-density approximation (LDA) calculations for $Nd_{2-x}Ce_xCuO_{4-\delta}$ [13, 30]. Due to the simple shape of the FS the curves just coincide. We note also that the canonical equation (1.8) would formally correspond to 1-band TB Hamiltonian of a 2D square lattice of the form

$$\epsilon(\mathbf{p}) = -2t(\cos p_x + \cos p_y) + 4t' \cos p_x \cos p_y,$$

with strong energy dependence of the hopping parameters, where t' is the anti-bonding hopping between the sites along the diagonal, cf. [31–33].

1.3.1 *Effective Cu Hamiltonian*

Studies of the electronic structure of the layered cuprates have unambiguously proved the existence of a large hole pocket — a rounded square centred at the (π, π) point. This observation is indicative for a Fermi level located in a single band of dominant $Cu3d_{x^2-y^2}$ character. To address this band

and the related wave functions, it is therefore convenient for an effective Cu-Hamiltonian to be derived by Löwdin downfolding of the oxygen orbitals. This is equivalent to expressing the oxygen amplitudes from the third and fourth rows of Eq. (1.5)

$$\begin{aligned} X &= \frac{1}{\eta_p}\left[t_{pd}s_x\left(1+\frac{t_{pp}}{\varepsilon_p}s_y^2\right)D + t_{sp}s_x\left(1-\frac{t_{pp}}{\varepsilon_p}s_y^2\right)S\right] \\ Y &= \frac{1}{\eta_p}\left[-t_{pd}s_y\left(1+\frac{t_{pp}}{\varepsilon_p}s_x^2\right)D + t_{sp}s_y\left(1-\frac{t_{pp}}{\varepsilon_p}s_x^2\right)S\right], \end{aligned} \tag{1.13}$$

where $\eta_p = \varepsilon_p - \frac{t_{pp}^2}{\varepsilon_p}s_x^2 s_y^2$, and substituting back into the first and the second rows of the same equation. Such a downfolding procedure results in the following energy-dependent copper Hamiltonian

$$H_{\rm Cu}(\epsilon) = \begin{pmatrix} \epsilon_d + \frac{(2t_{pd})^2}{\eta_p}\left(x+y+\frac{8t_{pp}}{\varepsilon_p}xy\right) & \frac{(2t_{pd})(2t_{sp})}{\eta_p}(x-y) \\ \frac{(2t_{pd})(2t_{sp})}{\eta_p}(x-y) & \epsilon_s + \frac{(2t_{pd})^2}{\eta_p}\left(x+y-\frac{8t_{pp}}{\varepsilon_p}xy\right) \end{pmatrix}, \tag{1.14}$$

which enters the effective Schrödinger equation $H_{\rm Cu}\binom{D}{S} = \epsilon\binom{D}{S}$. Thus, from Eq. (1.13) and Eq. (1.14) one can easily obtain an approximate expression for the eigenvector corresponding to a dominant Cu3$d_{x^2-y^2}$ character. Taking $D \approx 1$, in the lowest order with respect to the hopping amplitudes $t_{ll'}$ one has

$$|{\rm Cu}3d_{x^2-y^2}\rangle = \begin{pmatrix} D \\ S \\ X \\ Y \end{pmatrix} \approx \begin{pmatrix} 1 \\ (t_{sp}t_{pd}/\varepsilon_s\varepsilon_p)(s_x^2 - s_y^2) \\ (t_{pd}/\eta_p)s_x \\ -(t_{pd}/\eta_p)s_y \end{pmatrix}, \tag{1.15}$$

i.e., $|X|^2 + |Y|^2 + |S|^2 \ll |D|^2 \approx 1$. We note that within this Cu scenario the Fermi level location and the CEC shape are not sensitive to t_{pp}. Therefore one can neglect the oxygen-oxygen hopping as was done, for example, by Andersen *et al.* [25–27] (the importance of the t_{pp} parameter has been considered by Markiewicz [34]) and the band structure of the Hamiltonian (1.14) for the same set of energy parameters as used in Ref. [27] is shown in Fig. 1.3 (a). In this case, the Fermi surface can be fitted by its diagonal alone, i.e., using D alone as a reference point. Hence an equation for the Fermi energy follows, $\mathcal{A}(\epsilon_{\rm F})x_d^2 + 2\mathcal{B}(\epsilon_{\rm F})x_d + \mathcal{C}(\epsilon_{\rm F}) = 0$, which yields $\epsilon_{\rm F} = 2.5$ eV. As seen in Fig. 1.3 (b), the deviation from the two-parameter fit is almost vanishing, thus justifying the neglect of t_{pp}

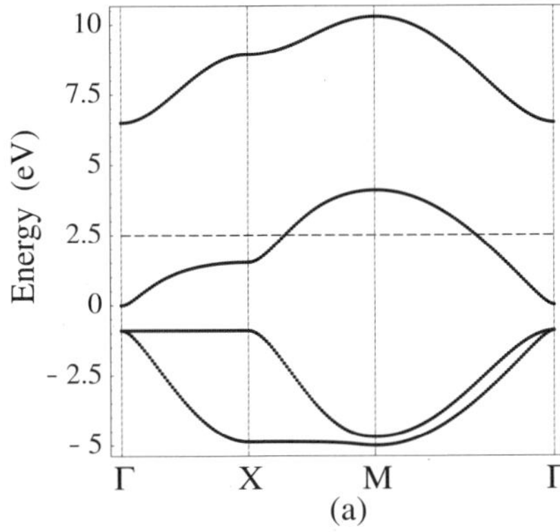

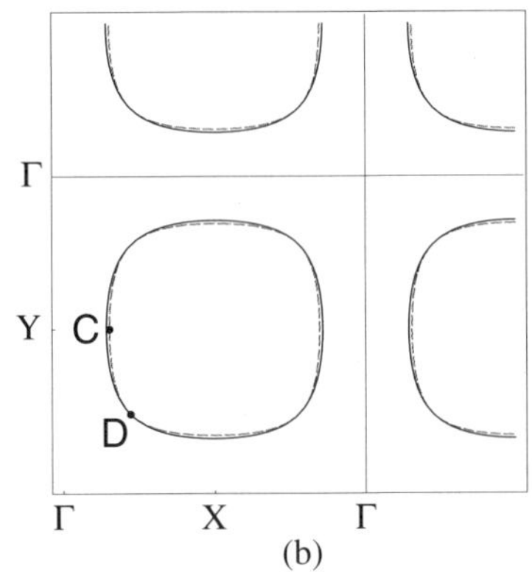

Fig. 1.3 (a) Electron band structure of the generic for the CuO_2 plane 4σ-band Hamiltonian using the parameters from Ref. [27] and the Fermi level $\epsilon_F = 2.5$ eV fitted from the LDA calculation by Yu and Freeman [13]; (b) The LCAO Fermi contour (solid line) fitted to the LDA Fermi surface (dashed line) for $Nd_{2-x}Ce_xCuO_{4-\delta}$ [13] using only D as a reference point. The deviation of the fit at the C point is negligible.

and using the one-parameter fit. However, despite the excellent agreement between the LDA calculations, the LCAO fit and the ARPES data regarding the FS shape, the theoretically calculated conduction bandwidth w_c in the layered cuprates is overestimated by a factor of 2 or even 3 [12]. Equation (1.11) describes acceptably also the experimental ARPES data, e.g., for $Nd_{2-x}Ce_xCuO_{4-\delta}$, material with single CuO_2 plane and no other complicating structural details. In Fig. 1.4 we compare the ARPES data from Ref. [12] and the Fermi contour calculated for $x = 0.15$.

There exist a tremendous number of ARPES/ARUPS data for layered cuprates which makes the reviewing of all those spectra impossible. To further illustrate our TB model we have chosen the data for Pb substitution for Bi in $Bi_2Sr_2CaCu_2O_8$. The ARUPS Fermi surface of $Pb_{0.42}Bi_{1.73}Sr_{1.94}Ca_{1.3}Cu_{1.92}O_{8+x}$ [17–21] is shown in Fig. 1.5 (a). In this case the CuO_2 planes are quite flat and the ARPES data are not distorted by structural details. When present, distortions were eventually misinterpreted as a manifestation of strong antiferromagnetic correlations. We believe, however, that the experiment by Schwaller *et al.* [21] reveals the main feature of the CuO_2 plane band structure — the large hole pocket found to be in agreement with the one-particle band calculations. Strong support to this view comes from a paper by Campuzano *et al.* [35] where the ARPES Fermi surface of pure $Bi_2Sr_2CaCu_2O_{8+\delta}$ was mapped [cf. the inset of Fig. 1 (a) therein]. This experimental finding is

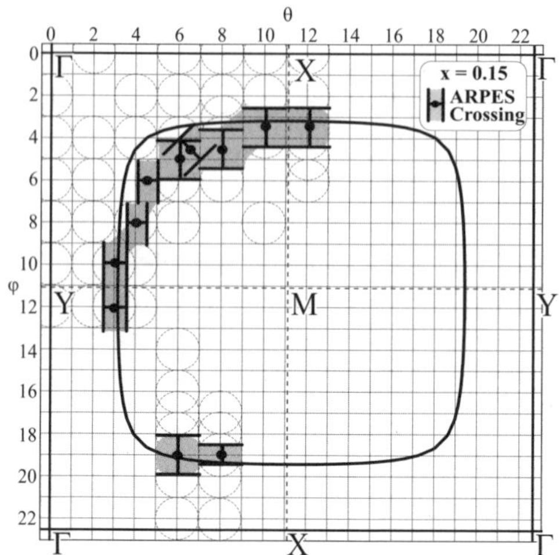

Fig. 1.4 The Fermi surface of $Nd_{2-x}Ce_xCuO_{4-\delta}$ (solid line) determined by equation Eq. (1.11) for $x = 0.15$ and compared with experimental data (points with error bars) for the same value of x after King *et al.* [12]. θ and φ denote the polar and azimuthal emission angles, respectively, measured in degrees. The empty dashed circles show **k**-space locations where ARPES experiments have been performed (cf. Fig. 2 in Ref. [12]) and their diameter corresponds to 2° experimental resolution.

in excellent agreement with our tight-binding fit to the Fermi surface of $Pb_{0.42}Bi_{1.73}Sr_{1.94}Ca_{1.3}Cu_{1.92}O_{8+x}$, studied by Schwaller and co-workers in Ref. [21], Fig. 1.5 (b). The remarkable coincidence of the Fermi surfaces of these two compounds is a nice confirmation that Pb substitution for Bi is irrelevant for the band structure of the CuO_2 plane and the Fermi surface of the latter is therefore revealed to be a common feature.

1.4 Conduction bands of the RuO_2 plane

Sr_2RuO_4 is the first coper-free perovskite superconductor isostructural to the high-T_c cuprates [23]. The layered ruthenates, just like the layered cuprates, are strongly anisotropic and in first approximation the nature of the conduction band(s) can be understood by analysing the bare RuO_2 plane. One should repeat the same steps as in the previous section but now having Ru instead of Cu and the Fermi level located in the metallic bands of Ru4$d\pi$ character. To be specific, the conduction bands arise form the hybridisation between the Ru4d_{xy}, Ru4d_{yz}, Ru4d_{zx} and O_a2p_y, O_b2p_x, $O_{a,b}2p_z$ π-orbitals. The LCAO wave function spanned over the four

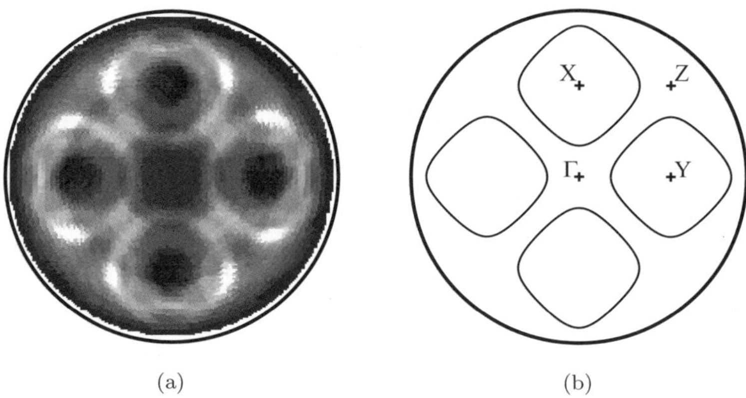

(a) (b)

Fig. 1.5 (a) ARUPS Fermi surface of $Pb_{0.42}Bi_{1.73}Sr_{1.94}Ca_{1.3}Cu_{1.92}O_{8+x}$ by Schwaller *et al.* [21]; (b) LCAO fit to (a) according to Eq. (1.11) using the D reference point with $p_d = 0.171 \times 2\pi$.

perpendicular to the RuO_2 plane orbitals reads

$$\begin{aligned}\Psi^{(z)}_{\rm LCAO}(\mathbf{r}) =& \frac{1}{\sqrt{N}}\sum_{\mathbf{p}}\sum_{\mathbf{n}}\Big[D_{zx,\mathbf{n}}\psi_{\mathrm{Ru4}d_{zx}}(\mathbf{r}-a_0\mathbf{n}) \\ &+ D_{zy,\mathbf{n}}\psi_{\mathrm{Ru4}d_{zy}}(\mathbf{r}-a_0\mathbf{n}) + \mathrm{e}^{\mathrm{i}\varphi_a}Z_{a,\mathbf{n}}\psi_{\mathrm{O}_a2p_z}(\mathbf{r}-\mathbf{R}_{\mathrm{O}_a}-a_0\mathbf{n}) \\ &+\mathrm{e}^{\mathrm{i}\varphi_b}Z_{b,\mathbf{n}}\psi_{\mathrm{O}_b2p_z}(\mathbf{r}-\mathbf{R}_{\mathrm{O}_b}-a_0\mathbf{n})\Big]\,\mathrm{e}^{\mathrm{i}\mathbf{p}\cdot\mathbf{n}}\end{aligned} \tag{1.16}$$

Hence, the π-analog of Eq. (1.5) takes the form

$$\left(H^{(z)}_p - \epsilon\mathbf{1}\right)\psi^{(z)}_p = \begin{pmatrix} -\varepsilon_{zx} & 0 & t_{z,zx}s_x & 0 \\ 0 & -\varepsilon_{zy} & 0 & t_{z,zy}s_y \\ t_{z,zx}s_x & 0 & -\varepsilon_{za} & -t_{zz}c_xc_y \\ 0 & t_{z,zy}s_y & -t_{zz}c_xc_y & -\varepsilon_{zb}\end{pmatrix}\begin{pmatrix} D_{zx} \\ D_{zy} \\ Z_a \\ Z_b\end{pmatrix} = 0, \tag{1.17}$$

where

$$\begin{aligned}\varepsilon_{zx} &= \epsilon - \epsilon_{zx}, & \varepsilon_{za} &= \epsilon - \epsilon_{za}, & c_x &= 2\cos(p_x/2), \\ \varepsilon_{zy} &= \epsilon - \epsilon_{zy}, & \varepsilon_{zb} &= \epsilon - \epsilon_{zb}, & c_y &= 2\cos(p_y/2),\end{aligned} \tag{1.18}$$

and ϵ_{zx}, ϵ_{zy}, ϵ_{za}, and ϵ_{zb} are the single site energies of the $\mathrm{Ru}4d_{zx}$, $\mathrm{Ru}4d_{zy}$ and O_a2p_z, O_b2p_z orbitals, respectively. t_{zz} denotes the hopping between the latter two orbitals and, assuming a negligible orthorhombic distortion, the metal-oxygen π-hopping parameters are equal, $t_{z,zy} = t_{z,zx}$ and also $\epsilon_z = \epsilon_{za} = \epsilon_{zb}$. The phase factors $\mathrm{e}^{\mathrm{i}\varphi_{a,b}}$ in Eq. (1.16) are chosen in compliance with Ref. [27], see Eq. (1.4).

Similarly, writing the LCAO wave function spanned over the three in-plane π-orbitals Ru4d_{xy}, O$_a$2p_y, and O$_b$2p_x in the way in which Eq. (1.16) is designed one has for the "in-plane" Schrödinger equation

$$\left(H_p^{(xy)} - \epsilon\mathbf{1}\right)\psi_p^{(xy)} = \begin{pmatrix} -\varepsilon_{xy} & t_{pd\pi}s_x & t_{pd\pi}s_y \\ t_{pd\pi}s_x & -\varepsilon_{ya} & t'_{pp}s_xs_y \\ t_{pd\pi}s_y & t'_{pp}s_xs_y & -\varepsilon_{xb} \end{pmatrix} \begin{pmatrix} D_{xy} \\ Y_a \\ X_b \end{pmatrix} = 0, \qquad (1.19)$$

where $t_{pd\pi}$ and t'_{pp} denote the Ru4d_{xy} $\rightarrow$ O$_{a,b}$2$p\pi$ and O$_a$2p_y $\rightarrow$ O$_b$2p_x hoppings, respectively. The definitions for the other energy parameters are in analogy to Eq. (1.18) (for negligible orthorhombic distortion $\epsilon_{ya} = \epsilon_{xb} \neq \epsilon_z$). Thus, the π-Hamiltonian of the RuO_2 plane takes the form

$$H^{(\pi)} = \sum_{p,\alpha=\uparrow,\downarrow} \psi_{p,\alpha}^{(z)\dagger} H_p^{(z)} \psi_{p,\alpha}^{(z)} + \psi_{p,\alpha}^{(xy)\dagger} H_p^{(xy)} \psi_{p,\alpha}^{(xy)}. \qquad (1.20)$$

The derivation of the corresponding secular equations was announced in a couple of short publications [36, 37]. Here we shall only provide the final expressions in terms of the notation used here

$$\det(H_p^{(z,xy)} - \epsilon\mathbf{1}) = \mathcal{A}^{(z,xy)}xy + \mathcal{B}^{(z,xy)}(x+y) + \mathcal{C}^{(z,xy)} = 0,$$

$$\begin{aligned} \mathcal{A}^{(z)} &= 16(t_{z,zx}^4 - t_{zz}^2\varepsilon_{zx}^2) & \mathcal{A}^{(xy)} &= 32t'_{pp}t_{pd\pi}^2 - 16\varepsilon_{xy}t'^2_{pp} \\ \mathcal{B}^{(z)} &= -16t_{zz}^2\varepsilon_{zx}^2 - 4t_{z,zx}^2\varepsilon_{zx}\varepsilon_z & \mathcal{B}^{(xy)} &= -t_{pd\pi}^2\varepsilon_{ya} \\ \mathcal{C}^{(z)} &= \varepsilon_{zx}^2(\varepsilon_z^2 - 16t_{zz}^2) & \mathcal{C}^{(xy)} &= \varepsilon_{xy}\varepsilon_{ya}^2. \end{aligned} \qquad (1.21)$$

The three sheets of the Fermi surface in Sr_2RuO_4 fitted to the ARPES data by Lu *et al.* [22] are shown in Fig. 1.6 (b). To determine the Hamiltonian parameters we have made use of the eigenvalues at the high-symmetry points of the Brillouin zone. To the best of our knowledge, the TB analysis of the Sr_2RuO_4 band structure was first performed in Refs. [36, 37]. Subsequently, the latter results were reproduced in Ref. [38] without referring to Refs. [36, 37]. The RuO_2-plane band structure resulting from the set of parameters

$$\begin{aligned} t_{zz} &= t'_{pp} = 0.3\,\text{eV} & \varepsilon_z &= -2.3\,\text{eV} & \varepsilon_{xy} &= -1.62\,\text{eV} \\ t_{pd\pi} &= t_{z,zx} = 1\,\text{eV} & \varepsilon_{zx} &= -1.3\,\text{eV} & \varepsilon_{ya,xb} &= -2.62\,\text{eV} \end{aligned} \qquad (1.22)$$

is shown in Fig. 1.6 (a). This fit is subjected to the requirement of providing as good as possible a description of the narrow energy window around ϵ_F whereas the filled bands, far below the Fermi level, might only qualitatively match the LDA calculations by Oguchi [39] and Singh [40]. Furthermore, the Sr_2RuO_4 Fermi surface mapping, based on the de Haas–van Alphen

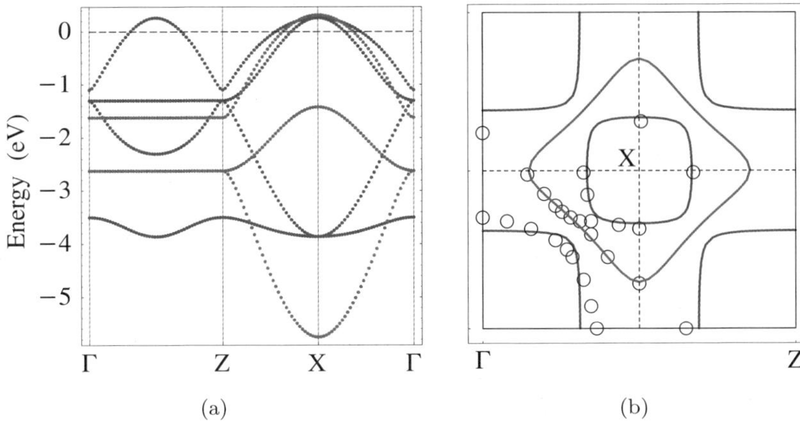

Fig. 1.6 (a) LCAO band structure of Sr_2RuO_4 according to Eq. (1.20). The Fermi level (dashed line) crosses the three Ru4$d\varepsilon$ bands of the RuO_2 plane; (b) LCAO fit (solid lines) to the ARPES data (circles) by Lu *et al.* [22], cf. also Ref. [36, 37].

(dHvA) effect [41], differs from the ARPES results [22]. Thus, fitting the dHvA data by using modified TB parameters is a natural refinement of the proposed model. We note that the diamond-shaped hole pocket, centred at the X point, Fig. 1.6 (b), is very sensitive to the "game of parameters". For that band the van Hove energy is fairly close to the Fermi energy. As a result, a minor change in the parameters could drive a van Hove transition transforming this hole pocket into an electron one, centred at the Γ point. Indeed, such a band arrangement has also been observed in the ARPES revision of the Sr_2RuO_4 Fermi surface [42] and reflects the topology of the energy surfaces $\epsilon(\mathbf{p})$ derived in Refs. [36,37]. The comparison of the ARPES data with TB energy surfaces could be the subject of a separate study.

1.5 Discussion

The LCAO analysis of the layered perovskites band structure, performed in the preceding sections, manifests a good compatibility with the experimental data and the band calculations as well. Due to the strong anisotropy of these materials, their FS within a reasonable approximation is determined by the properties of the bare CuO_2 or RuO_2 planes.

Despite these planes having identical crystal structure, their electronic structures are quite different. While for the RuO_2 plane the Fermi level crosses metallic π-bands, the conduction band of the CuO_2 plane is described by a σ-Hamiltonian (1.5). The latter gives for the CuO_2 plane a large hole pocket centered at the (π, π) point. Its shape, if no additional sheets exist, is well-described by the exact analytic results within the LCAO model, Eq. (1.8), as found for $Nd_{2-x}Ce_xCuO_{4-\delta}$ [12, 13] and $Pb_{0.42}Bi_{1.73}Sr_{1.94}Ca_{1.3}Cu_{1.92}O_{8+x}$ [17–21]. For a number of other cuprates,[1] this large hole pocket is easily identified. For all of the above compounds, however, its shape is usually deformed due to appearance of additional sheets of the Fermi surface originating from accessories of the crystal structure.

The applicability of the LCAO approximation to the electronic structure of the layered cuprates can be considered as being proved. The basis functions of the LCAO Hamiltonian can be included in a realistic one-electron part of the lattice Hamiltonians for the layered perovskites. This is an indispensable step preceding the inclusion of the two electron exchange (Chapter 2), electron-phonon interaction or any other interactions between conducting electrons.

1.6 Determining the density of states of thin high-T_c films by field-effect-transistor type microstructures

The importance of the density of states (DOS) for the physics of high-T_c cuprates has been discussed in many papers [34, 54–62]. In this section we shall suggest a simple electronic method for determining the DOS. The proposed experiment requires (i) preparation of a field-effect transistor (FET) type microstructure and (ii) standard electronic measurement. The FET controls the current between two points but does so differently than the bipolar transistor. The FET relies on an electric field to control the shape and hence the conductivity of a "channel" in a semiconductor material. The shape of the conducting channel in a FET is altered when a potential difference is applied to the gate terminal (potential relative to either source or drain). It causes the electrons flow to change its width and thus controls the voltage between the source and the drain. If the negative voltage applied to the gate is high enough, it can remove all the electrons from the gate and thus close the conducting channel in which the electrons flow. Thus, the FET gets blocked.

[1] $YBa_2Cu_3O_{7-\delta}$ [43], $YBa_2Cu_4O_8$ [44], $Bi_2Sr_2CaCu_2O_8$ [45–47], $Bi_2Sr_2CuO_6$ [48], the infinite-layered superconductor $Sr_{1-x}Ca_xCuO_2$ [49], $HgBa_2Ca_2Cu_3O_{8+\delta}$ [50], $HgBa_2CuO_{4+\delta}$ [51], $HgBa_2Ca_{n-1}Cu_nO_{2n+2+\delta}$ [52], $Tl_2Ba_2Ca_{n-1}Cu_nO_{4+2n}$, [9], $Sr_2CuO_2F_2$, $Sr_2CuO_2Cl_2$, $Ca_2CuO_2Cl_2$ [53].

The system, considered in this section is in hydrodynamic regime, which means low frequency regime where the temperature of the superconducting film adiabatically follows the dissipated Ohmic power. All working frequencies of the lock-in's, say up to 100 kHz, are actually low enough. The investigations of superconducting bolometers show that only in the MHz range it is necessary to take into account the specific heat of the superconducting film. As an example there is a publication, corresponding to this topic [63]. In Ref. [64] we have proposed an experiment with a FET, for which we need to measure the second harmonic of the source-gate voltage and the third harmonic of the source-drain voltage. Other higher harmonics will be present in the measurements (e.g., from the leads), but in principle they can also be used for determining the DOS. An analogous experimental research has already been performed for investigation of thermal interface resistance [65]. The suggested experiment can be conducted using practically the same experimental setup, only the gate electrodes should be added to the protected by insulator layer superconducting films.

Here we suggest a simple electronic experiment to determine the logarithmic derivative of the density of states by electronic measurements using a thin film of $Tl_2Ba_2CuO_{6+\delta}$. The thickness of the samples should be typical for the investigation of high-T_c films, say 50–200 nm. Such films demonstrate already the properties of the bulk phase. The numerical value of this parameter

$$\nu'(E_F) = \left.\frac{d\nu(\epsilon)}{d\epsilon}\right|_{\epsilon=E_F}, \tag{1.23}$$

will ensure the absolute determination of hopping integrals.

We propose a field effect transistor (FET) of $Tl_2Ba_2CuO_{6+\delta}$ Fig. 1.7 to be investigated electronically with lock-in at second and third harmonics. Imagine a strip of $Tl_2Ba_2CuO_{6+\delta}$ and between the ends of the strip, i.e., between the source (S) and the drain (D) is applied an AC current

$$I_{SD}(t) = I_0\cos(\omega t). \tag{1.24}$$

For low enough frequencies the ohmic power P increases the temperature of the film T above the ambient temperature T_0

$$P = RI_{SD}^2 = \alpha(T - T_0), \tag{1.25}$$

where the constant α determines the boundary thermo-resistance between the $Tl_2Ba_2CuO_{6+\delta}$ film and the substrate, and $R(T)$ is the temperature dependent source-drain (SD) resistance. We suppose that for thin film the

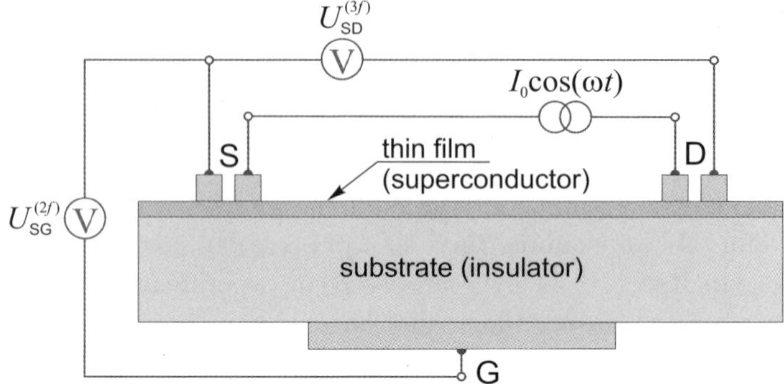

Fig. 1.7 A field effect transistor (FET) is schematically illustrated. The current $I(t)$, applied between the source (S) and the drain (D) has frequency ω. Running through the transistor the electrons create voltage U_{SG} with double frequency 2ω between the source (S) and the gate (G). The source-drain voltage U_{SD} is measured on the triple frequency 3ω.

temperature is almost homogeneous across the thickness of the film. In such a way we obtain for the temperature oscillations

$$T - T_0 = \frac{RI_{SD}^2}{\alpha} = \frac{RI_0^2}{\alpha}\cos^2(\omega t). \tag{1.26}$$

As the resistance is weakly temperature dependent

$$R(T) = R_0 + (T - T_0)R_0', \quad R_0'(T_0) = \left.\frac{dR(T)}{dT}\right|_{T_0}. \tag{1.27}$$

A substitution here of the temperature oscillations from Eq. (1.26) gives a small time variations of the resistance

$$R(t) = R_0\left(1 + \frac{R_0'}{\alpha}I_0^2\cos^2(\omega t)\right). \tag{1.28}$$

Now we can calculate the source-drain voltage as

$$U_{\mathrm{SD}}(t) = R(t)I_{\mathrm{SD}}(t). \tag{1.29}$$

Substituting here the SD current from Eq. (1.24) and the SD resistance from Eq. (1.28) gives for the SD voltage

$$U_{\mathrm{SD}}(t) = U_{\mathrm{SD}}^{(1f)}\cos(\omega t) + U_{\mathrm{SD}}^{(3f)}\cos(3\omega t). \tag{1.30}$$

The coefficient in front of the first harmonic $U_{\mathrm{SD}}^{(1f)} \approx R_0 I_0$ is determined by the SD resistance R_0 at low currents I_0, while for the third harmonic signal using the elementary formula $\cos^3(\omega t) = (3\cos(\omega t) + \cos(3\omega t))/4$ we obtain

$$U_{\mathrm{SD}}^{(3f)} = \frac{U_{\mathrm{SD}}^{(1f)}}{4\alpha}I_0^2R_0'. \tag{1.31}$$

From this formula we can express the boundary thermo-resistance by electronic measurements

$$\alpha = \frac{U_{\rm SD}^{(1f)}}{4U_{\rm SD}^{(3f)}} I_0^2 R_0'. \tag{1.32}$$

The realization of the method requires fitting of $R(T)$ and numerical differentiation at working temperature T_0; the linear regression is probably the simplest method if we need to know only one point.

Given α, we can express the time oscillations of the temperature substituting in Eq. (1.26)

$$T = T_0 + \frac{RI_0^2}{2\alpha}\left[1+\cos(2\omega t)\right] \approx T_0\left(1+\frac{R_{\rm SD}I_0^2}{2\alpha T_0}\cos(2\omega t)\right). \tag{1.33}$$

In this approximation terms containing I_0^4 are neglected and also we consider that shift of the average temperature of the film is small.

The variations of the temperature lead to variation of the work function of the film according to the well-known formula from the physics of metals

$$W(T) = -\frac{\pi^2}{6e}\frac{\nu'}{\nu}k_{\rm B}^2 T^2, \quad \nu'(E_{\rm F}) = \left.\frac{d\nu}{d\epsilon}\right|_{E_{\rm F}}, \tag{1.34}$$

where the logarithmic derivative of the density of states $\nu(\epsilon)$ taken for the Fermi energy $E_{\rm F}$ has dimension of inverse energy, the work function W has dimension of voltage, T is the temperature in °K and $k_{\rm B}$ is the Boltzmann constant. For an introduction see the standard text books on statistical physics and physics of metals. [66, 67] Substituting here the temperature variations from Eq. (1.33) gives

$$W = -\frac{\pi^2 k_{\rm B}^2}{6e}\frac{\nu'}{\nu}T_0^2\left[1+\frac{R_0 I_0^2}{\alpha T_0}\cos(2\omega t)\right] + O(I_0^4), \tag{1.35}$$

where $\mathcal{O}$-function again marks that the terms having I_0^4 are negligible.

The oscillations of the temperature creates AC oscillations of the source-gate (SG) voltage. We suppose that a lock-in with a preamplifier, having high enough internal resistance is switched between the source and the gate. In these conditions the second harmonics of the work function and of the SG voltage are equal

$$U_{\rm SG}^{(2f)} = -\frac{\pi^2 k_{\rm B}^2}{6e}\frac{\nu'}{\nu}T_0^2\frac{R_0 I_0^2}{\alpha T_0}, \tag{1.36}$$

$$U_{\rm SG}(t) = U_{\rm SG}^{(2f)}\cos(2\omega t) + U_{\rm SG}^{(4f)}\cos(4\omega t) + \dots \tag{1.37}$$

Substituting α from Eq. (1.32) we have

$$U_{\mathrm{SG}}^{(2f)} = -\frac{4\pi^2 k_{\mathrm{B}}^2}{6e}\,\frac{\nu'}{\nu}\,\frac{U_{\mathrm{SD}}^{(3f)}}{I_0}\,\frac{T_0}{R_0'}. \tag{1.38}$$

From this equation we can finally express the pursued logarithmic derivative of the density of states

$$\left.\frac{d\ln\nu(\epsilon)}{d\epsilon}\right|_{E_{\mathrm{F}}} = \frac{\nu'}{\nu} = -\frac{3e}{2\pi^2 k_{\mathrm{B}}^2}\,\frac{I_0}{T_0}\,\frac{U_{\mathrm{SG}}^{(2f)}}{U_{\mathrm{SD}}^{(3f)}}\,\frac{dR}{dT}. \tag{1.39}$$

In such way the logarithmic derivative of the density of states can be determined by fully electronic measurements with a FET. This important energy parameter can be used for absolute determination of the hopping integrals in the generic LCAO model. The realization of the experiment can be considered as continuation of already published detail theoretical and experimental investigations and having a set of complementary studies we can reliably determine the LCAO parameters.

We predict largest DOS logarithmic derivative values, associated with a sign change as well, for $La_{2-x}Sr_xCuO_4$, $0.09 < x < 0.22$, as the Fermi contour topology changes, see Fig. 3.44 in Ref. [2] and Refs. [68–71].

Chapter 2

The pairing mechanism of overdoped cuprates

What is the hardest thing of all?
That which seems the easiest
For your eyes to see,
That which lies before your eyes
– Goethe

2.1 Introduction

The discovery of high-temperature superconductivity [1,24] in cuprates and the subsequent "research rush" have led to the appearance of about 100,000 papers to date [72] (cf. Fig. 0.1 on page vii). Virtually every fundamental process known in condensed matter physics was probed as a possible mechanism of this phenomenon. Nevertheless, none of the theoretical efforts resulted in a coherent picture [72]. For the conventional superconductors the mechanism was known to be the interaction between electrons and crystal-lattice vibrations, but the development of its theory lagged behind the experimental findings. The case of cuprate high-T_c superconductivity appears to be the opposite: we do not convincingly know which mechanism is to be incorporated in the traditional Bardeen–Cooper–Schrieffer (BCS) theory [73–75]. Thus the path to high-T_c superconductivity in cuprates, perhaps carefully hidden or well-forgotten, has turned into one of the long-standing mysteries in physical science.

Features of the electronic spectrum of the CuO_2 plane, Fig. 2.1 (a), the structural detail responsible for the superconductivity of the cuprates, have become accessible from the angle-resolved photoemission spectroscopy (ARPES) [76,77]. Thus, any theory which pretends to explain the cuprate superconductivity is bound to include these features and account for them

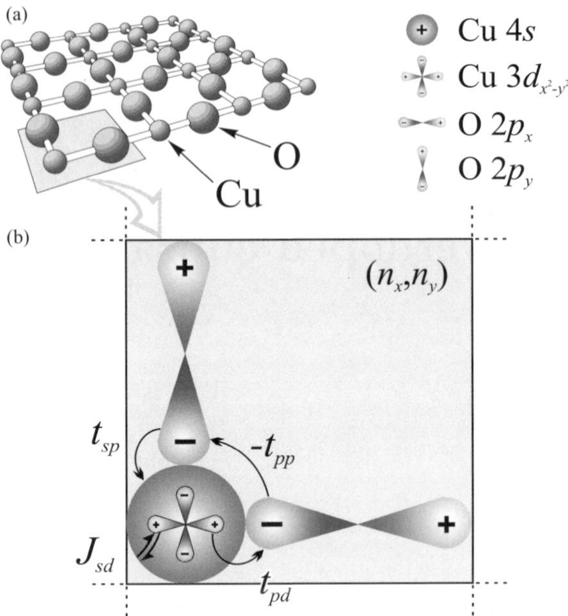

Fig. 2.1 (a) Ball-and-stick model of the CuO_2 plane. The shaded square is the unit cell indexed by $\mathbf{n} = (n_x, n_y)$, $n_{x,y} = 0, \pm 1, \pm 2, \ldots$. (b) The LCAO basis set: A single electron hops from the $3d$ atomic orbital to $2p_x$ with amplitude t_{pd}, contained in $\hat{\mathcal{H}}_{\rm BH}$. From $2p_x$ to $2p_y$ the hopping amplitude is t_{pp}, and from there to $4s$ the hopping amplitude is t_{sp}. Correlated hopping of two electrons in opposite directions between $3d$ and $4s$ with amplitude J_{sd} is depicted as a double arrow (see the discussion in Secs. 2.4 and 2.5).

consistently. A number of extensive reviews over the past years have been devoted to that theoretical problem [34, 78–96]. For further related discussion we also refer the reader to the review [97] on NMR-NQR spectroscopies in high-$T_{\rm c}$ superconductors.

In contrast with all previous proposals, we have advanced in Ref. [98] the *intra-atomic* exchange [99–112] of two electrons between the $4s$ and $3d_{x^2-y^2}$ states of the Cu atom as the origin of high-$T_{\rm c}$ superconductivity in the layered cuprates and have shown that the basic spectroscopic and thermodynamic experiments can be explained by it. Previously only *inter-atomic* Heitler–London-type [113] two-electron exchange [114–116] has been discussed. Thus, the present chapter offers a unabridged version of our theory announced in Ref. [98]. It builds upon the standard Bloch-Hückel [117–122] (tight-binding) approximation to the electronic band structure of the CuO_2

plane [123], developed in Chapter 1. We derive an analytical expression for the BCS pairing kernel, or pairing potential $V_{\mathbf{pp}'}$. For the case of the *s*-*d* pairing the analytical solution is compared to the ARPES data. Extensive discussion is also provided to help the juxtaposition of our theory with other models. Exact within the *s*-*d* model expressions for the specific heat, London penetration depth, Cooper pair effective mass and Hall constant of the vortex-free Meissner–Ochsenfeld phase will be derived later in Chapter 3.

2.2 Lattice Hamiltonian

The electronic properties of materials are strongly influenced by the local environment and in this sense the electronic features are local physics. The simplest possible model for high-T_c superconductivity contains single-particle and correlated two-electron hoppings between nearest neighbours and next-nearest neighbours. Formally, this is an expansion of the many-particle Hamiltonian containing two- and four-fermion operators. The two-fermion Hamiltonian determines the band structure, briefly considered in subsection 2.2.1, while the four-fermion terms (subsection 2.2.2) determine the pairing interaction, and lead to the gap equations considered in Sec. 2.3.

2.2.1 *The four-band model in a nutshell*

Every high-T_c superconductor has its specific properties. It is strongly believed, however, that the main features of the electronic band structure of the CuO_2 plane are adequately described by the four-band model spanning the Cu3$d_{x^2-y^2}$, Cu4s, O2p_x and O2p_y orbitals, Fig. 2.1 (b). In the spirit of the Bloch-Hückel (BH) model, using Jordan's second quantization language, we introduce Fermi annihilation operators for an electron with spin projection α at a particular orbital, respectively, $\hat{D}_{\mathbf{n}\alpha}, \hat{S}_{\mathbf{n}\alpha}, \hat{X}_{\mathbf{n}\alpha}$, and $\hat{Y}_{\mathbf{n}\alpha}$ in the unit cell with index $\mathbf{n} = (n_x, n_y)$. It is convenient to introduce a multicomponent Fermi creation operator in momentum space, $\hat{\Psi}^\dagger_{\mathbf{p}\alpha} = (\hat{D}^\dagger_{\mathbf{p}\alpha}, \hat{S}^\dagger_{\mathbf{p}\alpha}, \hat{X}^\dagger_{\mathbf{p}\alpha}, \hat{Y}^\dagger_{\mathbf{p}\alpha})$. In this notation, the one-electron BH Hamiltonian reads

$$\hat{\mathcal{H}}'_{\mathrm{BH}} = \hat{\mathcal{H}}_{\mathrm{BH}} - \mu\hat{\mathcal{N}} = \sum_{\mathbf{p},\alpha} \hat{\Psi}^\dagger_{\mathbf{p}\alpha}(H_{\mathrm{BH}} - \mu\mathbf{1}_{4\times4})\hat{\Psi}_{\mathbf{p}\alpha}, \tag{2.1}$$

where μ is the chemical potential, and using (1.6), (1.7) (cf. Ref. [123])

$$H_{\rm BH} = \begin{pmatrix} \epsilon_d & 0 & t_{pd}s_x & -t_{pd}s_y \\ 0 & \epsilon_s & t_{sp}s_x & t_{sp}s_y \\ t_{pd}s_x & t_{sp}s_x & \epsilon_p & -t_{pp}s_x s_y \\ -t_{pd}s_y & t_{sp}s_y & -t_{pp}s_x s_y & \epsilon_p \end{pmatrix}. \tag{2.2}$$

Note, that because of the orbital orthogonality $t_{sd} = 0$.

From a classical point of view, the Cu$3d_{x^2-y^2}$ state corresponds to a circular electron rotation in the CuO_2 plane while the Cu$4s$ state corresponds to a classical ensemble of electrons of zero angular momentum continuously falling to the nucleus. Pictorially, the s-electrons fall to the nuclei like comets, but after the impact the turning point of their motion is very far from the nucleus. This is the reason why t_{sp} is considerably larger than t_{pd}. The transfer amplitude t_{pp} is the smallest one since the hopping to the next-nearest neighbour requires a tunneling through free space. As a rule, the electron band calculations significantly overestimate t_{pp}, but the latter can be reliably calculated using the surface integral method, cf. Ref. [123]. Even for the largest transfer integrals t_{sp} and t_{pd}, which determine the bandwidth of the conduction band, the *ab initio* calculations give a factor 2 or even 3 "overbinding". Nonetheless, the band calculations substantiate this choice for the LCAO (linear combination of atomic orbitals) basis set and provide an adequate language for interpretation. In the end, these parameters should be determined by fitting to the spectroscopy data and be treated in the model lattice Hamiltonian as phenomenological parameters of the microscopic many-body theory. We shall briefly recall some basic properties of the four-band model as derived in Ref. [123].

Let $\epsilon_{b,\mathbf{p}}$ and $\Psi_{b,\mathbf{p}}$ be the eigenvalues and the corresponding eigenvectors of the BH Hamiltonian, $H_{\rm BH}\Psi_{b,\mathbf{p}} = \epsilon_{b,\mathbf{p}}\Psi_{b,\mathbf{p}}$, where $b = 1,\ldots,4$ is the band index. For the "standard model", $\epsilon_p < \epsilon_d < \epsilon_s$, the lowest energy band, $b = 1$, is an oxygen bonding band having a minimum at the (π,π) point. The next band, $b = 2$, is a narrow "nonbonding" oxygen band with an exactly (within the framework of the model) zero dispersion along the $(0,0)$-$(\pi,0)$ direction, i.e., this band is characterized by an extended Van Hove singularity. The conduction band, $b = 3$, is a nearly half-filled Cu$3d_{x^2-y^2}$ band with the analytical eigenvector

$$\tilde{\Psi}_{3,\mathbf{p}} = \begin{pmatrix} D_{3,\mathbf{p}} \\ S_{3,\mathbf{p}} \\ X_{3,\mathbf{p}} \\ Y_{3,\mathbf{p}} \end{pmatrix} = \begin{pmatrix} -\varepsilon_s\,\varepsilon_p^2 + 4\varepsilon_p\,t_{sp}^2\,(x+y) - 32t_{pp}\,\tau_{sp}^2\,xy \\ -4\varepsilon_p\,t_{sp}\,t_{pd}\,(x-y) \\ -(\varepsilon_s\,\varepsilon_p - 8\tau_{sp}^2\,y)\,t_{pd}\,s_x \\ (\varepsilon_s\,\varepsilon_p - 8\tau_{sp}^2\,x)\,t_{pd}\,s_y \end{pmatrix}, \tag{2.3}$$

where the ε's denote the energies measured relative to their respective atomic levels, Eq. (1.6), and $\tau_{sp}^2 = t_{sp}^2 - \varepsilon_s t_{pp}/2$. The topmost band, $b = 4$, is an empty Cu4s band. In elemental metals like Cu and Fe, the 4s band is a wide conduction band, but for the CuO_2 plane it is completely "oxidized". Having the analytical eigenvector we can calculate the corresponding eigenvalue:

$$\epsilon_{3,\mathbf{p}} = \frac{\langle \tilde{\Psi}_{3,\mathbf{p}} | H_{\mathrm{BH}} | \tilde{\Psi}_{3,\mathbf{p}} \rangle}{\langle \tilde{\Psi}_{3,\mathbf{p}} | \tilde{\Psi}_{3,\mathbf{p}} \rangle}. \tag{2.4}$$

If necessary, the nonorthogonality of the atomic orbitals at neighbouring atoms can be easily taken into account. In this case the normalizing denominator in the above equation reads (for arbitrary band index)

$$\begin{aligned} \langle \tilde{\Psi}_{\mathbf{p}} | \tilde{\Psi}_{\mathbf{p}} \rangle = D_{\mathbf{p}}^2 + S_{\mathbf{p}}^2 + X_{\mathbf{p}}^2 + Y_{\mathbf{p}}^2 + 2g_{\mathrm{pd}} s_x D_{\mathbf{p}} X_{\mathbf{p}} - 2g_{\mathrm{pd}} s_y D_{\mathbf{p}} Y_{\mathbf{p}} \\ +2g_{\mathrm{sp}} s_x S_{\mathbf{p}} X_{\mathbf{p}} + 2g_{\mathrm{sp}} s_y S_{\mathbf{p}} Y_{\mathbf{p}} - 2g_{\mathrm{pp}} s_x s_y X_{\mathbf{p}} Y_{\mathbf{p}}, \end{aligned} \tag{2.5}$$

where the "metric tensor" g_{ij} is given by the integral

$$g_{ij} = \int \psi_i^*(\mathbf{r}) \psi_j(\mathbf{r} - \mathbf{R}) d\mathbf{r}, \tag{2.6}$$

where $\psi_i^*(\mathbf{r})$ and $\psi_j(\mathbf{r} - \mathbf{R})$ are the atomic wave functions, and $\mathbf{R}$ is the inter-atomic distance. The phases are chosen such that all overlap integrals g_{pd}, g_{sp}, and g_{pp} be positive parameters, like the hopping integrals t_{pd}, t_{sp}, and t_{pp}. Note that these provisions apply only to the single-particle spectrum. As long as one deals with a single conduction band, all Bloch states are orthogonal and the further treatment of the second-quantized Hamiltonian proceeds in the standard way.

Thus, using the Rayleigh quotient iteration for equations Eqs. (2.2)–(2.4) one can obtain numerically the eigenvalue and the eigenvector. The band energies $\epsilon \equiv \epsilon_{b,\mathbf{p}}$ satisfy the secular equation (1.8) for $H_{\mathrm{BH}} \equiv H_p^{(4\sigma)}$. Furthermore we introduce the normalized eigenvector $\Psi_{b,\mathbf{p}} = \tilde{\Psi}_{b,\mathbf{p}} / \| \tilde{\Psi}_{b,\mathbf{p}} \|$ and write the noninteracting Hamiltonian in diagonal form,

$$\hat{\mathcal{H}}'_{\mathrm{BH}} = \sum_{b,\mathbf{p},\alpha} (\epsilon_{b,\mathbf{p}} - \mu) \hat{c}_{b,\mathbf{p}\alpha}^{\dagger} \hat{c}_{b,\mathbf{p}\alpha}. \tag{2.7}$$

The Fermi operators in real-space representation can be easily expressed using the band representation,

$$\hat{\Psi}_{\mathbf{n}\alpha} \equiv \begin{pmatrix} \hat{D}_{\mathbf{n}\alpha} \\ \hat{S}_{\mathbf{n}\alpha} \\ \hat{X}_{\mathbf{n}\alpha} \\ \hat{Y}_{\mathbf{n}\alpha} \end{pmatrix} = \frac{1}{\sqrt{N}} \sum_{b,\mathbf{p}} \mathrm{e}^{\mathrm{i}\mathbf{p}\cdot\mathbf{n}} \begin{pmatrix} D_{b,\mathbf{p}} \\ S_{b,\mathbf{p}} \\ \mathrm{e}^{\mathrm{i}\varphi_x} X_{b,\mathbf{p}} \\ \mathrm{e}^{\mathrm{i}\varphi_y} Y_{b,\mathbf{p}} \end{pmatrix} \hat{c}_{b,\mathbf{p}\alpha}, \tag{2.8}$$

where N is the number of unit cells, and the two phases in the right-hand side of the equation read $\varphi_x = \frac{1}{2}(p_x - \pi)$ and $\varphi_y = \frac{1}{2}(p_y - \pi)$. This transformation will be used in the next subsection for deriving the interaction Hamiltonian.

2.2.2 *The Heitler–London and Schubin–Wonsowsky–Zener interactions*

The Heitler–London (HL) interaction Hamiltonian describes the (intra- and inter-atomic) two-electron exchange. It comprises four parts [114, 115] corresponding to Cu4s $\leftrightarrow$ O2$p\sigma$, O2$p\sigma$ $\leftrightarrow$ Cu3$d_{x^2-y^2}$, O2p_x $\leftrightarrow$ O2p_y, and Cu3$d_{x^2-y^2}$ $\leftrightarrow$ Cu4s exchanges with transition amplitudes J_{sp}, J_{pd}, J_{pp}, and J_{sd}, respectively:

$$\begin{aligned}\hat{\mathcal{H}}_{\rm HL} = &-J_{sd}\sum_{\mathbf{n},\alpha\beta}\hat{S}^\dagger_{\mathbf{n}\alpha}\hat{D}^\dagger_{\mathbf{n}\beta}\hat{S}_{\mathbf{n}\beta}\hat{D}_{\mathbf{n}\alpha} - J_{sp}\sum_{\mathbf{n},\alpha\beta}\Big[\hat{S}^\dagger_{\mathbf{n}\alpha}\hat{X}^\dagger_{\mathbf{n}\beta}\hat{S}_{\mathbf{n}\beta}\hat{X}_{\mathbf{n}\alpha}\\ &+\hat{S}^\dagger_{\mathbf{n}\alpha}\hat{Y}^\dagger_{\mathbf{n}\beta}\hat{S}_{\mathbf{n}\beta}\hat{Y}_{\mathbf{n}\alpha} + \hat{S}^\dagger_{(n_x+1,n_y)\alpha}\hat{X}^\dagger_{\mathbf{n}\beta}\hat{S}_{(n_x+1,n_y)\beta}\hat{X}_{\mathbf{n}\alpha}\\ &+\hat{S}^\dagger_{(n_x,n_y+1)\alpha}\hat{Y}^\dagger_{\mathbf{n}\beta}\hat{S}_{(n_x,n_y+1)\beta}\hat{Y}_{\mathbf{n}\alpha}\Big]\\ &-J_{pd}\sum_{\mathbf{n},\alpha\beta}\Big[\hat{D}^\dagger_{\mathbf{n}\alpha}\hat{X}^\dagger_{\mathbf{n}\beta}\hat{D}_{\mathbf{n}\beta}\hat{X}_{\mathbf{n}\alpha} + \hat{D}^\dagger_{\mathbf{n}\alpha}\hat{Y}^\dagger_{\mathbf{n}\beta}\hat{D}_{\mathbf{n}\beta}\hat{Y}_{\mathbf{n}\alpha}\\ &+\hat{D}^\dagger_{(n_x+1,n_y)\alpha}\hat{X}^\dagger_{\mathbf{n}\beta}\hat{D}_{(n_x+1,n_y)\beta}\hat{X}_{\mathbf{n}\alpha}\\ &+\hat{D}^\dagger_{(n_x,n_y+1)\alpha}\hat{Y}^\dagger_{\mathbf{n}\beta}\hat{D}_{(n_x,n_y+1)\beta}\hat{Y}_{\mathbf{n}\alpha}\Big]\\ &-J_{pp}\sum_{\mathbf{n},\alpha\beta}\Big[\hat{X}^\dagger_{\mathbf{n}\alpha}\hat{Y}^\dagger_{\mathbf{n}\beta}\hat{X}_{\mathbf{n}\beta}\hat{Y}_{\mathbf{n}\alpha} + \hat{X}^\dagger_{\mathbf{n}\alpha}\hat{Y}^\dagger_{(n_x+1,n_y)\beta}\hat{X}_{\mathbf{n}\beta}\hat{Y}_{(n_x+1,n_y)\alpha}\\ &+\hat{X}^\dagger_{(n_x,n_y+1)\alpha}\hat{Y}^\dagger_{\mathbf{n}\beta}\hat{X}_{(n_x,n_y+1)\beta}\hat{Y}_{\mathbf{n}\alpha}\\ &+\hat{X}^\dagger_{(n_x,n_y+1)\alpha}\hat{Y}^\dagger_{(n_x+1,n_y)\beta}\hat{X}_{(n_x,n_y+1)\beta}\hat{Y}_{(n_x+1,n_y)\alpha}\Big].\end{aligned} \tag{2.9}$$

Let us now analyze the structure of the total electron Hamiltonian $\hat{\mathcal{H}}' = \hat{\mathcal{H}}'_{\rm BH} + \hat{\mathcal{H}}_{\rm HL}$. In terms of the Fermi operators $\hat{\Psi}_{i\alpha}$, corresponding to the atomic orbitals, $\hat{\mathcal{H}}'$ reads:

$$\begin{aligned}\hat{\mathcal{H}}' = &\sum_{i,\alpha}(\epsilon_i - \mu)\hat{\Psi}^\dagger_{i\alpha}\hat{\Psi}_{i\alpha} - \sum_{i<j,\alpha}\left(\tilde{t}_{ji}\hat{\Psi}^\dagger_{j\alpha}\hat{\Psi}_{i\alpha} + \tilde{t}^*_{ji}\hat{\Psi}^\dagger_{i\alpha}\hat{\Psi}_{j\alpha}\right)\\ &-\sum_{i<j,\alpha\beta} J_{ij}\hat{\Psi}^\dagger_{i\beta}\hat{\Psi}^\dagger_{j\alpha}\hat{\Psi}_{i\alpha}\hat{\Psi}_{j\beta},\end{aligned} \tag{2.10}$$

where $\tilde{t}_{ji} = t_{ji}{\rm e}^{{\rm i}\phi_{ji}}$, $t_{ji} = t_{ij}$ and $\phi_{ji} = \phi_j - \phi_i$ is the phase difference between the ith and jth atomic orbitals in the overlapping domain. Roughly speaking, onto every single-electron hopping amplitude t_{ij} one can map a corresponding two-electron hopping amplitude J_{ij}. The case of a strong electron correlation implies that J_{ij} could be of the order of t_{ij}. Thus, one can expect that the following inequalities hold true $J_{pp} < J_{pd} < J_{sp} < J_{sd}$.

In fact, the *s-d* exchange is the basic process responsible for the magnetism of transition metals; see for example Ref. [99–112]. It was understood since the dawn of quantum physics that the mechanism of ferromagnetism [124, 125] is the two-electron exchange owing to the electron correlations [126–135].

Here we shall add a few words in retrospect concerning the two-electron correlation parameterized by J_{ij} in Eq. (2.10). Probably the first two-electron problem was Bohr's consideration of the He atom [136–139] (cf. references [140, 141]) in which two electrons have opposite coordinates $\mathbf{r}_2 = -\mathbf{r}_1$ and momenta $\mathbf{p}_2 = -\mathbf{p}_1$. For a purely radial motion, such a fall to the nucleus is stable and many years after Bohr's prediction double Rydberg states, with an effective $\mathrm{Ry}_{\mathrm{eff}} = (2 - 1/4)\,\mathrm{Ry}$, were discovered by electron energy loss spectroscopy [142–145]. These double Rydberg states with opposite electron momenta can be considered as proto-forms of the Cooper pairs. Interestingly, in 1914, Sir J. J. Thomson proposed [146] (cf. also the textbook [147]) that electric charge can propagate as electron doublets—another proto-form of the local (Ogg–Schafroth) pairs [148,149]. Before the appearance of quantum mechanics, Lewis [150] and Langmuir [151, 152] introduced the idea of electron doublets in order to explain the nature of the chemical bond. Nearly at the same time Parson [153] came to the conclusion that "an electron is not merely an electron charge but a small magnet" or in his terminology "a magneton", cf. Ref. [150]. Later, in 1926, Lewis introduced also the notion of a photon [154] without any reliable theoretical background at the time.

In the era of new quantum mechanics, Heitler and London [113] realized the idea of electron doublets [155] and convincingly demonstrated how the two-particle correlation owing to a strong Coulomb repulsion can lead to a decrease of the energy, and by virtue of the Hellmann–Feynman theorem, to inter-atomic attraction for the singlet state of the electron doublet. The original Heitler–London calculation, which is nowadays interpreted in every textbook in quantum mechanics and/or quantum chemistry, gives indeed a wrong sign of the exchange energy for very large inter-atomic distances but, in principle, there are no conceptual difficulties in the Heitler–London theory. The exchange energy J was represented [156–158] as a surface integral in the two-electron six-dimensional space $(\mathbf{r}_1, \mathbf{r}_2)$ and this was shown to be an asymptotically exact result, cf. also reference [159]. The surface integral method gives amazingly accurate results (cf. the excellent monograph by Patil and Tang [160]) even if the exchange energy is of the order of the energies typical for solid state phenomena. Unfortunately, this method, that ought to be applied to *ab initio* calculated (e.g., from density functional theory (DFT) [161–163]) wave functions, is barely known in the solid state community (although a very recent work by Gor'kov and Krotkov [164] indicates that it is not completely forgotten).

This is one of the reasons why the t and J transfer integrals have been treated phenomenologically just as fitting parameters of the theory. A valuable discussion on a similar scope of ideas has recently been given by Brovetto, Maxia and Salis [165] but it may well not be the only case. In order to ease comparison of the HL Hamiltonian with the other types discussed in the search of a theory of high-T_c superconductivity we shall rewrite it in terms of spin variables.

The grounds for our theory have been set first by Schubin and Wonsowsky and later in more clear notions and notation by Zener [100–102]. The *s-d* two-electron exchange is the intra-atomic version of the HL interaction. Both of those 4-fermion interactions due to Heitler–London & Schubin–Wonsowsky–Zener can in principle mediate superconductivity and magnetism.

2.2.2.1 *Spin variables*

Let us introduce the spin operator $\hat{\mathbf{S}}_i$ and particle number operator $\hat{n}_i$ for each atomic orbital,

$$\hat{\mathbf{S}}_i = \hat{\Psi}_i^\dagger \frac{\boldsymbol{\sigma}}{2} \hat{\Psi}_i, \qquad \hat{n}_i = \hat{\Psi}_i^\dagger \sigma_0 \hat{\Psi}_i, \qquad \hat{\Psi}_i^\dagger = \left(\hat{\Psi}_{i\uparrow}^\dagger, \hat{\Psi}_{i\downarrow}^\dagger\right), \tag{2.11}$$

where $\sigma_0 = \mathbf{1}_{2\times 2}$ and $\boldsymbol{\sigma}$ are the Pauli sigma matrices, and the first two formulae imply summation over the spin indices. Introducing also the spin exchange operator $\hat{P}_{ij}$,

$$P\hat{\Psi}_{i\alpha}\hat{\Psi}_{j\beta} = \hat{\Psi}_{i\beta}\hat{\Psi}_{j\alpha}, \qquad \hat{P}_{ij} = \sum_{\alpha\beta}(\hat{\Psi}_{i\alpha}\hat{\Psi}_{j\beta})^\dagger P\,\hat{\Psi}_{i\alpha}\hat{\Psi}_{j\beta}, \tag{2.12}$$

we can rewrite the HL Hamiltonian *per bond* as [166–170]

$$-J\sum_{\alpha\beta}\hat{\Psi}_{i\beta}^\dagger\hat{\Psi}_{j\alpha}^\dagger\hat{\Psi}_{i\alpha}\hat{\Psi}_{j\beta} = J\hat{P}_{ij} = 2J\left(\hat{\mathbf{S}}_i\cdot\hat{\mathbf{S}}_j + \frac{1}{4}\hat{n}_i\hat{n}_j\right). \tag{2.13}$$

We should stress that in the t-J model the term $\propto \hat{n}_i\hat{n}_j$ enters with negative sign [171, 172]. Let us also provide the "mixed" representation:

$$\begin{aligned} 2\hat{\mathbf{S}}_i \cdot \hat{\mathbf{S}}_j =& \hat{S}_{i,x}\left(\hat{\Psi}^\dagger_{j\uparrow}\hat{\Psi}_{j\downarrow} + \hat{\Psi}^\dagger_{j\downarrow}\hat{\Psi}_{j\uparrow}\right) \\ &+ \hat{S}_{i,y}\left(-i\hat{\Psi}^\dagger_{j\uparrow}\hat{\Psi}_{j\downarrow} + i\hat{\Psi}^\dagger_{j\downarrow}\hat{\Psi}_{j\uparrow}\right) + \hat{S}_{i,z}\left(\hat{n}_{j\uparrow} - \hat{n}_{j\downarrow}\right) \\ =& \hat{S}_{i,+}\hat{\Psi}^\dagger_{j\downarrow}\hat{\Psi}_{j\uparrow} + \hat{S}_{i,-}\hat{\Psi}^\dagger_{j\uparrow}\hat{\Psi}_{j\downarrow} + \hat{S}_{i,z}\left(\hat{n}_{j\uparrow} - \hat{n}_{j\downarrow}\right), \end{aligned} \tag{2.14}$$

where $\hat{n}_{j\uparrow} \equiv \hat{\Psi}^\dagger_{j\uparrow}\hat{\Psi}_{j\uparrow}$, and $\hat{S}_{i,+} = \hat{\Psi}^\dagger_{i\uparrow}\hat{\Psi}_{i\downarrow} = \hat{S}^\dagger_{i,-}$. Note that (2.13) implies a purely orbital motion without spin flip: two electrons exchange their orbitals and only the spin indices reflect this correlated hopping. For $J > 0$, the HL Hamiltonian has a singlet ground state

$$\begin{aligned} |\mathrm{S}\rangle &= \frac{1}{\sqrt{2}}(\hat{\Psi}^\dagger_{i\uparrow}\hat{\Psi}^\dagger_{j\downarrow} - \hat{\Psi}^\dagger_{i\downarrow}\hat{\Psi}^\dagger_{j\uparrow})|\mathrm{vac}\rangle \\ \hat{\Psi}_{i\alpha}|\mathrm{vac}\rangle &= 0, \qquad \langle\mathrm{vac}|\mathrm{vac}\rangle = 1, \end{aligned} \tag{2.15}$$

with eigenvalue $-J$. The lowering in energy of the singlet state, having a symmetric orbital wave function, is of purely kinetic origin related to the delocalization of the particles at different orbitals. Symbolically, the "location" of the (approximately) localized electron doublet(s) in the structure signature of a molecule is designated by a colon, e.g., H:H for the H_2 molecule. This Lewis notation for the valence bond with energy $-J$ (or four-Fermion terms in the second quantization language) is an important ingredient of the chemical intuition. In principle, such an exchange lowering is expected to exist for Bose particles as well. For electrons, however, we have triplet excited states

$$\begin{aligned} |\mathrm{T}{+1}\rangle &= \hat{\Psi}^\dagger_{i\uparrow}\hat{\Psi}^\dagger_{j\uparrow}|\mathrm{vac}\rangle, \\ |\mathrm{T}\,0\rangle &= \frac{1}{\sqrt{2}}(\hat{\Psi}^\dagger_{i\uparrow}\hat{\Psi}^\dagger_{j\downarrow} + \hat{\Psi}^\dagger_{i\downarrow}\hat{\Psi}^\dagger_{j\uparrow})|\mathrm{vac}\rangle, \\ |\mathrm{T}{-1}\rangle &= \hat{\Psi}^\dagger_{i\downarrow}\hat{\Psi}^\dagger_{j\downarrow}|\mathrm{vac}\rangle, \end{aligned} \tag{2.16}$$

with eigenvalue J. In the present work we consider the parameter J to be *positive if it corresponds to antiferromagnetism, or pairing in the singlet channel.* Thus the singlet-triplet splitting for the single-bond HL Hamiltonian (2.13) is $2J$. Similarly, the bonding-antibonding splitting for the single-particle hopping Hamiltonian $-t\sum_\alpha(\hat{\Psi}^\dagger_{j\alpha}\hat{\Psi}_{i\alpha} + \hat{\Psi}^\dagger_{i\alpha}\hat{\Psi}_{j\alpha})$ is $2t$, and the energy threshold for creation of a pair of normal carriers, considered in the next section, is 2Δ. Besides stemming from bare inter- and intra-atomic processes, two-electron hopping amplitudes J can be created by strong correlations [173] within the Hubbard model. For a nice review on this subject the reader is referred to the work by Spalek and Honig [172].

2.3 Reduced Hamiltonian and the BCS gap equation

Substituting the Fermi operators $\hat{\Psi}_{\mathbf{n}\alpha}$, Eq. (2.8), into Eq. (2.9) one obtains the HL interaction Hamiltonian in a diagonal band representation. For the case of zero electric current, cf. Sec. 26, p. 203 and Sec 33, p. 243 in Ref. [174], solely the reduced Hamiltonian $\hat{\mathcal{H}}_{\text{HL-R}}$, including creation and annihilation operators with opposite momenta only, has to be taken into account:

$$\hat{\mathcal{H}}_{\text{HL-R}} = \frac{1}{2N} \sum_{b,\mathbf{p}} \sum_{b',\mathbf{p}'} \sum_{\alpha\beta} \hat{c}^{\dagger}_{b,\mathbf{p}\beta} \hat{c}^{\dagger}_{b,-\mathbf{p}\alpha}\, V_{b,\mathbf{p};b',\mathbf{p}'}\, \hat{c}_{b',-\mathbf{p}'\alpha} \hat{c}_{b',\mathbf{p}'\beta}. \tag{2.17}$$

For singlet superconductors it is necessary to take into account the pairing with opposite spins, thereby the *total reduced* Hamiltonian reads

$$\hat{\mathcal{H}}'_{\text{R}} = \sum_{b,\mathbf{p},\alpha} \eta_{b,\mathbf{p}}\, \hat{c}^{\dagger}_{b,\mathbf{p}\alpha} \hat{c}_{b,\mathbf{p}\alpha} + \frac{1}{N} \sum_{b,\mathbf{p}} \sum_{b',\mathbf{p}'} V_{b,\mathbf{p};b',\mathbf{p}'}\, \hat{c}^{\dagger}_{b,\mathbf{p}\uparrow} \hat{c}^{\dagger}_{b,-\mathbf{p}\downarrow} \hat{c}_{b',-\mathbf{p}'\downarrow} \hat{c}_{b',\mathbf{p}'\uparrow}, \tag{2.18}$$

where $\eta_{b,\mathbf{p}} \equiv \epsilon_{b,\mathbf{p}} - \mu$ are the band energies measured from the chemical potential [174]. Hence the BCS equation [73–75] for the superconducting gap takes the familiar form

$$\Delta_{b,\mathbf{p}} = \frac{1}{N} \sum_{b',\mathbf{p}'} (-V_{b,\mathbf{p};b',\mathbf{p}'}) \frac{1 - 2n_{b',\mathbf{p}'}}{2E_{b',\mathbf{p}'}} \Delta_{b',\mathbf{p}'}, \tag{2.19}$$

where $E_{b,\mathbf{p}} = (\eta^2_{b,\mathbf{p}} + |\Delta_{b,\mathbf{p}}|^2)^{1/2}$ are the quasiparticle energies and $n_{b,\mathbf{p}} = [\exp(E_{b,\mathbf{p}}/k_{\text{B}}T) + 1]^{-1}$ the Fermi filling factors. The summation over the band index b' should be restricted to the partially filled (metallic) bands, comprising sheets of the Fermi surface. Applying this standard procedure to the HL Hamiltonian (2.9), and after some algebra, we obtain the desired BCS pairing kernel,

$$\begin{aligned} V_{b,\mathbf{p};b',\mathbf{p}'} = &- 2J_{sd}\, S_{\mathbf{p}} S_{\mathbf{p}'} D_{\mathbf{p}} D_{\mathbf{p}'} - J_{pp}\, \gamma_x X_{\mathbf{p}} X_{\mathbf{p}'}\, \gamma_y Y_{\mathbf{p}} Y_{\mathbf{p}'} \\ &+ 2\,(J_{sp} S_{\mathbf{p}} S_{\mathbf{p}'} + J_{pd} D_{\mathbf{p}} D_{\mathbf{p}'})\,(\gamma_x X_{\mathbf{p}} X_{\mathbf{p}'} + \gamma_y Y_{\mathbf{p}} Y_{\mathbf{p}'}), \end{aligned} \tag{2.20}$$

where

$$\gamma_x = 4\cos\left(\frac{p_x + p'_x}{2}\right), \qquad \gamma_y = 4\cos\left(\frac{p_y + p'_y}{2}\right). \tag{2.21}$$

As the band indices b and b' enter implicitly in the band energies $\epsilon_{b,\mathbf{p}}$ in the equation for the eigenvectors $\Psi_{\mathbf{p}}(\epsilon_{b,\mathbf{p}})$, we will suppress them hereafter. The layered cuprates, admittedly, have a single conduction band and their

Fermi surface has the shape of a rounded square. In this simplest case one has to solve numerically the nonlinear integral equation

$$\Delta_{\mathbf{p}} = \int_{-\pi}^{\pi} \frac{dq_x}{2\pi} \int_{-\pi}^{\pi} \frac{dq_y}{2\pi} (-V_{\mathbf{pq}}) \frac{\Delta_{\mathbf{q}}}{2E_{\mathbf{q}}} \tanh\left(\frac{E_{\mathbf{q}}}{2k_{\mathrm{B}}T}\right). \tag{2.22}$$

The solution to this general gap equation, depending on the J_{ij} values, can exhibit s, p, or d-type symmetry. It has been shown previously that a purely p-p model [115,175] ($J_{pp} > 0$) results in a d_{xy} (B_{2g}) gap anisotropy. However, we found that an agreement with the experimentally observed $d_{x^2-y^2}$ (B_{1g}) gap anisotropy (for a review see for example Ref. [176]) can be achieved only in the simplest possible case of a dominant s-d exchange. This separable Hamiltonian deserves special attention and we will analyze it in the next sections.

2.4 Separable *s*-*d* model

For the special case of a purely s-d model, $J_{sp} = J_{pd} = J_{pp} = 0$, representing the spin exchange operator $\hat{P}$ as a (4×4)-matrix, cf. Eqs. (2.12)–(2.13), the reduced pairing Hamiltonian takes the form

$$\hat{\mathcal{H}}_{\text{HL-R}} = \frac{J_{sd}}{N} \sum_{\mathbf{p},\mathbf{q}} \begin{pmatrix} \hat{S}_{-\mathbf{p}\uparrow}\hat{D}_{\mathbf{p}\uparrow} \\ \hat{S}_{-\mathbf{p}\uparrow}\hat{D}_{\mathbf{p}\downarrow} \\ \hat{S}_{-\mathbf{p}\downarrow}\hat{D}_{\mathbf{p}\uparrow} \\ \hat{S}_{-\mathbf{p}\downarrow}\hat{D}_{\mathbf{p}\downarrow} \end{pmatrix}^{\dagger} \begin{pmatrix} 1\,0\,0\,0 \\ 0\,0\,1\,0 \\ 0\,1\,0\,0 \\ 0\,0\,0\,1 \end{pmatrix} \begin{pmatrix} \hat{S}_{-\mathbf{q}\uparrow}\hat{D}_{\mathbf{q}\uparrow} \\ \hat{S}_{-\mathbf{q}\uparrow}\hat{D}_{\mathbf{q}\downarrow} \\ \hat{S}_{-\mathbf{q}\downarrow}\hat{D}_{\mathbf{q}\uparrow} \\ \hat{S}_{-\mathbf{q}\downarrow}\hat{D}_{\mathbf{q}\downarrow} \end{pmatrix}. \tag{2.23}$$

Carrying out an additional reduction for a spin-singlet pairing, the interaction Hamiltonian reads

$$\hat{\mathcal{H}}_{\text{HL-R}} = -\frac{J_{sd}}{N} \sum_{\mathbf{p},\mathbf{q},\alpha} \hat{S}^{\dagger}_{\mathbf{p},\alpha} \hat{D}^{\dagger}_{-\mathbf{p},-\alpha} \hat{D}_{-\mathbf{q},-\alpha} \hat{S}_{\mathbf{q},\alpha}, \tag{2.24}$$

where $-\alpha$ stands for the electron spin projection opposite to α. For comparison, we provide again the kinetic energy part of the Hamiltonian employing the same notation,

$$\hat{\mathcal{H}}'_{\text{BH}} = \sum_{\mathbf{p},\alpha} \begin{pmatrix} \hat{D}_{\mathbf{p},\alpha} \\ \hat{S}_{\mathbf{p},\alpha} \\ \hat{X}_{\mathbf{p},\alpha} \\ \hat{Y}_{\mathbf{p},\alpha} \end{pmatrix}^{\dagger} \begin{pmatrix} \epsilon_{\mathrm{d}} - \mu & 0 & t_{pd}s_x & -t_{pd}s_y \\ 0 & \epsilon_{\mathrm{s}} - \mu & t_{sp}s_x & t_{sp}s_y \\ t_{pd}s_x & t_{sp}s_x & \epsilon_{\mathrm{p}} - \mu & -t_{pp}s_x s_y \\ -t_{pd}s_y & t_{sp}s_y & -t_{pp}s_x s_y & \epsilon_{\mathrm{p}} - \mu \end{pmatrix} \begin{pmatrix} \hat{D}_{\mathbf{p},\alpha} \\ \hat{S}_{\mathbf{p},\alpha} \\ \hat{X}_{\mathbf{p},\alpha} \\ \hat{Y}_{\mathbf{p},\alpha} \end{pmatrix}. \tag{2.25}$$

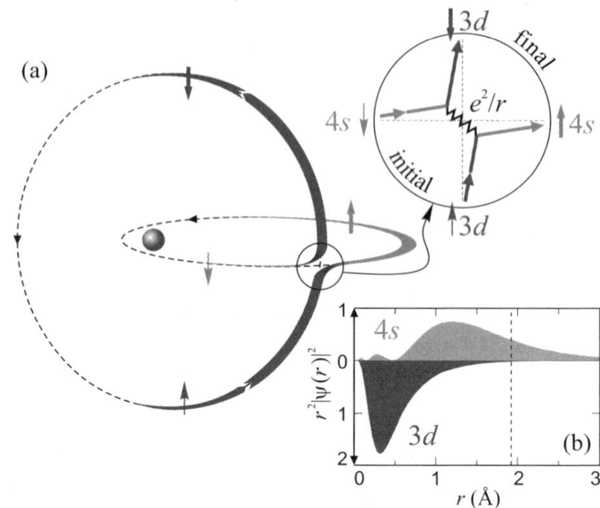

Fig. 2.2 Pairing two-electron exchange amplitude J_{sd} "hidden" in the Cu atom. (a) Classical Bohr-Sommerfeld representation of the s-d two-electron exchange process. The inset shows how the Coulomb scattering leads to an effective electron spin exchange. (b) Electron charge distribution for Cu 4s and Cu 3d orbitals: the dashed line marks the Cu-O distance in the CuO_2 plane.

Within the s-d model considered, the pairing kernel (2.20) factors into functions depending only on $\mathbf{p}$ or $\mathbf{q}$,

$$(-V_{\mathbf{pq}}) = 2J_{sd}\, S_{\mathbf{p}} D_{\mathbf{p}}\, S_{\mathbf{q}} D_{\mathbf{q}} \equiv 2J_{sd}\, \chi_{\mathbf{p}}\, \chi_{\mathbf{q}}. \tag{2.26}$$

A schematic representation of the J_{sd} exchange amplitude is given in Fig. 2.2. This factorizable Markowitz–Kadanoff [177] form of the pairing kernel is a direct consequence of the local intra-atomic character of the s-d exchange in the transition ion. Substituting in Eq. (2.22)

$$\Delta_{\mathbf{p}}(T) = \Xi(T)\, S_{\mathbf{p}} D_{\mathbf{p}} = \Xi(T)\, \chi_{\mathbf{p}}, \tag{2.27}$$

one obtains in a closed form, cf. Ref. [178],[1] a simple BCS equation for the

[1]Here the author writes "... the interlayer tunneling theory which I advocated for five years, and wrote a book about, turns out to be one of those which must be consigned to the dustbin". This voluntary concession marks the end of an epoch of unconventional theories of high-T_c superconductivity. Further, in the same paper, the assumption "without any justification" of the BCS gap equation for "one phonon which can be expected to couple rather strongly..." is a premonition of the return to conventional BCS interpretations of the high-T_c problem. Apropos, in order to derive an effective J_{sd} we need to take into account a local B_{1g} phonon mode of the oxygen ligands around the copper ion.

temperature dependence of the gap,

$$2J_{sd}\left\langle \frac{\chi_{\mathbf{p}}^2}{2E_{\mathbf{p}}}\tanh\left(\frac{E_{\mathbf{p}}}{2k_{\mathrm{B}}T}\right)\right\rangle = 1, \tag{2.28}$$

where

$$E_{\mathbf{p}} \equiv \left(\eta_{\mathbf{p}}^2 + \Delta_{\mathbf{p}}^2\right)^{1/2} = \left[(\epsilon_{\mathbf{p}} - E_{\mathrm{F}})^2 + (\Xi(T)\,\chi_{\mathbf{p}})^2\right]^{1/2}, \tag{2.29}$$

$$\langle f_p\rangle = \int_0^{2\pi}\int_0^{2\pi}\frac{dp_x\,dp_y}{(2\pi)^2}\,f(\mathbf{p}), \tag{2.30}$$

$E_{\mathrm{F}} \equiv \mu$. We wish to mention that separability of the order parameter Eq. (2.27) has been derived by Pokrovsky [179] in the general weak-coupling case and not only for factorizable pairing kernels.

According to Eq. (2.3) we have

$$\begin{aligned}\chi_{\mathbf{p}} \equiv& S_{\mathbf{p}}D_{\mathbf{p}} = 4\varepsilon_p\,t_{sp}\,t_{pd}\,(x-y)\left[\varepsilon_s\,\varepsilon_p^2 - 4\varepsilon_p\,t_{sp}^2\,(x+y) + 32t_{pp}\,\tau_{sp}^2\,xy\right]\\ &\times\Big\{\left[4\varepsilon_p\,t_{sp}\,t_{pd}\,(x-y)\right]^2 + \left[\varepsilon_s\,\varepsilon_p^2 - 4\varepsilon_p\,t_{sp}^2\,(x+y) + 32t_{pp}\,\tau_{sp}^2\,xy\right]^2\\ &+ 4x\left[(\varepsilon_s\,\varepsilon_p - 8\tau_{sp}^2\,y)\,t_{pd}\right]^2 + 4y\left[(\varepsilon_s\,\varepsilon_p - 8\tau_{sp}^2\,x)\,t_{pd}\right]^2\Big\}^{-1}.\end{aligned} \tag{2.31}$$

The gap symmetry is then easily made obvious in the narrow-band approximation. Formally, it is the asymptotic behaviour of the eigenvector (2.3) for vanishing hopping integrals $t \to 0$. In this limit case [123], we have $\epsilon_{3,\mathbf{p}} \approx \epsilon_{\mathrm{d}}$, and

$$\tilde{\Psi}_{3,\mathbf{p}} = \begin{pmatrix} D_{3,\mathbf{p}}\\ S_{3,\mathbf{p}}\\ X_{3,\mathbf{p}}\\ Y_{3,\mathbf{p}}\end{pmatrix} \approx \begin{pmatrix} 1\\ -(t_{sp}t_{pd}/\varepsilon_s\varepsilon_p)\,(s_x^2 - s_y^2)\\ (t_{pd}/\varepsilon_p)\,s_x\\ (t_{pd}/\varepsilon_p)\,s_y\end{pmatrix}. \tag{2.32}$$

Clearly, $D_{3,\mathbf{p}}$ exhibits A_{1g} symmetry, while $S_{3,\mathbf{p}}$ has the B_{1g} symmetry. Whence the product $S_{3,\mathbf{p}}D_{3,\mathbf{p}} \propto \cos p_x - \cos p_y$ "inherits" the B_{1g} symmetry, Fig. 2.3 (b), which is conserved even for realistic values of the hopping integrals, and from (2.27) it follows

$$\Delta_{\mathbf{p}} \propto S_{3,\mathbf{p}}D_{3,\mathbf{p}} \approx \frac{2t_{sp}t_{pd}}{(E_{\mathrm{F}}-\epsilon_s)\,(E_{\mathrm{F}}-\epsilon_p)}(\cos p_x - \cos p_y). \tag{2.33}$$

As can be seen in Fig. 2.3 (c) this small-t approximation fits the ARPES data for the gap anisotropy quite well. Similar experimental data have been previously reported, e.g., in Ref. [181–184]. Note, additionally, that close to

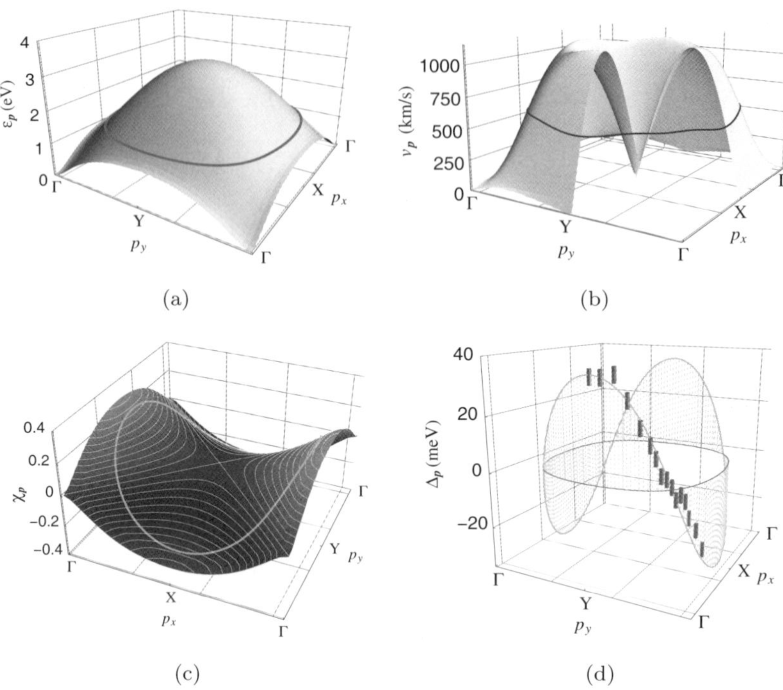

Fig. 2.3 Electronic properties of the superconducting CuO_2 plane. (a) Conduction band energy $\epsilon_{\mathbf{p}}$ as a function of the quasi-momentum $\mathbf{p}$. The red contour corresponding to the Fermi energy, $\epsilon_{\mathbf{p}} = E_F$, is in excellent agreement with the ARPES data [21]. (b) Quasiparticle velocity $v_{\mathbf{p}}$ as a function of quasimomentum. The velocity variation along the Fermi contour is less than 10%. The energy parameters are fitted to be in agreement with the typical *ab initio* calculations [27]. The significant overestimate disappears if the bandwidth is fitted to the experimental data, but the shape is conserved. (c) Momentum dependence of the gap-anisotropy function $\chi_{\mathbf{p}}$ within the *s*-*d* model. The functional values along the Fermi contour are indicated by a green line. (d) Superconducting gap at zero temperature $\Delta_{\mathbf{p}}$ (green line) according to our analytical result (2.27), plotted along the Fermi contour (red line). The ARPES data [180] for $Bi_2Sr_2CaCu_2O_{8+\delta}$, Fig. 2.4, are given as prisms with sizes corresponding to the experimental error bars. The gap function along the Fermi contour has the same qualitative behavior and symmetry as the $Cu3d_{x^2-y^2}$ electron wave function along the circular orbit sketched in Fig. 2.2 (a).

the (π, π)-point, where $(p_x - \pi)^2 + (p_y - \pi)^2 \ll 1$, the angular dependence of the gap can be written in the form

$$\Delta_{\mathbf{p}} \propto \cos p_x - \cos p_y \approx \left[(p_x - \pi)^2 + (p_y - \pi)^2\right] \cos 2\phi \qquad (2.34)$$
$$\tan\phi \equiv \frac{p_y - \pi}{p_x - \pi}.$$

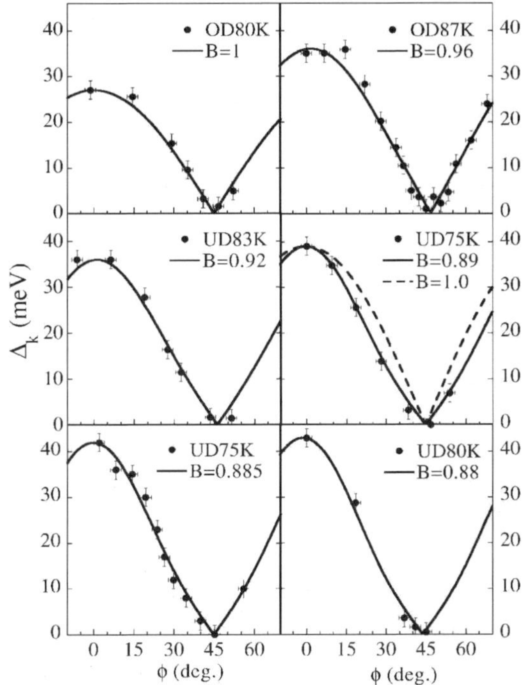

Fig. 2.4 Superconducting gap of $Bi_2Sr_2CaCu_2O_{8+\delta}$ as a function of the Fermi surface angle ϕ, Eq. (2.34) (reprinted with permission from Ref. [180]; Copyright © 1999 by the American Physical Society). The data for the "OD87K" sample were fitted in Fig. 2.3 (d).

The d-type angular dependence of both the gap anisotropy and the separable pairing kernel is often postulated in phenomenological model Hamiltonians to describe high-T_c superconductivity. The previous discussion thus provides a microscopic justification based on the fundamental exchange amplitudes.

Employing the analytical expression (1.8) for the constant-energy contours (CEC), one can implement an efficient numerical integration,

$$\int_0^{2\pi} dp_x \int_0^{2\pi} dp_y\, f(\epsilon_{\mathbf{p}}) = \int_{\epsilon_b}^{\epsilon_t} d\epsilon \oint \frac{dp_l}{v_{\mathbf{p}}}\, f(\epsilon), \tag{2.35}$$

where

$$v_{\mathbf{p}} = \left|\frac{\partial \epsilon_{\mathbf{p}}}{\partial \mathbf{p}}\right| = \frac{\left[(\mathcal{A}y+\mathcal{B})^2 x(1-x) + (\mathcal{A}x+\mathcal{B})^2 y(1-y)\right]^{1/2}}{|\mathcal{A}'xy + \mathcal{B}'(x+y) + \mathcal{C}'|}, \tag{2.36}$$

with $\mathcal{A}'$, $\mathcal{B}'$ and $\mathcal{C}'$ being the energy derivatives of the polynomials (1.9),

$$\begin{aligned} \mathcal{A}'(\epsilon) &= 16\left[2t_{sp}^2 t_{pp} - 2t_{pd}^2 t_{pp} - t_{pp}^2(\varepsilon_d + \varepsilon_s)\right], \\ \mathcal{B}'(\epsilon) &= -4(t_{sp}^2\varepsilon_d + t_{pd}^2\varepsilon_s) - 4\varepsilon_p(t_{sp}^2 + t_{pd}^2), \\ \mathcal{C}'(\epsilon) &= \varepsilon_s\varepsilon_p^2 + \varepsilon_d\varepsilon_p^2 + 2\varepsilon_d\varepsilon_s\varepsilon_p. \end{aligned} \tag{2.37}$$

Using these functions, the band spectrum, see Eq. (1.8), can be obtained by Newton iterations

$$\epsilon_{\mathbf{p}}^{[i]} = \epsilon_{\mathbf{p}}^{[i-1]} - \frac{\mathcal{A}xy + \mathcal{B}(x+y) + \mathcal{C}}{\mathcal{A}'xy + \mathcal{B}'(x+y) + \mathcal{C}'} \tag{2.38}$$

with initial approximation for the conduction band $\epsilon_{3,\mathbf{p}}^{[0]} = \varepsilon_d$.

The charge carrier velocity is $v_{\mathbf{p}}a_0/\hbar$, a_0 is the lattice constant, p_l the dimensionless momentum component along the CEC, and ϵ_{b} and ϵ_{t} are the bottom and the top of the conduction band, respectively, $\epsilon_{\mathrm{b}} \leqslant \epsilon_{\mathbf{p}} \leqslant \epsilon_{\mathrm{t}}$. The canonic equation for the CEC Eq. (1.8),

$$\mathcal{A}\cos p_x \cos p_y - (\mathcal{A} + 2\mathcal{B})(\cos p_x + \cos p_y) + \mathcal{A} + 4\mathcal{C} = 0, \tag{2.39}$$

can be cast in an explicit form

$$p_{y,1}(p_x) = 2\arcsin\sqrt{-\frac{\mathcal{B}x + \mathcal{C}}{\mathcal{A}x + \mathcal{B}}}, \qquad p_{y,2}(p_x) = 2\pi - p_{y,1}(p_x), \tag{2.40}$$

and for the length element dp_l we obtain

$$dp_l = \sqrt{1 + \left(\frac{dp_y}{dp_x}\right)^2}\, dp_x, \qquad \left(\frac{dp_y}{dp_x}\right)^2 = \frac{x(1-x)}{y(1-y)}\left(\frac{\mathcal{A}y + \mathcal{B}}{\mathcal{A}x + \mathcal{B}}\right)^2. \tag{2.41}$$

The contour integration along the hole pocket $\varepsilon_{\mathbf{p}} = \mathrm{const}$ centered at the (π, π) point needs to be performed only over one eight of the CEC

$$\oint dp_l f(p_x, p_y) = 8\int_{p_{\mathrm{d}}}^{\pi} f(p_x, p_y(p_x)) \left(\frac{dp_{\mathrm{d}}(p_x)}{dp_x}\right) dp_x, \tag{2.42}$$

where

$$x_{\mathrm{d}} = \sin^2(p_{\mathrm{d}}/2), \qquad \mathcal{A}x_{\mathrm{d}}^2 + 2\mathcal{B}x_{\mathrm{d}} + \mathcal{C} = 0. \tag{2.43}$$

2.5 Antiferromagnetic character of J_{sd}

Let us address now the atomic physics underlying the s-d pairing mechanism. Within the framework of the Hartree–Fock (HF) theory the exchange energy is given [185] as an integral of the Cu4s and Cu3$d_{x^2-y^2}$ atomic wave functions,

$$-J_{sd}^{(\mathrm{HF})} = \iint \psi_{\mathrm{s}}^*(\mathbf{r}_1)\psi_{\mathrm{d}}^*(\mathbf{r}_2)\frac{e^2}{|\mathbf{r}_1-\mathbf{r}_2|}\psi_{\mathrm{d}}(\mathbf{r}_1)\psi_{\mathrm{s}}(\mathbf{r}_2)\,d\mathbf{r}_1 d\mathbf{r}_2, \tag{2.44}$$

and its sign corresponds to repulsion and depairing for singlet Cooper pairs. Thus, one can formulate the following conceptual problem, emerging in fundamental physics:

(i) Is it possible, as in the case of a covalent bond, for two-electron correlations to trigger a change of the sign of the exchange amplitude?
(ii) How can one adapt the Heitler–London idea to a transition ion perturbed by ligands?

There is no doubt that the solution to this problem will illuminate other problems in the physics of magnetism as well. In brief, the enigma can be stated as to whether the Heitler–London approximation for the exchange energy may result in $J_{sd} > 0$, cf. Ref. [185]. Let us recall that already in 1962 Herring [156] was advocating that "antiferromagnetic J_{ij}'s should be the rule, ferromagnetic J_{ij}'s the exception". For the present, we can adopt the s-d model as a convenient microscopic phenomenology of superconductivity in the CuO_2 plane. On the other hand, the exchange amplitude J_{sd} is an important ingredient in the physics of magnetism as well.

Physics of magnetism certainly displays lots of subtleties, but for a qualitative comparison let us trace the "operation" of the s-d exchange amplitude J_{sd} in the case of the simplest model for a ferromagnetic metal. While for the CuO_2 plane the s-band is empty, for transition metals it is the widest conduction band. The width of the d-band is significantly smaller and thus, making a caricature of the ferromagnetic metals, we completely neglect the width of the d-band. In this "heavy Fermion" approximation the d-electrons are considered as localized, and without significant energy loss they can be completely spin polarized, $\langle \hat{n}_{\mathrm{d}\uparrow}\rangle \approx 1$, $\langle \hat{S}_{\mathrm{d},z}\rangle \approx \frac{1}{2} > 0$. In this case the self-consistent approximation applied to Eq. (2.14) gives

$$2\hat{\mathbf{S}}_d \cdot \hat{\mathbf{S}}_s \approx \hat{S}_{d,z}\,(\hat{n}_{s\uparrow} - \hat{n}_{s\downarrow}) \approx \langle \hat{S}_{d,z}\rangle\,(n_{s\uparrow} - n_{s\downarrow})\,. \tag{2.45}$$

Here $n_{s\alpha} \equiv \langle \hat{n}_{s\alpha}\rangle$ denotes the average number of s-electrons per atom with spin projection α. In order to calculate these variables, one has to take into account the different filling of the s-bands with different polarizations, and sum over the quasi-momenta. Finally, the exchange energy per atom reads

$$E_X = -\frac{1}{2} J_{sd}(n_{s\downarrow} - n_{s\uparrow}) < 0. \tag{2.46}$$

In the CuO_2 plane, positive values of the J_{sd} parameter lead to singlet superconductivity. For ferromagnetic metals, positive values of J_{sd} correspond to polarization of the s-band opposite to d-state polarization, $n_{s\downarrow} - n_{s\uparrow} > 0$. Thereby ferromagnetism could be brought about by an exchange amplitude with a sign corresponding to antiparallel spin polarization of s- and d-orbitals, cf. Figures 4–15 of Ref. [169]. Thus the same sign of the s-d exchange amplitude J_{sd} can be at the origin of ferromagnetism, e.g., in Fe and Ni, and superconductivity in the CuO_2 plane. This is perhaps the simplest scenario for cuprate superconductivity based on the two-electron exchange processes.

According to a naive interpretation of Hund's rule the Kondo effect should not exist. In the epoch-making paper [186, 187] on the resistance minimum in dilute magnetic alloys, Kondo concluded that in the s-d exchange model, due to Zener [100–102], Kasuya and Yosida [188–194], the sign of the direct exchange amplitude J_{sd} must be antiferromagnetic. And vice versa, the minimum disappears if J_{sd} is ferromagnetic. Such minimum exists for many magnetic metals and alloys and is another hint in favor of Herring's argument [156] mentioned earlier. In his analysis Kondo speculates that J_{sd} is a parameter whose sign and magnitude have to be determined so as to fit the experiment, and concluded that antiferromagnetic values of the order of eV are quite reasonable. For a review on the Kondo problem we refer the reader to Ref. [195].

On the other hand every textbook on atomic physics tells us that parallel electron spins and an antisymmetric wave function minimize the electrostatic energy. Put differently, the tendency toward ferromagnetism in Hund's rule is of electrostatic origin. As Kondo has pointed out [186, 187], the problem is to find the origin of an antiferromagnetic J_{sd}, or how to overcome the strong electrostatic repulsion. It is very plausible that it is not a single driving force, but instead one has to take into account several interfering electron scattering amplitudes.

2.5.1 *Intra-atomic correlations*

The self-consistent approximation has been known in celestial mechanics for ages. Accordingly, the motion of a planet is averaged over its orbit. One has to calculate then the potential created by this orbital-averaged motion and perform a sum over all particles. Where does this scheme fail? It fails in the case of a resonance when the periods for some planets are commensurate or just equal. This is nothing but the case of a transition ion for which the energies and classical periods are very close. Then the resonant repetitive electron scattering, symbolically presented in Fig. 2.2, leads to strong electron correlations like in the double Rydberg states of atoms [142–145]. For double Rydberg states in He it is necessary to solve a two-electron quantum problem but for other atoms we have to take into account the influence of the other electrons in some self-consistent approximation, the LDA, for example. For two $4s$ electrons the two-electron correlations are so strong that they have to be taken into account from the very beginning [142–145]. There are no doubts that the two-electron correlations between $4s$ and $3d$ electrons having almost equal energies cannot be neglected. In other words Hartree–Fock theory cannot be used directly. Hence, the Bohr picture is not merely a historical remark but rather an indispensable ingredient of the contemporary physics of magnetism: two-electron correlations can be important even in a *single* atom. We thus conclude that the two-electron correlations may overrule Hund's rule for the local s-d exchange. Note also that the single-particle orbitals (accessible, e.g., from DFT [161–163], HF and X_α methods, etc.) only form an adequate basis for a subsequent account of electronic correlations. A first step in this direction will be *ab initio* calculation of the J_{sd}.

2.5.2 *Indirect s-d exchange*

The antiferromagnetism of the insulating phase of the undoped cuprates is mediated by the Bloch-Kramers-Anderson indirect exchange [196, 197] between $3d$ electrons of nearest-neighbour Cu ions via O$2p$ electrons. It is unlikely that the numerical value of this J_{dd} exchange integral be dramatically changed in the metallic phase obtained by hole doping. In the metallic phase, however, the same indirect exchange mechanism will operate between 3d and 4s electrons at the same Cu atom via the $2p$ electrons of the O ligands. For illustration, let us compare the indirect s-d exchange amplitude $J_{sd}^{(\mathrm{ind})}$ with J_{dd}. There are three important factors: (i) Every Cu ion has four O ligands, Fig. 2.1 (a). (ii) The hopping amplitude between $4s$ and $2p$ orbitals is bigger than the $3d$-$2p$ transfer. (iii) The Cu on-site Coulomb repulsion between $4s$ and $3d$ electrons U_{sd} is much smaller than the $3d$-$3d$ Hubbard repulsion U_{dd}. Taking into account these factors one can expect that $J_{sd}^{(\mathrm{ind})}$ is an order of magnitude bigger than J_{dd}:

$$J_{sd}^{(\mathrm{ind})} \simeq 4\left(\frac{t_{sp}}{t_{pd}}\right)^2 \frac{U_{dd}}{U_{sd}} J_{dd}. \tag{2.47}$$

The relatively small J_{dd} ensures Néel temperature T_{N} of the order of room temperature. Hence we can conclude that the indirect exchange can contribute significantly to the total J_{sd} amplitude responsible for the pairing. However, only very detailed first-principles calculations can clarify the relative contributions of the direct and the indirect *s*-*d* exchange.

2.5.3 *Effect of mixing wave functions*

In an early paper [198], by analyzing the *g*-shift and the anomalous Hall effect in Gd metal, Kondo showed that an antiferromagnetic J_{sd} can result from the effect of mixing the wave functions of conduction and d electrons. We believe that this property is preserved if the *d* electrons also form a conduction band, or even in the case of a single *s*-*p*-*d* hybridized band. We should note that Kondo's argumentation for the need of a J_{sd} with an antiferromagnetic sign in the Kondo effect is related to Anderson's consideration of localized magnetic states in metals [199]. In the latter schematized model, based on the works of Friedel [200–202], Anderson shows that "any *g*-shift caused by free-electron polarization will tend to have antiferromagnetic sign."

As it was expected by various investigators the later numerical calculations confirmed that the striking features of negative hyperfine field with large amplitude comes mainly from the contact contribution of the core electrons [203–207]. The antipolarization between the *s*- and *d*-electrons in transition metal compounds is also well observed by Mössbauer spectroscopy, however the contribution of the core *s*-electrons and conduction band cannot be experimentally resolved. For the pairing, the amplitude of the *s*-*d* Kondo scattering is essential because in some terminology the CuO_2 plane can be considered to be a single-band Kondo lattice, cf. Ref. [208].

Given the above diversity of channels for *s*-*d* exchange it is not surprising that an adequate first-principles scheme to calculate J_{sd} is still sought. Furthermore, J_{sd} is involved in the theory of magnetism in an entangled way precluding so far a direct relation between *ab initio* calculations and formulae fitting the experiment [209].

2.5.4 *Cooper and Kondo singlet formation*

Although Kondo [186, 187] does cite Zener's paper [100], in many publications in the field of magnetism the *s-d* interaction is referred to as Kondo interaction. Often this term is used in a broader sense causing eventually terminological misunderstanding. Here we cite some works which could be related to our theory even though the relation may not be direct or immediately apparent. Analyzing the possible "interplay of Cooper and Kondo singlet formations in high-T_c cuprates" Sekitani *et al.* [210] point out that "In the 20th century, two significant many-body phenomena due to spin singlet formation were discovered in the field of solid state physics: superconductivity and the Kondo effect". These authors believe that the pseudogap in the normal state corresponds to the dissociation energy of the Kondo bound states and that superconductivity and the Kondo effect are competing in $La_{1.85}Sr_{0.15}CuO_4$. They speculate that the interplay between superconductivity and the Kondo effect has not be considered seriously for high-T_c cuprates (further references on the Kondo effect in cuprate superconductors can be found, e.g., in [211]). It would be premature for us to comment on these ideas; we note however that within this terminology our theory could be considered as a Kondo interaction mechanism for pairing in high-T_c cuprates.

It is unclear from microscopic point of view if the same "Kondo interaction" amplitude J_{sd} is responsible for the empirical Kadowaki–Woods ratio [212] but the location of $La_{1.7}Sr_{0.3}CuO_4$ on the A (the coefficient of the T^2 resistivity) versus γ_0 (the electronic specific-heat coefficient) plot, cf. figure 4 in [213], is a serious hint that J_{sd} in the CuO_2 plane is one of the largest exchange amplitudes in solid state physics, comparable with the uranium heavy-fermion compounds and $SrVO_3$. In this sense our theory requires a large, yet acceptable J_{sd} value, putting the cuprates among the most interesting materials with considerable exchange interaction.

2.6 Dogmatics and more

In a review on the history of studies of superconductivity and the prospects for further research in the field Ginzburg [72] conditionally divided the history into several periods:

(i) The "Day Before Yesterday" (1911–1941). This period starts with the discovery of superconductivity in Leiden by Gilles Holst and Heike Kamerlingh Onnes [72, 214].

(ii) "Yesterday" (1942–1986). This period embraces the appearance of the Ψ-Δ theories and the first significant technical applications.
(iii) "Today" (1987–?). This epoch emerged with the discovery of the high-T_c cuprates [1].
(iv) "Tomorrow" (?). The final landmark of "Today" must be some event.

Long ago, in the "Day Before Yesterday" high-T_c superconductivity was known as a "*blue dream*" of physicists. Considerable theoretical efforts were applied "Yesterday", attempting to predict possible realizations of this phenomenon [215–219]. At that time the problem of high-T_c superconductivity was "one of the most interesting and attractive problems from the purely scientific point of view" [215–219]. Intriguingly, the special role of *layered metallic systems* and *almost two-dimensional superconductivity* [220, 221] was mentioned already in this epoch, and a big variety of mechanisms of superconductivity were considered including the *s*-*d* exchange [222–230]. This exchange process was well known in the physics of magnetism since the dawn of quantum mechanics. Thus it is not surprising that the first work on the *s*-*d* pairing mechanism, by Akhiezer and Pomeranchuk [222], was accomplished about a year after the celebrated BCS paper [75]. These pioneering works, however, "have been ignored thus far" [231].

After Bednorz and Müller's work [1] the problem of high-T_c superconductivity soon came into fashion. "*After experiencing the 'smell of roast meat', yesterday skeptics or even critics can become zealous advocates of a new direction of endeavor. But this is another story—more in the realm of psychology and sociology than scientific and technical activity*" [232, 233]. All models of high-T_c superconductivity were revisited in great detail in the uncountable number of papers that have appeared in the epoch "Today".

2.6.1 *Aesthetics and frustrations of the central dogmas*

The common trends of some new theoretical models for cuprate superconductivity were systematized by Anderson [234] in six dogmas. We find it very instructive to compare our theory of high-T_c superconductivity with these dogmas.

"Dogma I: All the relevant carriers of both spin and electricity reside in the CuO_2 plane and derive from the hybridized O 2p–Cu3$d_{x^2-y^2}$ orbital which dominates the binding in these compounds. ...in summary look at the planes only (a great and welcome simplification.)"

The key ingredient of our pairing theory is the four-fermion *s-d* interaction between the Cu4*s* and Cu3$d_{x^2-y^2}$ orbitals. If we cut the Cu4*s* orbital off from the Hilbert space of the CuO_2 plane such a pairing interaction cannot exist. Although Cu4*s* is an empty band, it is an important component of the theory of high-T_c superconductivity. The O2*p* orbitals are the intermediaries between the Cu4*s* and Cu3$d_{x^2-y^2}$ orbitals that create the necessary *s-d* hybridization of the conduction Cu3$d_{x^2-y^2}$ band.

"Dogma II: Magnetism and high-T_c superconductivity are closely related, in a very specific sense: i.e., the electrons which exhibit magnetism are the same as the charge carriers. ... We must solve the old problem of doping of a single Mott–Hubbard band before we can begin the problem of high-T_c."

The incommensurate spin-density waves (SDW) observed in the superconducting phase of $La_2CuO_{4.11}$ and $La_{1.88}Sr_{0.12}O_4$ by neutron scattering [235] and muon spin relaxation [236], respectively, demonstrated that antiferromagnetism of the Cu site is innocuous for superconductivity in the cuprates. These antiferromagnetic correlations are not depairing and do not change significantly T_c and the electronic structure of CuO_2 plane. The observed correspondence between the magnetic and the superconducting order parameters is an additional hint that both phenomena have a common origin [237, 238]; see also the detailed theoretical works [239, 240]. Nevertheless the coexistence of SDW and superconductivity with a common critical temperature cannot be clearly observed in every high-T_c cuprate. As a result superconductivity can be considered, at least in first approximation, separately from a possible antiferromagnetism as is done in the present Chapter. In Cr metal the amplitude of the SDW shows also a BCS-like temperature dependence [241] and the SDW-theory is based on the conventional theory of metals based in turn on the Landau Fermi-liquid theory. We consider the quasiparticle picture as a reliable starting point for the theory of high-T_c cuprates as well.

"Dogma III: The dominant interactions are repulsive and their energy scales are all large. ... Restrict your attention to a single band, repulsive (not too big) U Hubbard Model."

Indeed, the dominant interactions are repulsive — "Nobody has abrogated the Coulomb's law", as Landau used to emphasize [72]. However, something subtle occurs when the atomic orbitals are analyzed. The strong electron repulsion leads to Heitler–London type correlations: two electrons cannot occupy simultaneously the same orbital, not even if they have opposite spins. The exchange of electrons between two orbitals decreases the electron kinetic energy and thereby the total energy of the whole system.

In molecular physics, according to the Hellmann–Feynman theorem such a decrease in energy drives an inter-atomic attraction for large inter-atomic distances. Thus, the valence attraction is the final result of the dominant Coulomb repulsion between electrons. In this way the Heitler–London-type exchange between itinerant electrons gives rise to electron-electron attraction and conventional Cooper pairing. The *s*-*d* exchange, "residing" in the Cu atom, can be considered as an "intra-atomic-valence bond" — an attraction-sign scattering amplitude due to the Coulomb repulsion between the correlated electrons. The *s*-*d* exchange in the transition ions is one of the most intensive exchange processes in solid state physics. Such a high-frequency process is described by the exchange amplitude J_{sd} in the lattice models for the electronic structure and its sign is determined by the inter-electronic Coulomb repulsion. The Heitler–London interaction is a result of strong electron repulsion and survives even for infinite Hubbard U. This interaction is lost when starting with the infinite-U Hubbard model, however. Thus, not a single-band Hubbard model but a single-band *s*-*d* model with antiferromagnetic exchange amplitude is the adequate starting point for a realistic treatment of CuO_2 superconductivity.

"Dogma IV: The 'normal' metal above T_c... is not a Fermi liquid... but retains a Fermi surface satisfying Luttinger's theorem at least in the highest-T_c materials. We call this a Luttinger Liquid."

Very recently, the crucial experiment has finally been conducted. After 15 years of intensive investigations of the cuprates it is now experimentally established [242] that the overdoped cuprates obey the 150-years-old Wiedemann–Franz law within a remarkable 1% accuracy. After this experimental clarification the theoretical comprehension will hardly keep us waiting long. This experiment has also solved the old problem of the nature of charge carriers created by doping of a single Mott-Hubbard band, cf. *Dogma II*. Now we know that charge carriers of the normal state are standard Landau quasi-particles [243, 244] for which we have conventional Cooper pairing in the superconducting phase. "Holons", "spinons" and spin-charge separation are unlikely to occur and behave so as to emulate the properties of the ideal Fermi gas. As a function of the hole doping per Cu atom, $\tilde{p}$, the critical temperature is a smooth parabola [245],

$$T_c/T_c^{\max} \approx 1 - 82.6(\tilde{p} - 0.16)^2. \tag{2.48}$$

Thus, it is improbable that the nature of the carriers and pairing mechanism can be dramatically changed in the optimal and underdoped regimes although a number of new and interesting phenomena complicate the physics of the underdoped cuprates.

In short, in our opinion the experimental validation of the Wiedemann-Franz law in overdoped cuprates [242] is a triumph of the Landau [243, 244] and Migdal concept of Fermi quasi-particles (and Landau spirit of *trivialism* in general) and provides a refutation of the spin-charge separation in cuprates [246]. Hence, the problem of deriving the Wiedemann-Franz law for strongly correlated electrons in the CuO_2 plane has just been set in the agenda. According to the Fermi liquid theory [247] interactions between the particles create an effective self-consistent Hamiltonian. As Kadanoff [248] has pointed out, this idea was much developed by Landau [249] and Anderson [250]. Unfortunately, for high-T_c cuprates a link is still missing between the Landau quasiparticle concept and the one due to Slater that even scattering matrix elements can be calculated from first principles.

"Dogma V: Nonetheless, enough directions have been probed to indicate strongly that this odd-even splitting of CuO_2 planar states does not exist. ... The impact of Dogma V, then, is that the two-dimensional state has separation of charge and spin into excitations which are meaningful only within their two-dimensional substrate; to hop coherently as an electron to another plane is not possible, since the electron is a composite object, not an elementary excitation."

Within the single-particle approximation (Sec. 2.2) the bilayer band splitting is readily obtained from Eq. (1.8) and Eq. (2.40) by the replacements

$$\epsilon_i \to \epsilon_i \pm t_{\perp,ii}, \quad i = s,\, p,\, d, \tag{2.49}$$

where $t_{\perp,ii}$ is the hopping amplitude between the ith orbitals in the adjacent CuO_2 planes. In other words, the two constant energy curves due to the bilayer splitting are described by the same equation Eq. (1.8). Since it is plausible that $t_{\perp,ss}$ dominates, from Eq. (2.32) one finds

$$\Delta E_{\mathrm{bilayer}} \approx 2t_{\perp,ss}|S_{3,\mathbf{p}}|^2 \approx 22\ \mathrm{meV}\ (\cos p_x - \cos p_y)^2, \tag{2.50}$$

in agreement with references [27, 251]. The numerical value of 22 meV has been reported for heavily overdoped $Bi_2Sr_2CaCu_2O_{8+\delta}$ (BSCCO) [251]. This experiment, crucial for *Dogma V*, cf. reference [252], is another piece of evidence in favor of the conventional behaviour of the electron excitations in the $(CuO_2)_2$ slab. Since $\Delta E_{\mathrm{bilayer}}$ is relatively small in comparison with the width of the conduction band, it is another hint that even for bilayer superconductors like BSCCO and $YBa_2Cu_3O_{7-\delta}$ (YBCO) the analysis of a single CuO_2 plane is an acceptable initial approximation. Similarly, for fitting the three-dimensional Fermi surface of $Tl_2Ba_2CuO_{6+\delta}$ determined by angle magnetoresistance oscillations [253] one can start with the simplest possible tight-binding approximation

$$\epsilon_i \rightarrow \epsilon_i + t_{\perp,ii} \cos p_z, \quad i = s,\, p,\, d. \tag{2.51}$$

"Dogma VI: Interlayer hopping together with the "confinement" of Dogma V is either the mechanism of or at least a major contributor to superconducting condensation energy."

The interlayer hopping which is understood as a single-electron process definitely cannot be considered as a two-electron pairing interaction creating the condensation energy. It is only one of the details when one concentrates on the material-specific effects in high-T_c superconductors. The inter-slab hopping between double $(CuO_2)_2$ layers is a coherent Josephson tunneling responsible for the long-living plasma oscillations with frequency $\omega_{pl} < \Delta$. These plasma oscillations along with far infrared transparency of the superconducting phase were theoretically predicted [4] for BSCCO — one of the few predictions made for high-T_c cuprates, cf. the postdiction [254]. After the experimental observation [255], the plasma resonances associated with the Cooper-pair motion soon turned into a broad research field [256]. Subgap plasmons were predicted [5] for conventional superconducting thin films as well, and shortly after experimentally confirmed [257, 258] for thin Al films on $SrTiO_3$ substrate. The relatively lagged development of the physics of this effect was partially due to the false neglect of the longitudinal current response in the classical works on microscopic theory. Concluding, let us note that the London penetration depth λ can be considered as the Compton wave length of the Higgs boson of mass $m_H c^2 = \hbar\omega_{pl}$, but the overall contribution of the interlayer hopping to the condensation energy is negligible.

2.6.2 *Discussion*

The band structure of the CuO_2 plane is now believed to be understood. However, after 15 years of development a mismatch of a factor of two or three between the *ab initio* and the experimental spectroscopic estimates for the single-electron hopping amplitudes t, or the bandwidth, tends to be interpreted rather as a state-of-the-art "coincidence". The Heitler–London approach is well-known in quantum chemistry [259–261], and has been successfully used for a long time in the physics of magnetism [262]. We hope that realistic first-principles calculations aiming at the exchange integrals J of the CuO_2 plane can be easily carried out. Should they validate the correct (antiferromagnetic) sign and the correct order of magnitude of J_{sd}, we can consider the theory of high-T_c superconductivity established. We stress that the two-electron exchange, analyzed here, is completely different from the double exchange considered in reference [263, 264].[2]

In order to compare the derived results with the experiment, it is necessary that the tight-binding conduction band energy be fitted to the available ARPES data. In doing so a few parameters have to be properly taken into account: the Fermi energy E_F, as determined from the total area of the hole Fermi contour, the difference between the Fermi energy and the Van Hove singularity, $E_F - \epsilon(\pi, 0)$, the difference between the Van Hove singular point and the bottom of the conduction band at the Γ point, $\epsilon(\pi, 0) - \epsilon(0, 0)$. If the superconducting gap has a B_{1g}-type symmetry, its maximum value along the Fermi contour, $\Delta_{max} = \max|\Delta_{\mathbf{p}}(T = 0)|$, determines the J_{sd} exchange integral in the *s*-*d* model. Thus, the temperature dependence of the gap, described by the function $\Xi(T)$, and the overall thermodynamic behaviour and low frequency electrodynamic response will be determined without free fitting parameters.

The fit to the extended Van Hove singularity as observed, e.g., in Ref. [44] also points to a relatively low-lying Cu4*s* level and one needs to consider the minimum value of ϵ_s. Although the 4*s* band is completely empty (Cu4*s* level is above the Fermi level), a very close location is not "harmless" and would necessarily lead to some prediction for the optical behaviour. With some risk of opening the Pandora's box, we should mention that the lowest position of the Cu4*s* level is determined by the mid-infrared (MIR) response. According to this possible interpretation, the broadly discussed maximum of the absorption in the MIR range is due to 3*d*-4*s* interband transition: one electron in the conduction band is excited by the light to the empty Cu4*s* band. It seems that, up to now, there is no natural explanation of this MIR optical adsorbtion (for a review see [81]).

The derived gap anisotropy function (2.31) and its interpolation (2.33) compared to the ARPES experiment showed that the "standard" four-band model spanned on the Cu3$d_{x^2-y^2}$, Cu4*s*, O2p_x, and $O2p_y$ orbitals, with an antiferromagnetic *s*-*d* pairing interaction, successfully describes the main

---

[2]On p. 27 in [263], de Gennes writes: "Indeed one can generate an interesting attraction electron-electron coupling via magnetic excitations. But the double exchange picture predicts a ferromagnetic state at strong doping, which has never been seen in the cuprates – hence this idea failed". And further: "A third group, led by P. Anderson, believes that electrons in the cuprate form a 'Luttinger liquid' which is not a simple Fermi liquid. I like the aesthetics of this approach, but I do not have a deep understanding of it".

features of the ARPES data: the rounded-square-shaped Fermi surface, small energy dispersion along the $(0,0)$-$(2\pi,0)$ line, and the d-type (B_{1g}) symmetry of the energy gap $\Delta_{\mathbf{p}}$ along the Fermi contour. According to the pairing scenario proposed here, strong electron correlations "drive" the electron exchange amplitudes. These inter- and intra-atomic processes occur on energy scales unusually large for solid state physics. However, the subsequent treatment of the lattice Hamiltonian can be performed completely within the framework of the traditional BCS theory. The criterion for applicability of the BCS scheme is not given by the J versus t, but rather by the $T_{\rm c}$ versus $E_{\rm F} - \epsilon_{\rm b}$ relation. Taking into account the typical ARPES-derived bandwidths, which are much bigger than $T_{\rm c}$ we come to the conclusion that the BCS trial wave function [265] is applicable for the description of superconductivity in the layered cuprates with an acceptable accuracy if $T_{\rm c}$ does not significantly exceed room temperature.

It is worth adding also a few remarks on the normal properties of the layered cuprates. Among all debated issues in the complex physics of the cuprates, the most important one is perhaps that of the normal-phase kinetics. The long-standing problem is whether the paring interaction dominates (or totally determines) the mechanism of Ohmic resistance in the normal phase, as is the case for conventional superconductors. Within the present theory this question can be formulated as follows: does the s-d exchange interaction dominate the scattering of the normal-state charge carriers above $T_{\rm c}$? This is a solvable kinetic problem whose rigorous treatment will be given elsewhere. Here we shall restrain ourselves in providing only a qualitative discussion.

In electron-electron scattering, just like in traffic accidents, the crucial effect comes from the backscattering in "head-on" collisions. For backscattering (i.e., $\vartheta = \pi$) in the case of s-d interaction, it turns out that the matrix elements entering the pairing amplitude are also important. It can be easily realized that this amplitude vanishes along the diagonals of the Brillouin zone $(0,0)$-(π,π) (the Γ-M direction). Thereby the cold spots on the Fermi contour correspond to the zeros of $\Delta_{\mathbf{p}}$. And vice versa, the hot spots are associated with a maximum gap along (π,π)–$(0,\pi)$ (the M-X direction). In this sense, cuprates repeat the qualitative feature of the conventional superconductors, with a maximal gap corresponding to maximal scattering on the Fermi surface.

All layered cuprates are strongly anisotropic and two-dimensional models give a reasonable starting point to analyze the related electronic processes. Most importantly, the picture of a layered metal brings in something qualitatively new which does not exist for a bulk metal — the "interstitial" electric field between the layers, like the one in any plane capacitor. The thermodynamic fluctuations of this electric field and related fluctuations of the electric potential and charge density constitute an intensive scattering mechanism analogous to the blue-sky mechanism of light scattering by density fluctuations. It has recently been demonstrated [266] that the experimentally observed linear resistance can be rationalized in terms of the plane capacitor scenario; density fluctuations in the layered conductors are more important than the nature of the interaction. In such a way the linear normal-state resistivity is an intrinsic property [266] of the "layered" electron gas and cannot be used as an argument in favor of non-Fermi-liquid behavior. The resistance of the normal phase may not be directly related to the pairing mechanism and these problems can be solved separately. Nevertheless it will be interesting to check whether the anisotropic scattering in cuprates [267–270] can be explained within the framework of the *s*-*d* pairing Hamiltonian.

The present theory can also predict a significant isotope effect in the cuprates. Even though the J_{sd} pairing amplitude does not depend on the atomic mass, the charge carriers reside the ionic CuO_2 2D lattice, thereby rendering polaron effects, as in any ionic crystal, possible. For the lighter oxygen isotope the lattice polarization is more pronounced, leading to enhanced effective mass and density of states, and reducing the transfer integrals. Overall, the isotope effect in the CuO_2 plane is due to the isotope effect of the density of states. This rationale can be quantitatively substantiated. In our theory, upon isotope substitution at $T = 0$, e.g., the change of the penetration depth would be mainly driven by the Cooper pair effective mass that could be determined by means of the Bernoulli effect. At that the superfluid density remains unchanged. The calculation of the isotope effect on T_c requires an evaluation of the polaron effects on the conduction band. Although this is a feasible problem, it is beyond the scope of the present work.

The proposed mechanism for pairing in the CuO_2 plane can be handled much like an "Alice-in-Wonderland" toy-model, but we find it fascinating that all ingredients of our theory are achievements of quantum mechanics dating back to the memorable 1920s, presently described in every physics textbook, and constituting the fundamentals of solid state physics [271,272]. It would be worthwhile attempting to apply the approach, used in this chapter, for modeling triplet and heavy-fermion superconductivity as well. It is tempting to speculate about the relevance of the *s*-*d* exchange to the pairing mechanism even of the iron-based superconductors [273]. The discussion so far points to a simple recipe for testing the viability of such a "speculation" — apply the formalism of this and previous chapters to the relevant states and analyze the corresponding gap anisotropy function on the Fermi surface.

2.6.3 The reason for the success of the CuO_2 plane

We find it very instructive to analyze qualitatively the reasons for the success of the realization of high-T_c superconductivity in the CuO_2 plane:

(i) Because of the relatively narrow quasi-two-dimensional conduction d-band, due to p-d hybridization, the density of states is rather high. The wide s-band resulting from s-p hybridization is completely empty, which is somewhat unusual for compounds containing transition ions.

(ii) The pairing s-d exchange process was known since the first years of quantum physics. It is omnipresent in the physics of the transition ions but in order for it to become the pairing mechanism in perovskites it is necessary that the s- and d-levels be close. In other words, a virtual population of the s-level is at least needed in order to make the J_{sd} amplitude operative. Indeed, the conduction d-band is, actually, a result of the s-p-d hybridization in the two-dimensional CuO_2 plane.

With the above remarks, one can speculate that among the perovskites the layered ones are more favorable for achieving higher T_c(see discussion in the next subsection). The transition ion series ends with Cu^{2+} and the $Cu3d_{x^2-y^2}$ and $4s$ levels are too close. One should keep in mind that the filling of the electron shells finishes with a "robbery" in Cu [169]: $3d^{10}4s^1$ instead of $3d^9 4s^2$ as one could expect from the electron configuration of the Ni atom ($3d^8 4s^2$). However, the energy difference between these two Cu shell configurations is very small.

Another favorable factor is the proximity of the $O2p$ and $Cu3d_{x^2-y^2}$ levels. Thus, *post factum* the success of Cu and O looks quite deterministic: the CuO_2 plane is a tool to realize a narrow d-band with a strong s-p-d hybridization. It was mentioned earlier that J_{sd} is one of the largest exchange amplitudes, but the $4s$ and $3d$ orbitals are orthogonal and necessarily require an intermediary whose role is played by the $O2p$ orbital. Hence this theory can be nicknamed "*the 3d-to-4s-by-2p highway to super-*

conductivity" [98]. The J_{sd} amplitude is omnipresent for all transition ion compounds, the hybridization of $3d$, $4s$ and $2p$ is however specific only for the CuO_2 plane.

How this qualitative picture can be employed to predict new superconducting compounds is difficult to assess immediately. We believe, however, that this picture, working well for the overdoped regime, is robust enough against the inclusion of all the accessories inherent to the physics of optimally doped and underdoped cuprates: cohabitation of superconductivity and magnetism [237, 238], stripes [274], pseudo-gap [275, 276], interplay of magnetism and superconductivity at individual impurity atoms [277], apex oxygen, CuO_2 plane dimpling, doping in chains [278], the 41 meV resonance [279], etc. Perhaps some of these ingredients can be used in the analysis of triplet superconductivity in the copper-free layered perovskite Sr_2RuO_4 [280]. It is also likely that the superconductivity of the RuO_2 plane is a manifestation of a ferromagnetic exchange integral J. The two-electron exchange mediates superconductivity and magnetism in heavy Fermion compounds [281–283] as well. We suppose that lattice models similar to the approach here will be of use in revealing the electronic processes in these interesting materials. Two-electron exchange may even contribute to the 30 K T_c of the cubic perovskite $Ba_{0.6}K_{0.4}BiO_3$ but so far it is difficult to separate the exchange contribution from the phonon part of the pairing interaction. However, the strange doping behaviour of $Tl_2Ba_2CuO_{6\pm\delta}$ in comparison with YBCO requires more detailed investigation [86].

2.6.4 T_c–ϵ_s correlations: a crucial test for the pairing mechanism in cuprates

In a study by Pavarini *et al.* [284], a strong correlation is observed, Fig. 2.5, between $T_{c\,\max}$ and a single parameter

$$s(\epsilon) = (\epsilon_s - \epsilon)(\epsilon - \epsilon_p)/(2t_{sp})^2. \tag{2.52}$$

It is unfortunate that theorists have not so far paid any attention to this observation because it is an important correlation between the *ab initio* calculated parameter $r = \frac{1}{2}/(1+s)$ and the experimentally measured $T_{c\,\max}$ which can reveal the subtle link between the experiment and the theory and finally solve the long-standing puzzle of the mechanism of HTS. The significance of this correlation to the physics of HTS was, however, emphasized in a handbook on high-T_c superconductivity edited by Shrieffer and Brooks [2] (see Fig. 3.18 on p. 105 and discussion on p. 531 therein).

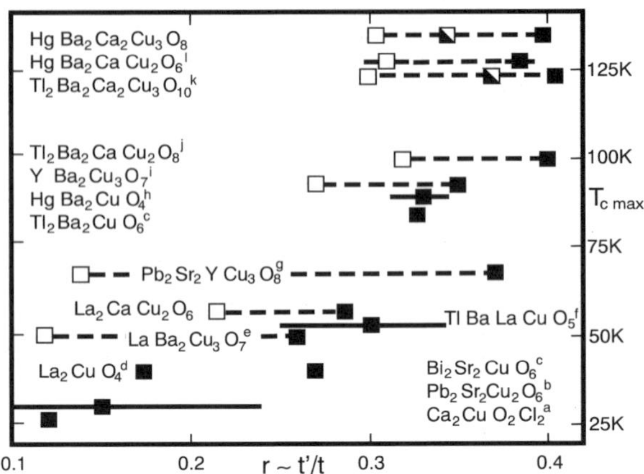

Fig. 2.5 Correlation between calculated r and observed $T_{c\ max}$ (reprinted with permission from [284]; Copyright © 2001 by the American Physical Society). □ – most antibonding subband. ■ – single-layer materials and most bonding subband for multilayers. Half-filled square – nonbonding subband. The explanation of this correlation is the crucial test for the theory of HTS.

Here we shall emphasize that the missing link, in fact, has already been found [285], and the work by Pavarini *et al.* can be used as a crucial test for theoretical models of HTS. Perhaps the simplest possible interpretation, though one could search for alternatives, is given within the framework of the present theory. In order for the Schubin–Zener–Kondo exchange amplitude J_{sd} to operate as a pairing interaction of the charge carriers, the Cu4s orbital needs to be significantly hybridized with the conduction band. The degree of this hybridization depends strongly on the proximity of the Cu4s level to the Fermi level ϵ_F. Thus, it is not surprising that ϵ_s controls the maximal critical temperature $T_{c\ max}$, being the only parameter of the CuO_2 plane which is essentially changed for different cuprate superconductors.

Cu3d and Cu4s are orthogonal orbitals and their hybridization is indirect. First, the Cu3$d_{x^2-y^2}$ orbital hybridizes with the O2p_x and O2p_y orbitals, then the O2p orbitals hybridize with Cu4s. As a result we have a "3d-to-4s-by-2p" hybridization of the conduction band of HTS cuprates which makes it possible the strong *antiferromagnetic* amplitude J_{sd} to create pairing in a relatively narrow Cu3$d_{x^2-y^2}$ conduction band. The hybridization "filling" of the Cu4s orbitals can be seen in cluster calculations as well [286]. The $s(\epsilon)$ parameter introduced in Ref. [284] reflects the proximity of all 3 levels in the generic 4-band Hamiltonian of the CuO_2 plane—their "random coincidence" for the Cu-O combination. Suppose that those levels are not so close to each other. In this case the slightly modified parameter

$$\tilde{s}(\epsilon) = (\epsilon_s - \epsilon)(\epsilon - \epsilon_p)/(4t_{sp}t_{pd}) \equiv -\varepsilon_s\varepsilon_p/(4t_{sp}t_{pd}) \tag{2.53}$$

is simply the energy denominator of the perturbation theory which describes the hybridization filling of the axial orbital, see (2.32). Whence $[\tilde{s}(\epsilon)]^2$ is a denominator of the pairing amplitude in the BCS equation (2.28). Hence, we conclude that the correlations reported in Ref. [284] *are simply the correlations between the critical temperature* T_c *and the dimensionless BCS coupling constant* $\rho(\epsilon_\mathrm{F})J_{sd}/[\tilde{s}(\epsilon_\mathrm{F})]^2$. Of course, for coupling constants ~ 1 the BCS trial wavefunction can be used only for qualitative estimates, but knowing the Hamiltonian the mathematical problem may somehow be solved. In any case, even qualitatively, we are sure that the stronger pairing amplitude J_{sd} and hybridization $1/\tilde{s}(\epsilon_\mathrm{F})$ enhance T_c.

With first-principles electronic structure calculations available for many cuprates, it is worthwhile performing a LCAO fit to them [27] and using experimental values of T_c to extract the pairing amplitude J_{sd} for all those compounds. The *ab initio* calculation of the Kondo scattering amplitude parameterized by J_{sd} is an important problem which has to be set in the agenda of computational solid state physics. We expect that it will be a weakly material dependent parameter of the order of the s-d exchange amplitude in Kondo alloys, but perhaps slightly bigger as for the Cu ion the $3d$ and $4s$ levels are closer compared to many other ions. Closer energy levels, from classical point of view, imply closer classical periods of orbital motion which leads to some "resonance" enhancement of the exchange amplitude due to intra-atomic two-electron correlations.

The final qualitative conclusion that can be extracted from the correlations reported by Pavarini *et al.* is the explanation why only the CuO_2 plane renders HTS possible, whereas hundreds other similar compounds are not even superconducting, or have only a "conventional" value of T_c. The natural explanation is: because its s-parameter is not small enough below its critical value. Even among the cuprates one can find compounds with "conventional" values of T_c having relatively large value of the s-parameter. For other transition metal compounds the parameter

$$\tilde{s}(\epsilon_d) = (\epsilon_s - \epsilon_d)(\epsilon_d - \epsilon_p)/(4t_{sp}t_{pd}) \tag{2.54}$$

is much bigger than its critical value s_c which can be reached probably only for the Cu-O combination. Thereby, the correlation reported by Pavarini *et al.* is a crucial test which of the models for HTS is still on the arena. We also note that the $4s$ hybridization is responsible for the three-dimensional coherent Fermi surface of $Tl_2Ba_2CuO_{6+\delta}$ [287].

2.6.5 *Perspectives: if "Tomorrow" comes...*

The technological success in preparing the second generation of high-T_c superconducting cables by depositing thin-layer superconducting ceramics on a flexible low-cost metallic substrate is crucial for the future energy applications. The USA Department of Energy suggests global superconducting energy products would command an annual market of 30 G$ by about 2020. High-T_c superconductor power cables, transformers, motors and generators could grab a 50% market share by 2013, 2015, 2016 and 2021, respectively [288]. On the other hand atomic-layer engineering of superconducting oxides will trigger progress in materials science and electronics. One can envision multi-functional all-oxide electronics, e.g., sensors, processing and memory devices, all monolithically integrated within a single chip [289]. In spite of the technological progress and tens of thousands of publications the theoretical "*picture in early 2000 remains fairly cloudy on the whole*" [72]. The landmark of "Today" must be some event. "What event will it be? It is desirable that this landmark be the insight into the mechanism of superconductivity in high-T_c cuprates" [72].

In this chapter we presented a traditional theory for superconductivity in overdoped, and possibly also optimally doped cuprates. All of its ingredients can be found in the textbooks and there is a considerable chance that we witness the victory of traditionalism, as it was in the history of quantum electrodynamics (QED) half a century ago, but it may well be just a personal viewpoint "*brainwashed by Feynman*" [290]. Nonetheless, let us use the example of QED to illustrate the essence of our contribution. QED appeared as a synthesis between perturbation theory and relativity. Both components had been known well before the QED conception. Similarly, both the BCS theory and the exchange interaction have been known for ages, so the point in the agenda was how to conceive out of them the theory of high-T_c cuprates. Such a theory contains necessarily a big number of energy parameters (E_F, ϵ_s, ϵ_p, ϵ_d, t_{sp}, t_{pp}, t_{pd}, J_{sd}, J_{pd}, J_{sp}, J_{pp}) which are difficult to determine simultaneously[3] (for the current status of the problem see for example Ref. [291,292]). The first step will definitely be

[3]The gap-anisotropy fit in Fig. 2.3 (d) is quite robust against the choice of the parameters. To illustrate and emphasize the capability of the model we have used, for example, unrealistically big values of the hopping integrals: $\varepsilon_d = 0$, $\varepsilon_s = 5$, $\varepsilon_p = -0.9$, $t_{pd} = 1.13$, $t_{sp} = 1.63$, $t_{pp} = 0.2$ eV. This set of parameters corresponds to band calculations but gives a factor 2–3 wider conduction band. If the band is fitted to the ARPES data J_{sd} can be less than 1 eV. A realistic fit is deemed to be a subject of a collaboration with experimentalists.

to use ARPES data in which the spectrum is clearly seen and to neglect in a first approximation the "irrelevant" inter-atomic exchange integrals J_{pd}, J_{sp} and J_{pp}. In this case, for a known normal spectrum one can determine J_{sd} from T_c or from the maximum gap at $T = 0$.

Having a big variety of calculated variables the parameters of the theory can be reliably fitted. Another research direction is the first-principles calculation of the transfer amplitudes and two-electron exchange integrals. The level of agreement with the fitted values will be indicative for the completeness of our understanding. In addressing more realistic problems, the properties of a single space-homogeneous CuO_2 plane will be a reasonable starting point. Concluding, we believe that there is a true perspective for the theoretical physics of cuprate superconductors to become an important ingredient of their materials science.

Magnetism and superconductivity are among the most important collective phenomena in condensed matter physics. And, remarkably, magnetism of transition metals and high-T_c superconductivity of cuprates seem to be two faces of the same ubiquitous two-electron exchange amplitude.

Chapter 3

Specific heat and penetration depth

3.1 Specific heat

Virtually all recently studied superconductors exhibit considerable anisotropy of the superconducting gap $\Delta_p(T)$ over the Fermi surface $\varepsilon_p = E_{\mathrm{F}}$. Despite the strong coupling effects and influence of disorder, which are all essential as a rule, for a qualitative analysis it is particularly useful to start with the weak-coupling BCS approximation for clean superconductors. In this case, very often model factorizable pairing potentials give an acceptable accuracy for the preliminary analysis of the experimental data.

The aim of the present chapter is twofold. Firstly, we shall derive an explicit interpolation formula for the temperature dependence of the specific heat $C(T)$. The formula is formally exact for factorizable pairing kernels which are consequence of the approximative separation in superconducting order parameter derived in BCS weak-coupling approximation by Pokrovskii [179,293]. Our formula reproduces the specific heat jump derived by Pokrovskii [179,293] for arbitrary weak coupling kernels and Gor'kov and Melik-Barkhudarov [294] results for the Ginzburg–Landau (GL) coefficients of an anisotropic superconductor. That is why we believe that the suggested formula can be useful for the analysis of experimental data when only gap anisotropy and band structure are known. Secondly, within the same system of notions and notation we present the recent results by Kogan [295] for the penetration depth $\lambda(T)$, and propose for the zero-scattering case new formulas which may be used for experimental data processing.

We begin with the entropy of a Fermi system per unit volume divided by the Boltzmann's constant k_{B}

$$S(T) = -2\,\overline{n_p \ln n_p + (1-n_p)\ln(1-n_p)}, \tag{3.1}$$

where the factor 2 takes into account the spin degeneracy and the overline denotes integration over the D-dimensional momentum space

$$\overline{f_p} = \int_{-\infty}^{\infty} \cdots \int_{-\infty}^{\infty} \frac{d^D p}{(2\pi\hbar)^D} f(\mathbf{p}). \tag{3.2}$$

The Fermi filling factors of independent Fermions

$$n_p = \frac{1}{\exp(2z_p) + 1}, \qquad z_p \equiv \frac{E_p}{2T}, \tag{3.3}$$

where T is the temperature times k_{B}, are expressed by spectrum of superconductor

$$E_p = \sqrt{\xi_p^2 + |\Delta_p|^2}, \qquad \xi_p = \varepsilon_p - E_{\mathrm{F}}. \tag{3.4}$$

Here we have to emphasize that for a model factorizable pairing potential $V_{p,q} \propto \chi_p \chi_q$ the gap function is always separable as a product of a temperature dependent function which can be associated with the GL order parameter $\Xi(T)$ and a rigid temperature independent function of the momentum χ_p. The nontrivial results [293] is that this separation of the variables is asymptotically correct in the BCS weak-coupling limit for an arbitrary kernel which is generally non factorizable. In fact, a factorizable kernel is a fairly unnatural property which, however, can occur if the pairing interaction is local, intra-atomic and located in a single atom in the unit cell. This is the special case of the *s*-*d* interaction at the copper site(s) in the CuO_2 plane [60], considered in the previous chapter; The separability ansatz, though, shall be employed here to obtain a general interpolation formula formally exact for factorizable kernels. We assume that the gap anisotropy function χ_p is known, either as a result of solving the general BCS equation at T_c, inferred from experimental data processing, or merely postulated within some model Hamiltonian, which is often the case for the high-temperature and exotic superconductors.

With the above remarks, we will derive $C(T)$ for the separable gap

$$\Delta_p(T) = \Xi(T)\chi_p \tag{3.5}$$

and a factorizable kernel [296]. We apply the ansatz (3.5) to the BCS gap equation [75]

$$\Delta_p(T) = \int \frac{d^D q}{(2\pi\hbar)^D} V_{p,q} \frac{1 - 2n_q}{2E_q} \Delta_q(T), \tag{3.6}$$

and use the convention that a positive sign of $V_{p,q}$ corresponds to attraction of charge carriers and a negative potential energy of interaction. Substituting here

$$V_{p,q} \approx G\chi_p\chi_q \tag{3.7}$$

and introducing $\eta \equiv |\Xi|^2$ we obtain a transcendental equation for the temperature dependence of the gap $\Xi(T)$

$$GA(\eta, T) = 1, \qquad A(\eta, T) \equiv \overline{(\chi_p^2/2E_p)\tanh z_p}, \tag{3.8}$$

where we have used the identity $1-2n_p = \tanh z_p$ and the coupling constant is defined by $G \equiv 1/A(0, T_c)$. Details on the derivation of the trial function approximation Eq. (3.7) and the numerical solution of Eq. (3.8) for $T_c \ll \omega_D$ are given in Sec. 3.2.

For the specific heat of the superconducting phase per unit volume divided by k_B we have

$$C(T) = T d_T S(\eta(T), T) = 2\overline{E_p\, d_T n_p} = C_\nu + C_\Delta, \tag{3.9}$$

where $d_T = d/dT$. Here C_ν is the "normal" part of the specific heat

$$C_\nu(T) \equiv T(\partial_T S)_\eta = \frac{\pi^2}{3}\overline{g_c(z_p)}, \tag{3.10}$$

where

$$g_c(z) \equiv \frac{6}{\pi^2}\frac{z^2}{\cosh^2 z}, \qquad \int_{-\infty}^{\infty} g_c(z)dz = 1, \tag{3.11}$$

and $(\partial_T \ldots)_\eta$ is the temperature differentiation for constant order parameter. For zero order parameter, $\eta = 0$ at T_c and above, C_ν is just the specific heat of the normal phase $C_N(T) = C_\nu(T, \eta = 0)$.

Introducing

$$\alpha(\eta, T) \equiv -(\partial_T A)_\eta = -(\partial_\eta S)_T = \frac{\overline{\chi_p^2\, g_a(z_p)}}{2T^2}, \tag{3.12}$$

where

$$g_a(z) \equiv \frac{1}{2\cosh^2 z}, \qquad \int_{-\infty}^{\infty} g_a(z)dz = 1, \tag{3.13}$$

the other term of the specific heat

$$C_\Delta \equiv T\partial_\eta S(\eta, T) d_T \eta(T) \tag{3.14}$$

can be written as

$$C_\Delta = \alpha(\eta, T)[-d_T\eta(T)]\,\theta(T_c - T). \tag{3.15}$$

Eq. (3.12) is actually a Maxwell-type equation $\partial_\eta \partial_T F = \partial_T \partial_\eta F$, where F is the free energy: $S = -(\partial_T F)_\eta$, $A = -(\partial_\eta F)_T$; cf. Refs. [297–299].

Differentiating Eq. (3.8) we obtain $dA = 0$ and

$$-d_T\eta(T) = \left.\frac{(\partial_T A)_\eta}{(\partial_\eta A)_T}\right|_{\eta(T)} = \left.\frac{\alpha}{b}\right|_{\eta(T)}, \tag{3.16}$$

where the functions α and b represent a generalization of the GL coefficients for arbitrary temperature and order parameter

$$b(\eta, T) \equiv -(\partial_\eta A)_T = \frac{7\zeta(3)}{16\pi^2 T^3}\,\overline{\chi_p^4\, g_b(z_p)}, \tag{3.17}$$

$$g_b(z) \equiv \frac{\pi^2}{14\zeta(3)}\frac{1}{z^2}\left(\frac{\tanh z}{z} - \frac{1}{\cosh^2 z}\right), \tag{3.18}$$

$$\int_{-\infty}^{\infty} g_b(z)dz = 1, \tag{3.19}$$

and ζ is the Riemann zeta function. Then

$$C_\Delta = T\frac{\alpha^2}{b} = \frac{4\pi^2}{7\zeta(3)}\,\theta(T_c - T)\left[\overline{\chi_p^2\, g_a(z_p)}\right]^2 / \;\overline{\chi_p^4\, g_b(z_p)} \tag{3.20}$$

and

$$\frac{C_\Delta}{C_\nu} = \frac{12}{7\zeta(3)}\,\theta(T_c - T)\left[\overline{\chi_p^2\, g_a(z_p)}\right]^2 / \left[\overline{\chi_p^4\, g_b(z_p)}\;\overline{g_c(z_p)}\right]. \tag{3.21}$$

The functions $g_i(z_p)$, $i = a, b, c$, introduced in Refs. [297–300], have sharp maximum at the Fermi surface and in a good approximation we have

$$\overline{\chi_p^n g_i(z_p)} \approx 2T\nu_F\,\langle\chi_p^n\, r_i\,(y_p)\rangle, \qquad y_p \equiv \frac{\Delta_p}{2T}, \tag{3.22}$$

where

$$r_i(y) \equiv \int_{-\infty}^{\infty} g_i(\sqrt{x^2+y^2})dx, \qquad x = \frac{\xi_p}{2T}, \tag{3.23}$$

$$r_i(0) = 1, \qquad r_i(\infty) = 0, \qquad i = a,\ b,\ c.$$

We define averaging over the Fermi surface

$$\langle f_p\rangle = \frac{\overline{f_p\,\delta(\xi_p)}}{\nu_F}, \qquad \nu_F = \nu(E_{\rm F}) = \overline{\delta(\xi_p)}, \tag{3.24}$$

where ν_F is the density of electron states per unit energy, volume and spin at the Fermi level. In such a way we obtain

$$C_\nu(T) = \frac{2}{3}\pi^2 T\nu_F\,\langle r_c(y_p)\rangle \tag{3.25}$$

and

$$\frac{C_\Delta}{C_\nu} = \frac{12}{7\zeta(3)}\,\frac{\langle\chi_p^2\, r_a(y_p)\rangle^2\;\;\theta(T_c - T)}{\langle\chi_p^4\, r_b(y_p)\rangle\;\;\langle r_c(y_p)\rangle}. \tag{3.26}$$

At T_c, where the gap is small and $r_i(0) = 1$ this formula gives the Pokrovskii [293] result for the reduced specific heat jump

$$\frac{\Delta C}{C_N(T_c)} = \frac{12}{7\zeta(3)}\,\frac{\langle\chi_p^2\rangle^2}{\langle\chi_p^4\rangle}. \tag{3.27}$$

For the GL coefficient Eq. (3.12) and Eq. (3.17) the approximation (3.22) gives

$$\alpha(\eta, T) = \frac{\nu_F}{T} \left\langle \chi_p^2 \, r_a \left(\Delta_p / 2T \right) \right\rangle , \tag{3.28}$$
$$b(\eta, T) = \frac{7\zeta(3)\nu_F}{8\pi^2 T^2} \left\langle \chi_p^4 \, r_b \left(\Delta_p / 2T \right) \right\rangle .$$

Then the specific heat takes the simple GL form for arbitrary temperatures

$$C(T) = C_\nu(\eta, T) + T \frac{\alpha^2(\eta, T)}{b(\eta, T)} \theta(T_c - T). \tag{3.29}$$

Here, for the functions on the right-hand side we have substituted the thermal equilibrium value of the order parameter $\eta(T) = |\Xi(T)|^2$, obtained from the solution of Eq. (3.8). This BCS formula (3.29) is an example how good the physical intuition was in the phenomenology of superconductivity. According to the Gorter-Casimir [301] model the specific heat is a sum of a "normal" part and another term, governed by the temperature dependence of the order parameter and having exactly the GL form. The Gorter-Casimir two fluid model has very simple physical grounds. In the self-consistent approximation, the entropy $S(T, \Delta(T))$ is a function of the temperature and a temperature dependent order parameter $\Delta(T)$. The temperature differentiation $C(T) = T(dS/dT)$ inevitably gives two terms in Eq. (3.9). According to the general idea by Landau [302], the order parameter is an adequate notion for description of second order phase transitions, regardless of the concrete particle dynamics. The ϵ-expansion by Wilson and Fisher is only an ingenious realization of the same Landau's idea when the influence of fluctuations is essential.

Again, at T_c the general formulas Eq. (3.28) give the Gor'kov and Melik-Barkhudarov [294] result for the GL coefficients

$$\alpha(0, T_c) = \frac{\nu_F}{T_c} \langle \chi_p^2 \rangle, \qquad b(0, T_c) = \frac{7\zeta(3)\nu_F}{8\pi^2 T_c^2} \langle \chi_p^4 \rangle. \tag{3.30}$$

This result can be directly derived [297] from the variational free energy $F(\eta, T)$ of the superconductor which close to T_c has the GL form

$$F_{GL}(\eta, T) \approx \alpha(0, T_c) \, (T - T_c) \, |\Xi|^2 + \frac{1}{2} b(0, T_c) \, |\Xi|^4. \tag{3.31}$$

The derivation of the Ginzburg–Landau coefficients for *disordered* anisotropic-gap superconductors is given in our work with Indekeu and Pokrovsky [303].

The simplest method to calculate the GL coefficients is to differentiate [297–299] the free energy after a u-v transformations $F(\eta, T) = \overline{H} - TS$. Then

$$\alpha(0, T_\mathrm{c}) = \left.\frac{\partial F}{\partial \eta}\right|_{\eta=0, T=T_\mathrm{c}}, \qquad b(0, T_\mathrm{c}) = \left.\frac{\partial^2 F}{\partial \eta^2}\right|_{\eta=0, T=T_\mathrm{c}}. \tag{3.32}$$

If a Van Hove singularity (VHS) is close to the Fermi level the formulas for GL coefficients are slightly modified [300]

$$\begin{aligned} \alpha(0, T_\mathrm{c}) &= \frac{\langle \chi_p^2 \rangle}{T_\mathrm{c}} \int_{-\infty}^{+\infty} \nu(E_\mathrm{F} + 2T_\mathrm{c}\, x)\, g_a(x)\, dx, \\ b(0, T_\mathrm{c}) &= \frac{7\zeta(3)\langle \chi_p^4 \rangle}{8\pi^2 T_\mathrm{c}^2} \int_{-\infty}^{+\infty} \nu(E_\mathrm{F} + 2T_\mathrm{c}\, x)\, g_b(x)\, dx, \\ C_\nu(T_\mathrm{c}) &= \frac{2}{3}\pi^2 T_\mathrm{c} \int_{-\infty}^{+\infty} \nu(E_\mathrm{F} + 2T_\mathrm{c}\, x)\, g_c(x)\, dx. \end{aligned} \tag{3.33}$$

Some important references on the influence of the VHS on the properties of superconductors, and pioneering works on the two-band model are given in Ref. [300]. Let us evaluate the upper limit which can give a VHS. Let us take 1D density of states $\nu(E) \propto 1/\sqrt{E - E_\mathrm{VHS}}$ and $E_\mathrm{F} = E_\mathrm{VHS} = 0$; there is no doubt that this mathematical illustration is unphysical. In this case we have for the reduced specific heat jump $\Delta C / C_N(T_\mathrm{c})$, Eq. (3.27), an additional factor

$$\frac{\left[\int_0^\infty g_a(\tilde{x}^2) d\tilde{x}\right]^2}{\int_0^\infty g_c(\tilde{x}^2) d\tilde{x} \int_0^\infty g_b(\tilde{x}^2) d\tilde{x}} = 2.51, \quad \tilde{x} \propto \sqrt{E}. \tag{3.34}$$

Although this mathematical example is not realistic, it can be seen that the VHS emulates qualitatively strong coupling corrections to the BCS theory: an enhancement of $\Delta C / C_N(T_\mathrm{c})$ and $2\Delta_\mathrm{max}(0)/T_\mathrm{c}$. Another simulation of strong coupling effects can be demonstrated by simple model density of states, corresponding to the case of layered cuprates

$$\nu(\xi) = 1 + k \ln \frac{1}{|\xi - E_\mathrm{VHS}|}. \tag{3.35}$$

For illustration, we solve the equation

$$\int_{-\omega_\mathrm{D}}^{\omega_\mathrm{D}} \frac{\tanh(\sqrt{\xi^2 + \Delta^2(T)}/2T)}{2\sqrt{\xi^2 + \Delta^2(T)}}\, \nu(\xi)\, d\xi = G^{-1} \tag{3.36}$$

taking $\omega_\mathrm{D} = 10$, $G = 1/2$, and $k = 10$. The $Z \equiv (2\Delta(0)/T_\mathrm{c})/(2\pi/\gamma)$ versus $E_\mathrm{VHS}/T_\mathrm{c}$ plot is given in Fig. 3.1. It can be seen that 7% enhancement

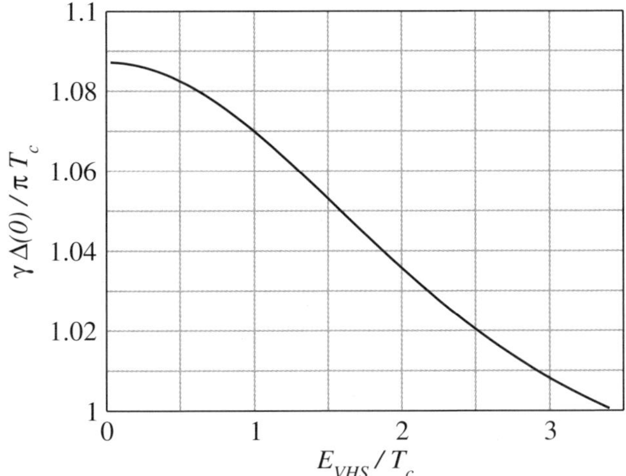

Fig. 3.1 $Z \equiv (2\Delta(0)/T_c)/(2\pi/\gamma)$ versus E_{VHS}/T_c computed for the model density of states Eq. (3.35). Note that 7% enhancement corresponds to $E_{VHS} = T_c$ and the maximum enhancement is $\approx 9\%$.

corresponds to $E_{VHS} = T_c$. Thus, the influence of the VHS on the specific heat is much stronger than on the $\Delta(0)/T_c$ ratio.

Let us also recall the general GL formula for the specific heat jump at T_c

$$\Delta C = T_c \frac{a^2(0, T_c)}{b(0, T_c)}. \tag{3.37}$$

The two-band model provides probably the simplest possible illustration of the derived formula for the specific heat; for pioneering references on the two-band model see Ref. [300]. The model is applicable with a remarkable accuracy [304] to MgB_2—a material which is in the limelight in the physics of high-T_c superconductivity over the past years.

For the normal specific heat we have

$$C_\nu(T) = \frac{2}{3}\pi^2 T \nu_F \left[c_1 r_c(y_1) + c_2 r_c(y_2)\right], \tag{3.38}$$

where

$$y_1 = \frac{\Delta_1}{2T}, \qquad y_2 = \frac{\Delta_2}{2T}, \qquad c_1 + c_2 = 1, \tag{3.39}$$

and $c_1\nu_F$ and $c_2\nu_F$ are the densities of states for the 2 bands of the superconductor. Above T_c or in the case of strong magnetic fields $B > B_{c2}$ we have

$$C_N(T) = \frac{2}{3}\pi^2 \nu_F T. \tag{3.40}$$

As pointed out earlier, within the weak-coupling BCS approximation Pokrovskii [293] has proved the general separation of the variables Eq. (3.5) which for a two-band superconductor results in a weakly temperature dependent gap ratio $\delta = \Delta_1/\Delta_2 = \chi_1/\chi_2$. For MgB_2 determination of the two gaps has been carried out by directional point-contact spectroscopy [305] in single crystals. One can see that for model evaluations the temperature dependence of the gap ratio could be neglected.

For the moments of the gap we have

$$\langle \chi_p^n \, r_i \rangle = \frac{c_1 \delta^n r_i(y_1) + c_2 r_i(y_2)}{(c_1 \delta^2 + c_2)^{n/2}}, \qquad i = a, b, c, \tag{3.41}$$

where the normalization is irrelevant in further substitution in the GL coefficients. Finally for the second, GL-order-parameter term of the specific heat below the T_c we obtain

$$C_\Delta(T) = \frac{8\pi^2}{7\zeta(3)} \nu_F T \frac{\left[c_1 \delta^2 r_a(y_1) + c_2 r_a(y_2)\right]^2}{c_1 \delta^4 r_c(y_1) + c_2 r_c(y_2)}. \tag{3.42}$$

For the jump of the specific heat this formula reduces to the Moskalenko [306] result

$$\frac{\Delta C}{C_N(T_c)} = \frac{12}{7\zeta(3)} \frac{\left(c_1 \chi_1^2 + c_2 \chi_2^2\right)^2}{c_1 \chi_1^4 + c_2 \chi_2^4}, \tag{3.43}$$

which is, in fact, a special case of the Pokrovskii [293] formula Eq. (3.27) applied to the two-band model. For application of the two-band model to the specific heat of MgB_2 the reader is referred to Ref. [307].

The analysis of the specific heat for MgB_2 gives perhaps the best corroboration of the BCS results due to Pokrovskii [293] and Moskalenko [306]. Solving the Eliashberg equation and performing first-principle calculations for the specific heat of MgB_2 Golubov *et al.* [Ref. [308], Fig. 3] derived 65% reduction of the specific heat jump at T_c. On the other hand, Eqs. (3.27) (3.43), using the parameters from Ref. [308], gives $\langle \chi^2 \rangle^2 / \langle \chi^4 \rangle = 58\%$ reduction of the $\Delta C / C_N(T_c)$ ratio. The 7% difference between those two estimates is in the range of the experimental accuracy and the Eliashberg corrections to the BCS result is difficult to extracted. Unfortunately, the groups solving the Eliashberg equation have not compared their results to the classical results of the BCS theory for anisotropic superconductors [293] in order to analyze several percent strong-coupling corrections to the specific heat jump for MgB_2.

In the single band case $c_1 = 1$ and Eq. (3.42) gives a simple relation between the specific heat and the BCS isotropic gap

$$\frac{C(T)}{C_N(T)} = r_c(y) + \frac{12}{7\zeta(3)} \frac{r_a^2(y)}{r_b(y)}, \tag{3.44}$$

where $y(T) = \Delta(T)/2T$. For anisotropic superconductors, functions of the gap have to be averaged independently on the Fermi surface; this is the interpretation of the general formulas Eq. (3.28) and Eq. (3.29). Thus, we have the natural generalization

$$\frac{C(T)}{C_N(T)} = \langle r_c(y_p)\rangle + \frac{12}{7\zeta(3)} \frac{\langle \chi_p^2 r_a(y_p)\rangle^2}{\langle \chi_p^4 r_b(y_p)\rangle}, \tag{3.45}$$

where $y_p(T) = \Delta_p(T)/2T = \chi_p\,\Xi(T)/2T$.

For illustration, we now apply this general formula to three typical cases and the results are shown in Fig. 3.2: (i) the isotropic-gap BCS model $\chi_p = 1$, familiar from a number of textbooks; [272, 309, 310] (ii) the two-dimensional (2D) d-wave superconductor $\chi_p = \cos 2\varphi$, $\tan\varphi = p_y/p_x$; and (iii) a two-band superconductor $c_1 = c_2 = 1/2$, for which the gap ratio parameter is taken to reproduce the same reduced specific heat jump of the d-wave superconductor ($\delta = \sqrt{3 \pm \sqrt{8}} = 2.41$ or 0.41).

The latter two models are often applied to analyze the behavior of CuO_2 or MgB_2 superconductors. Note also the qualitative difference: for a d-wave superconductor we have a quadratic specific heat at $T \ll T_c$, whereas

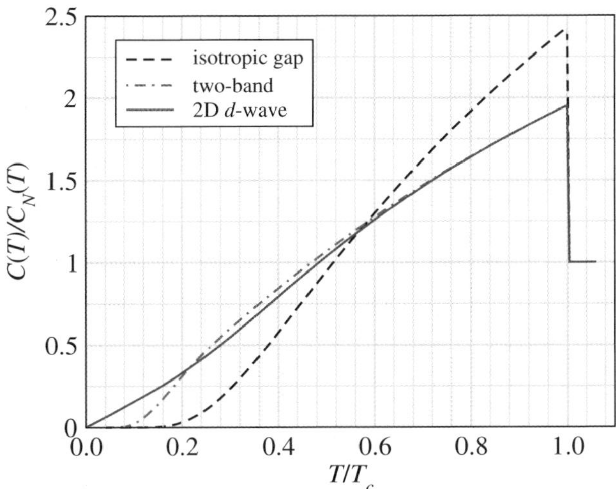

Fig. 3.2 Superconducting-to-normal specific heat ratio $C(T)/C_N(T)$ versus the reduced temperature $t = T/T_c$ according to Eq. (3.45) computed for: (i) an isotropic-gap BCS superconductor (dashed line), (ii) a two-band superconductor $c_1 = c_2 = 1/2$ with a gap ratio parameter $\delta = 2.41$ (dash-doted line) and (iii) 2D d-wave superconductor $\chi_p = \cos 2\varphi$, $\tan\varphi = p_x/p_y$ (solid line). Note that for $t > 0.2$ two of the curves would be experimentally indistinguishable.

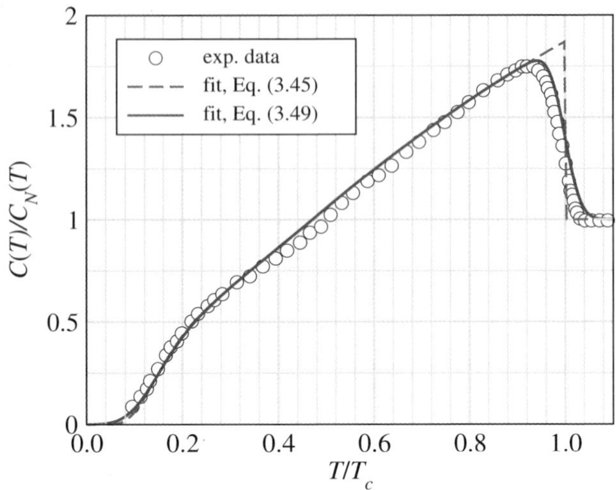

Fig. 3.3 Comparison between the superconducting-to-normal specific heat ratio $C(T)/C_N(T)$; the theoretical curve is computed following Ref. [307] with $c_1 = 0.49$, $\delta = 2.9$ (solid line) and the experimental data for MgB_2 are taken from Ref. [304] (circles). The theoretical curve is convoluted with a Gaussian kernel Eq. (3.49), chosen to fit best the experimental data ($\Delta t = 0.027$). The experimental data [304] are digitized from Ref. [307], Fig. 3.

for a two-band superconductor we have the exponential behavior $C(T) \propto \exp(-\Delta_2/2T)$; see also Fig. 3.3 below.

Consider now the low temperature behavior of the specific heat per unit area for a 2D d-wave superconductors. Close to a node the gap is proportional to the momentum component along the Fermi contour $\Delta_p(0) \approx v_\Delta p_l$. The corresponding superfluid velocity v_Δ is much smaller than the Fermi velocity v_F, which parameterizes the dependence of the normal excitations energy $\xi_p \approx v_F p_t$ as a function of the transversal to the Fermi contour momentum component. For the ground state quasiparticle spectrum we have $E_p \approx \sqrt{v_\Delta^2 p_l^2 + v_F^2 p_t^2}$. It is convenient to introduce the dimensionless variables $q_1 = v_\Delta p_l/2T$ and $q_2 = v_F p_t/2T$. In terms of the latter we have for the element of the area in momentum space

$$4\frac{dp_l dp_t}{(2\pi\hbar)^2} = 4\frac{(2T)^2}{v_\Delta v_F}\frac{2\pi q\, dq}{(2\pi\hbar)^2} = \frac{2EdE}{\pi\hbar^2 v_\Delta v_F}, \tag{3.46}$$

where $q = \sqrt{q_1^2 + q_2^2} = z_p = E_p/2T$, and for axial symmetric functions we can use polar coordinates; cf. Ref. [311]. Here we have taken into account

4 nodal points. In such a way Eq. (3.10) gives

$$C_\nu(T \ll T_c) = \frac{16}{\pi\hbar^2}\frac{T^2}{v_\Delta v_F}\int_0^\infty \frac{q^3\,dq}{\cosh^2 q} \approx 6.89\frac{T^2}{\hbar^2 v_\Delta v_F}, \tag{3.47}$$

where we used $18\zeta(3)/\pi \approx 6.89$; cf. Ref. [311], Eq. (2.9). This result together with Eq. (3.40) gives for the superconducting-to-normal specific heat ratio

$$\left.\frac{C_\nu}{C_N}\right|_{t\ll 1} = 1.047\,\frac{T_c}{\hbar^2\nu_F\,v_\Delta v_F}\,t, \tag{3.48}$$

where $t = T/T_c$ is the reduced temperature. The penetration depth has a similar linear low temperature behavior for d-wave superconductors.

Very often fluctuations of stoichiometry and crystal defects make the theory of homogeneous crystal inapplicable close to the critical region. Let $T_c(\mathbf{r})$ be a weakly fluctuating Gaussian field of the space vector $\mathbf{r}$. Hence, the simplest possible empirical model is to apply a Gaussian kernel to the theoretically calculated curve. Then for the heat capacity we have

$$C(t) = \int_{-\infty}^{+\infty} C_{\mathrm{theor}}(t)\,\exp\left\{-\frac{(t-t')^2}{2(\Delta t)^2}\right\}\frac{dt'}{\Delta t\sqrt{2\pi}}. \tag{3.49}$$

The philosophy of applying the convolution technique to all theoretical curves with singularities was advocated in the book by Migdal [312]. Such an empirically smeared curve with $\Delta t = 0.027$ describes better the experimental data for MgB_2 close to T_c; $T_c\Delta t \approx 1.1$ K, $B_{c2}(0) = 2.5$ T and $B_{c2}(0)\Delta t = 750$ G. The result is depicted at Fig. 3.3, where the smeared theoretical curve is compared with the experimental data [304]. In order to achieve a good fit of the theory to the experimental data we have treated c_1 and δ as fitting parameters (cf. Refs. [304, 307, 313, 314]). The values used $c_1 = 0.49$ and $\delta = 2.9$ are slightly different from the set of parameters used latter for computing the penetration depth, but are still in agreement with different spectroscopic evaluations. In order to reach the analogous quality of the fit of $C(T)$ for cuprates we have to take into account simultaneously the gap anisotropy and the VHS in the general expressions Eq. (3.12) and Eq. (3.17).

An analogous to Eq. (3.49) smearing of the fluctuation magnetization above T_c reads

$$M(B, T-T_c) = \int M_{\mathrm{theor}}(B, T-T_c')\,\exp\left\{-\frac{(T_c'-T_c)^2}{2(T_c\Delta t)^2}\right\}\frac{dT_c'}{\sqrt{2\pi}T_c\Delta t}. \tag{3.50}$$

However, for big fluctuations of T_c we have to take into account the appearance of superconducting domains. Such a precise investigation of fluctuations in the magnetization of Nb and Sn in the past led to the discovery of twinning plane superconductivity. For analytical GL results for twinning plane superconductivity see Ref. [315].

Here we wish to emphasize that a large body of experimental data for $B_{c2}(T)$ are strongly influenced by the disorder. It is imperative to cut off a region of width $T_c\Delta t$ or $B_{c2}(0)\Delta t$ close to $B_{c2}(T_c)$ if we wish to determine $B_{c2}(T)$ by extrapolation of properties from the superconducting phase or fluctuation behavior of the normal phase. Various spurious curvatures of $B_{c2}(T)$ have been reported merely as a result of disorder of the crystals.

3.2 Order parameter equation for anisotropic-gap superconductors

Following Ref. [179,293], let us scrutinize the derivation of and the solution to Eq. (3.8). The gap anisotropy function will have non-zero values only in a narrow region near the Fermi surface

$$\chi_p = \chi_p\,\theta(\omega_D - |\xi_p|), \qquad T_c \ll \omega_D \ll E_F. \tag{3.51}$$

Later, the differential volume in the momentum space can be separated to Fermi surface element dS and a normal element dp_t

$$d^D p = dp_t\,dS = \frac{d\varepsilon}{v_F}\,dS, \qquad v_F(\mathbf{p}) = \left|\frac{\partial\varepsilon_p}{\partial\mathbf{p}}\right|. \tag{3.52}$$

Returning to Eq. (3.8) we have

$$\frac{G}{(2\pi\hbar)^D}\oint\!\!\int \frac{\chi_p^2}{2E_p}\tanh(z_p)\,\theta(\omega_D - |\xi_p|)\,\frac{d\varepsilon\,dS}{v_F} = 1, \tag{3.53}$$

where $\oint$ denotes integration over the Fermi surface. With the account of the energy cutoff ω_D the last reads

$$\frac{G}{(2\pi\hbar)^D}\oint\frac{dS}{v_F}\,\chi_p^2\int_0^{\omega_D}\frac{\tanh(\sqrt{\xi^2+\Delta_p^2}/2T)}{\sqrt{\xi^2+\Delta_p^2}}\,d\xi = 1. \tag{3.54}$$

According to Eq. (3.24) we have for the density of states

$$\nu_F = \overline{\delta(\xi_p)} = \frac{1}{(2\pi\hbar)^D}\int\delta(\varepsilon - E_F)d\varepsilon\,\frac{dS}{v_F} = \frac{1}{(2\pi\hbar)^D}\oint\frac{dS}{v_F}.$$

Similarly, the averaging over the Fermi surface can be represented as a surface integral

$$\langle f(p)\rangle = \frac{1}{\nu_F}\oint \frac{dS}{(2\pi\hbar)^D\, v_F} f(p). \tag{3.55}$$

In these notation Eq. (3.54) reads

$$\left\langle \chi_p^2 \int\limits_0^{\omega_\mathrm{D}} \frac{\tanh(\sqrt{\xi^2+\Delta_p^2}/2T)}{\sqrt{\xi^2+\Delta_p^2}}\, d\xi \right\rangle = \frac{1}{G\nu_F} = \frac{1}{\lambda_\mathrm{BCS}}, \tag{3.56}$$

where $\lambda_\mathrm{BCS} \equiv G\nu_F$ is the dimensionless BCS coupling constant.

At $T = T_\mathrm{c}$, where $\Delta_p = 0$ and $E_p = |\xi_p|$, substituting $x = \xi/2T$ we obtain

$$\langle \chi_p^2\rangle \int_0^M \frac{\tanh x}{x}\, dx = \frac{1}{\lambda_\mathrm{BCS}}, \qquad M = \frac{\omega_\mathrm{D}}{2T_\mathrm{c}} \gg 1. \tag{3.57}$$

Now the identity

$$\int_0^M \frac{\tanh x}{x}\, dx = \ln\left(\frac{4\gamma}{\pi} M\right) \tag{3.58}$$

gives

$$T_\mathrm{c} = 2\omega_\mathrm{D}\, \frac{\gamma}{\pi}\, \exp\left(-\frac{1}{\langle\chi_p^2\rangle\, \lambda_\mathrm{BCS}}\right). \tag{3.59}$$

Analogously, at $T = 0$ we have

$$\left\langle \chi_p^2 \int_0^{\omega_\mathrm{D}} \frac{d\xi}{\sqrt{\xi^2+\Delta_p^2(0)}} \right\rangle = \frac{1}{\lambda_\mathrm{BCS}}. \tag{3.60}$$

Then taking into account that $\omega_\mathrm{D} \gg \Delta_p(0)$ we have

$$\int\limits_0^{\omega_\mathrm{D}} \frac{d\xi}{\sqrt{\xi^2+\Delta_p^2}} = \ln\left(\frac{\omega_\mathrm{D}}{|\Delta_p|} + \sqrt{1+\frac{\omega_\mathrm{D}^2}{|\Delta_p|^2}}\right) \approx \ln\frac{2\omega_\mathrm{D}}{|\Delta_p|}. \tag{3.61}$$

As we will see later, it is convenient to modify the normalization of the order parameter and gap anisotropy function:

$$\tilde\chi_p = \frac{\chi_p}{\chi_\mathrm{av}}, \quad \tilde\Xi = \Xi\,\chi_\mathrm{av}, \quad \chi_\mathrm{av} \equiv \exp\left\{\frac{\langle \chi_p^2 \ln|\chi_p|\rangle}{\langle\chi_p^2\rangle}\right\}. \tag{3.62}$$

The renormalizing multiplier χ_av is chosen in order for the renormalized gap anisotropy function to obey the relation

$$\langle \tilde\chi_p^2 \ln \tilde\chi_p^2\rangle = 0. \tag{3.63}$$

For the two-band model this gives

$$\chi_{\rm av} = \chi_1^{c_1\chi_1^2}\chi_2^{c_2\chi_2^2}, \tag{3.64}$$

and one can easily verify that

$$c_1\tilde{\chi}_1^2 \ln|\tilde{\chi}_1| + c_2\tilde{\chi}_2^2 \ln|\tilde{\chi}_2| = 0. \tag{3.65}$$

Similarly, using

$$\int_0^{\pi/2} \cos^2\varphi \ln|\cos\varphi|\, d\varphi = \frac{\pi}{8}\ln({\rm e}/4) \tag{3.66}$$

we obtain for a 2D d-wave superconductor

$$\tilde{\chi}_p(\varphi) = \frac{2}{\sqrt{\rm e}}\cos 2\varphi, \tag{3.67}$$

$$\int_0^{2\pi} \tilde{\chi}_p^2(\varphi)\ln|\tilde{\chi}_p(\varphi)|\, d\varphi = 0.$$

Using the approximation (3.61) with a renormalized order parameter and gap anisotropy function, from Eq. (3.60) we derive

$$\tilde{\Xi}(0) = 2\omega_{\rm D}\exp\left(-\frac{1}{\langle\chi_p^2\rangle\lambda_{\rm BCS}}\right). \tag{3.68}$$

This equation together with (3.59) gives the well-known BCS relation for the renormalized order parameter for anisotropic superconductors [179,293]

$$\frac{2\tilde{\Xi}(0)}{T_{\rm c}} = \frac{2\pi}{\gamma} \approx 3.53. \tag{3.69}$$

We assume that the density of states $\nu(E)$ is almost constant in the energy interval $E_{\rm F} \pm 2T_{\rm c}$.

The renormalization does not change the gap $\Delta_p(T) = \Xi\chi_p = \tilde{\Xi}\tilde{\chi}_p$, but in a sense $\tilde{Q}(T)$ is the "true" BCS gap for an anisotropic superconductor. For $T = 0$ the BCS model gives for d-wave superconductors $\Delta_p(0) = \Delta_{\rm max}\cos 2\varphi$, where

$$\frac{2\Delta_{\rm max}}{T_{\rm c}} = \frac{2\pi}{\gamma}\frac{2}{\sqrt{\rm e}} \approx 4.28. \tag{3.70}$$

However, for cuprates we have to take into account the influence of Van Hove singularity and strong coupling correlations. As we fitted from the temperature dependence of the penetration depth for $YBa_2Cu_3O_{7-\delta}$ we have 40% bigger gap $\Delta_{\rm max} = Z\Delta_{\rm max}^{\rm (BCS)}$ and $2\Delta_{\rm max}/T_{\rm c} \approx 6.0$. In such a way the thermodynamic behavior is in agreement with the spectroscopic data. This is a good hint in favor of the Landau–Bogoliubov quasiparticle picture applied to high-T_c cuprates. For MgB_2 taking $c_1 = 0.44$ and $\Delta_1(0) = 7.1$ meV and $\Delta_2(0) = 2.8$ we obtain $\tilde{\chi}_1 \approx 1.17$ and $\tilde{\chi}_2 \approx 0.46$. Then $\tilde{\Xi}(0) = \Delta_1(0)/\tilde{\chi}_1 = \Delta_2(0)/\tilde{\chi}_2 \approx 6.08$ meV $= 70.6$ K. For the critical temperature $T_c = 39$ K we obtain $2\tilde{\Xi}(0)/T_c \approx 3.62$ which agrees with the BCS ratio (3.69) within 3% accuracy as found in Ref. [307]

For arbitrary temperatures using the identity

$$\tanh\frac{x}{2} = 1 - \frac{2}{\mathrm{e}^x + 1} \tag{3.71}$$

Eq. (3.56) reads

$$\left\langle \chi_p^2 \int_0^{\omega_\mathrm{D}} \frac{d\xi}{\sqrt{\xi^2 + \Delta_p^2(T)}} \right\rangle - \frac{1}{\lambda_\mathrm{BCS}} \tag{3.72}$$

$$= 2\left\langle \chi_p^2 \int\limits_0^{\omega_\mathrm{D}} \frac{d\xi}{\sqrt{\xi^2 + \Delta_p^2(T)}\left[\exp\left(\frac{\sqrt{\xi^2+\Delta_p^2(T)}}{T}\right) + 1\right]} \right\rangle.$$

Substituting here $1/\lambda_\mathrm{BCS}$ from Eq. (3.60) and taking into account the $\omega_\mathrm{D} \gg |\Delta_p(0)|$ approximation, Eq. (3.61), we obtain the Pokrovskii equation

$$q := \exp\left\{-\frac{\langle \chi_p^2 F(2y_p)\rangle}{\langle \chi_p^2 \rangle}\right\}, \quad 2y_p = \frac{\pi}{\gamma}\frac{\chi_p}{\chi_\mathrm{av}}\frac{q}{t} = \frac{\Delta_p}{T}, \tag{3.73}$$

where

$$q(t) = \frac{\Delta_p(T)}{\Delta_p(0)} = \frac{\Xi(T)}{\Xi(0)} = \frac{\tilde{\Xi}(T)}{\tilde{\Xi}(0)} \tag{3.74}$$

is the reduced order parameter $0 \leqslant q \leqslant 1$ as a function of the reduced temperature $t = T/T_c$. In physical variables Pokrovskii [179, 293] equation reads

$$\ln\frac{\Delta_p(T)}{\Delta_p(0)} + \left\langle \chi_p^2\, F(\Delta_p(T)/T)\right\rangle_p = 0. \tag{3.75}$$

The function $F(x)$ associated with the right-hand side of Eq. (3.72) is defined by an integral, for which we have one integral and two different summation formulas, convenient for small and large arguments [316]

$$\begin{aligned} F(x) &\equiv \int_{-\infty}^{\infty} \frac{du}{\sqrt{u^2+x^2}\left[\exp\left(\sqrt{u^2+x^2}\right)+1\right]} \\ &= 2\int_0^{\infty} \frac{du}{\exp(x\cosh u) + 1} \\ &= \ln\frac{\pi}{\gamma x} + 2\pi\sum_{l=1}^{\infty}\left[\frac{1}{(2l-1)\pi} - \frac{1}{\sqrt{x^2+(2l-1)^2\pi^2}}\right] \\ &= -2\sum_{n=1}^{\infty}(-1)^n K_0(nx), \end{aligned} \tag{3.76}$$

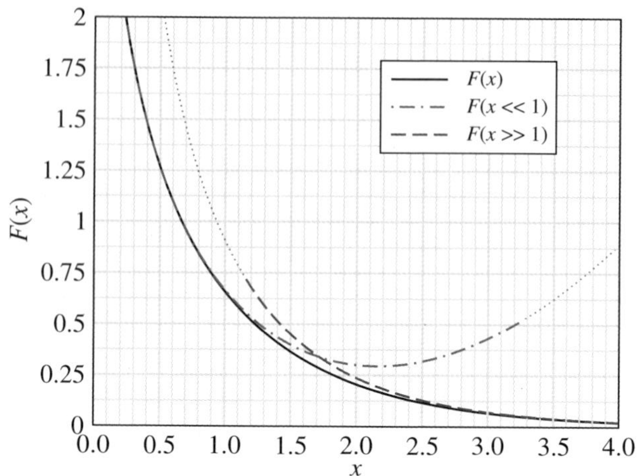

Fig. 3.4 Plot of the $F(x)$ function. The approximations to $F(x)$ for $x \ll 1$ and $x \gg 1$ are given by Eqs. (3.78) and (3.79), respectively.

where for large arguments we have the approximate formula

$$2K_0(x \gg 1) \approx \sqrt{\frac{2\pi}{x}}\, \mathrm{e}^{-x}\left(1 - \frac{1}{8x} + \frac{9}{128x^2} - \frac{225}{3972x^3}\right). \tag{3.77}$$

Physically, here $x = \Delta/T$, $u = \xi/T$ and the upper integration bound ω_{D}/T has been replaced by ∞. For this function we have the approximate formulas

$$F(x \ll 1) \approx \ln\frac{\pi}{\gamma x} + \frac{7}{8\pi^2}\,\zeta(3)\,x^2, \tag{3.78}$$

$$F(x \gg 1) \approx 2K_0(x). \tag{3.79}$$

The Euler constant is $\gamma = \mathrm{e}^C \approx 1.781072418$ and $\zeta(3) \approx 1.202056903$, where ζ is the Riemann zeta function. A plot of the function $F(x)$ is shown in Fig. 3.4. A simple C++ code for numerical evaluation of $F(x)$ is provided in the preprint version of Ref. [317]. For fast calculations one has to take only several terms of the expansions Eq. (3.76). The := sign in Eq. (3.73) represents an iterative assignment in which we use the initial approximation $q = 1$.

The BCS order parameter equation Eq. (3.73) is not specific for the physics of superconductivity. Recently, Abrikisov [318] has derived the same equation for the temperature dependence of the amplitude of spin-density waves in cuprates.

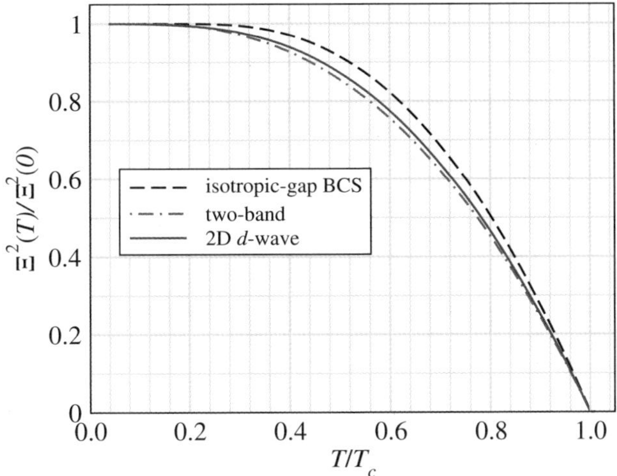

Fig. 3.5 Squared reduced order parameter $\Xi^2(T)/\Xi^2(0)$ versus reduced temperature $t = T/T_{\rm c}$. For the two-band model, the c_1 and δ parameters are chosen so as to simulating a d-wave $\mathrm{CuO_2}$: $c_1 = 1/2$, $\delta = 2.41$.

For 2D d-wave superconductors the Pokrovskii equation (3.73) reads

$$\ln q = -\int_0^{2\pi} 2\cos^2(2\varphi)\, F\left(\frac{2\pi}{\gamma\sqrt{\mathrm{e}}}\cos(2\varphi)\frac{q}{t}\right)\frac{d\varphi}{2\pi}. \tag{3.80}$$

The numerical solution for the squared reduced order parameter $q^2(t)$ is shown in Fig. 3.5. The linear dependence near the critical temperature $t = 1$ corresponds to the GL approximation. In Fig. 3.6 the squared reduced order parameter for $\mathrm{MgB_2}$ (two-band model with $c_1 = 0.44$, $\delta = 7.1/2.8$) is compared with the experimental data from Ref. [319].

As a last problem, let us derive the factorizable kernel (3.7) as a result from the BCS equation (3.6). For $\omega_{\rm D} \ll E_F$, Eq. (3.6) reads

$$\begin{aligned}\Delta_q(T) &= \oint_{\rm FS} V_{q,p}\,\Delta_p \int_0^{\omega_{\rm D}} \frac{\tanh(E_p/2T)}{E_p}\, d\xi_p \frac{dS_p}{(2\pi\hbar)^D v_F} \\ &= \nu_F \left\langle V_{q,p}\Delta_p \int_0^{\omega_{\rm D}} \frac{\tanh(E_p/2T)}{E_p}\, d\xi_p \right\rangle_p .\end{aligned} \tag{3.81}$$

At $T = T_{\rm c}$ [cf. Eqs. (3.57)–(3.59)] this formula gives

$$\Delta_q(T_{\rm c}) \approx \nu_F \ln\left(\frac{2\gamma\omega_{\rm D}}{\pi T_{\rm c}}\right)\langle V_{q,p}\Delta_p\rangle_p. \tag{3.82}$$

Let us also mention the dimensions of the variables. Since the integration $\int \cdots \frac{d^3p}{(2\pi\hbar)^3}$ has a dimension of 1/volume, then $V_{q,p}$, being a Fourier component of potential energy, has dimension of energy×volume. For example, the Coulomb potential e^2/r has dimension of energy and its Fourier transform

$$\frac{4\pi e^2}{k^2} = \int \frac{e^2}{r} \mathrm{e}^{-i\,\mathbf{k}\cdot\mathbf{r}}\, d^3r \tag{3.83}$$

has dimension of energy×volume. The same holds for the contact attraction in the BCS model potential $V(\mathbf{r}) = -G\delta(\mathbf{r})$ having a constant Fourier component $-G$. The density of states ν_F has dimension of (energy×volume)$^{-1}$, Δ_p and E_p have dimension of energy and the Fermi surface averaging brackets $\langle \ldots \rangle$ represent a dimensionless operation.

Let the dimensionless parameter V_0 denotes the maximum eigenvalue of the problem

$$\langle V_{q,p}\, \chi_p \rangle_p = V_0\, \chi_q, \tag{3.84}$$

and χ_p is the corresponding eigenvector, with normalization $\langle \chi_p^2 \rangle = 1$. The comparison of Eq. (3.84) and Eq. (3.82) gives

$$T_{\mathrm{c}} = \frac{2\gamma\omega_{\mathrm{D}}}{\pi} \exp\left(-\frac{1}{\nu_F V_0}\right), \tag{3.85}$$

which is identified with Eq. (3.59) and we obtain

$$G = V_0 = \frac{\langle \chi_q V_{q,p} \chi_p \rangle_{q,p}}{\langle \chi_p^2 \rangle_p}. \tag{3.86}$$

As the maximal eigenvalue is sought, one can apply in this case the Krilov iterations

$$\chi_q^{(n+1)} \propto \langle V_{q,p} \chi_p^{(n)} \rangle_p, \qquad \langle (\chi_p^{(n+1)})^2 \rangle = 1, \tag{3.87}$$

starting from some solution-like trial vector $\chi_p^{(0)}$. Then the gap anisotropy function χ_p is just the limit of the Krilov iterations $\chi_p^{(\infty)}$.

For $T = 0$, the gap equation (3.81) gives

$$\Delta_q(0) = \left\langle V_{q,p} \ln\left(\frac{2\omega_{\mathrm{D}}}{\tilde{\Xi}(0)|\tilde{\chi}_p|}\right) \Delta_p \right\rangle_p. \tag{3.88}$$

Within the weak-coupling BCS approximation, in the integrant

$$\ln\left(\frac{2\omega_{\mathrm{D}}}{|\Delta_p(0)|}\right) = \ln\left(\frac{2\omega_{\mathrm{D}}}{\tilde{\Xi}(0)}\right) - \ln|\tilde{\chi}_p| \tag{3.89}$$

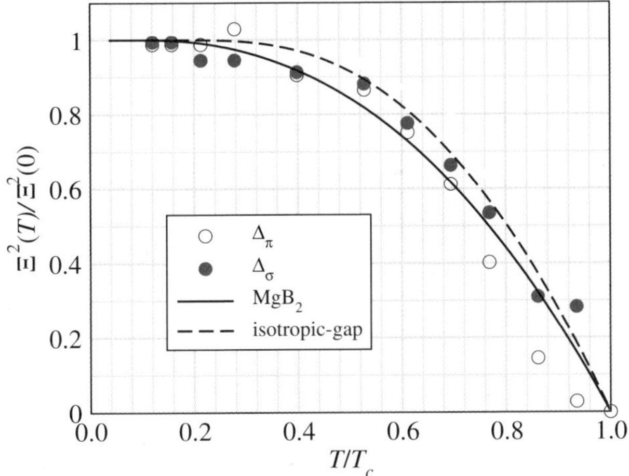

Fig. 3.6 Squared reduced order parameter $\Xi^2(T)/\Xi^2(0)$ versus the reduced temperature $t = T/T_c$ for MgB_2 (solid line) with $c_1 = 0.44$, $\delta = 7.1/2.8$. The experimental points for MgB_2 (circles) are digitized from Ref. [319].

the first term is much bigger than the second one. For details we refer to the original work by Pokrovsky [179, 293], but roughly speaking $\ln[2\omega_D/\Delta_p(0)] \approx \text{const} \gg 1$. Within the latter approximation for $\Delta_p(0)$ we obtain again the same eigenvalue problem and this constitutes the proof that the momentum dependence of the gap is rigid. Hence we derive the separation of the variables $\Delta_p(T) \approx \Xi(T)\chi_p$. When the term $\ln|\chi_p|$ in Eq. (3.89) is small it can be treated perturbatively, and according to the normalization Eq. (3.63) its influence diminish. The properties of this approximative separation of the variables can be simulated by a factorizable kernel

$$V_{q,p} = \sum_n V_n \Psi_q^{(n)} \Psi_p^{(n)} \to V_0 \chi_q \chi_p, \tag{3.90}$$

where V_n are the eigenvalues and $\Psi_p^{(n)}$ are the corresponding eigenvectors of the problem

$$\langle V_{q,p}\, \Psi_p^{(n)} \rangle_p = V_n\, \Psi_q^{(n)}, \qquad \langle |\Psi_p^{(n)}|^2 \rangle = 1. \tag{3.91}$$

In other words, the factorizable approximation, Eq. (3.90) and Eq. (3.7), works well when the influence of smaller eigenvalues is small.

Generally speaking, the separability ansatz is a low-T_c approximation; T_c should be much smaller than all other energy parameters: energy cutoff,

Debye frequency for phonon superconductors, exchange integrals for exchange mediated superconductivity, the Fermi energy and the bandwidths. Room temperature superconductivity is not yet discovered, but the good message is that we have still a simple approximation acceptably working for all superconductors. For theoretical models the accuracy of the separable approximation can be easily probed when investigating the angle between the order parameter at different temperatures, e.g.,

$$\arccos \frac{\langle \Delta_p^*(T)\Delta_p(T_{\rm c})\rangle}{\sqrt{\langle|\Delta_p(T)|^2\rangle\langle|\Delta_p(T_{\rm c})|^2\rangle}} \ll 1, \tag{3.92}$$

or

$$\arccos \frac{\overline{\Delta_p^*(T)\Delta_p(T_{\rm c})}}{\sqrt{\overline{|\Delta_p(T)|^2}\;\overline{|\Delta_p(T_{\rm c})|^2}}} \ll 1. \tag{3.93}$$

Those angles are just zero at $T_{\rm c}$ and the expressions for the specific heat jump and the GL coefficients is correct. Only for $T \to 0$ some small deviations can be observed, but in that case one can treat χ_p as a trial function in a variational approach.

The performed analysis shows that the separation of the variables Eq. (3.5) due to Pokrovsky [179, 293] and consequent factorizable kernel Eq. (3.7) are tools to apply the weak-coupling BCS approximation to anisotropic-gap superconductors. The factorizable kernel gives a simple solution to the gap equation, the nontrivial detail being that this separability can be derived by the BCS gap equation. The factorizable kernel has also been discussed by Markowitz and Kadanoff [296] and employed, e.g., by Clem [320] to investigate the effect of gap anisotropy in pure and superconductors with nonmagnetic impurities. Factorizable kernels are now used in many works on exotic superconductors. However, in none of them is mentioned that the separability of the superconducting order parameter is an immanent property of the BCS theory [179, 293]. The accuracy of the separable approximation is higher if the other eigenvalues of the pairing kernel are much smaller than the maximal one. This is likely to be the situation for the s-d model for layered cuprates [60], where the s-d pairing amplitude J_{sd} is much bigger than the phonon attraction and the other interatomic exchange integrals. In order for us to clarify this important approach to the theory of superconductivity, we have given here a rather methodical derivation of the Pokrovsky theory.

3.3 Electrodynamic behavior

An analysis of the London penetration depth tensor, similar to that carried out by Kogan in Ref. [295], gives

$$(\lambda^{-2}(T))_{\alpha\beta} = \frac{e^2}{\varepsilon_0 c^2} 2\nu_F \langle r_d(y_p) v_\alpha v_\beta \rangle, \quad \alpha, \beta = x, y, z, \tag{3.94}$$

where

$$\mathbf{v}_p = \frac{\partial \varepsilon_p}{\partial \mathbf{p}}, \qquad m_p^{-1} = \frac{\partial \mathbf{v}_p}{\partial \mathbf{p}} = \frac{\partial^2 \varepsilon_p}{\partial \mathbf{p}^2} \tag{3.95}$$

are the band velocity and effective mass and

$$r_d(y) \equiv (y/\pi)^2 \sum_{n=0}^{\infty} \left[(y/\pi)^2 + \left(n + \frac{1}{2} \right)^2 \right]^{-3/2}, \tag{3.96}$$
$$r_d(y) \approx 7\zeta(3)(y/\pi)^2 \ll 1, \qquad r_d(\infty) = 1.$$

For comparison, the conductivity tensor of the normal phase in τ_p-approximation reads

$$\sigma_{\alpha\beta} = 2\nu_F e^2 \langle \tau_p \, v_\alpha v_\beta \rangle. \tag{3.97}$$

For the penetration depths along the principal crystal axes we have in the two-band model

$$\lambda_\alpha^{-2}(T) = \lambda_{\alpha,1}^{-2}(0) \, r_d(y_1) + \lambda_{\alpha,2}^{-2}(0) \, r_d(y_2), \tag{3.98}$$

where for uniaxial crystals like MgB_2 there are only 4 constants: $\lambda_{x,1}(0) = \lambda_{y,1}(0)$, $\lambda_{x,2}(0) = \lambda_{y,2}(0)$, $\lambda_{z,1}(0)$ and $\lambda_{z,2}(0)$. These can be obtained from electron band calculations [321],

$$\begin{aligned}(\lambda^{-2}(T))_{\alpha\beta} &= \frac{e^2}{\varepsilon_0 c^2} 2\nu_F \sum_{b=1,2} c_b \, r_d(\Delta_b/2T) \langle v_\alpha v_\beta \rangle_b \\ &= \sum_{b=1,2} (\lambda_b^{-2}(0))_{\alpha\beta} \, r_d(\Delta_b/2T),\end{aligned} \tag{3.99}$$

where the band index b labels the leaf of the Fermi surface over which the averaging of the electron velocities is carried out. For a discussion and details see the review by Kogan and Bud'ko [295]. There is a natural "Eliashbergization" of this result (cf. Refs. [308, 314, 322–324]):

$$r_d(\Delta_p/2T) = \sum_{n=0}^{\infty} \frac{2\pi T \Delta_p^2}{\left(\Delta_p^2 + \omega_n^2\right)^{3/2}} \longrightarrow \sum_{n=0}^{\infty} \frac{2\pi T \tilde{\Delta}_p^2}{\left[\tilde{\Delta}_p^2(\omega_n) + \tilde{\omega}_{n,p}^2\right]^{3/2}}, \tag{3.100}$$

where $\omega_n = (2n+1)\,\pi T$ are the Matsubara frequencies, $\tilde{\omega}_{n,p} = Z_p(\omega_n)\,\omega_n$, $\tilde{\Delta}_p(\omega_n) = Z_p(\omega_n)\,\Delta_p(\omega_n)$ and $Z_p(\omega_n)$ is the normalization factor. Analogous expressions can be worked out for the specific heat.

For a heuristic consideration of the result by Kogan [295] at $T = 0$ see Ref. [321]. At $T = 0$ the Fermi surface is shifted as a rigid object in the momentum space under the influence of electromagnetic field. This shift of all conduction electrons explains why for the penetration depth the influence of VHS is less essential than the influence on the heat capacity. The increase of the kinetic energy of all conduction electrons is actually the increase of the Gibbs free energy density $\Delta G = \frac{1}{2\varepsilon_0 c^2}\lambda^2 j^2$. At finite temperatures the number of superfluid electrons is $r_d(\Delta_p/2T)$ times smaller.

The penetration depths at $T = 0$ can be also expressed by the optical masses and the Hall constant of the normal metal at high magnetic field

$$(\lambda^{-2}(0))_{\alpha\beta} = \frac{e}{\varepsilon_0 c^2}\frac{1}{\mathcal{R}_\infty}(m^{-1})_{\alpha\beta}, \tag{3.101}$$

$$\frac{1}{\mathcal{R}_\infty} = 2e\int_{\varepsilon_p<E_{\rm F}}\frac{d^3p}{(2\pi\hbar)^3},$$

$$m^{-1} = \frac{\displaystyle\int_{\varepsilon_p<E_{\rm F}}\frac{d^3p}{(2\pi\hbar)^3}m_p^{-1}}{\displaystyle\int_{\varepsilon_p<E_{\rm F}}\frac{d^3p}{(2\pi\hbar)^3}} = \frac{\displaystyle\oint_{\varepsilon_p=E_{\rm F}}\frac{dS_p}{(2\pi\hbar)^3 v_p}\mathbf{v}_p\otimes\mathbf{v}_p}{\displaystyle\int_{\varepsilon_p<E_{\rm F}}\frac{d^3p}{(2\pi\hbar)^3}},$$

the last equation being a consequence of the Gauss theorem $\int_{\varepsilon_p<E_{\rm F}} d^3p\frac{\partial}{\partial\mathbf{p}} = \oint_{\varepsilon_p=E_{\rm F}} d\mathbf{S}_p$, where $d\mathbf{S}_p$ is the element of the Fermi surface oriented along the outward normal. For an extensive discussion on galvanomagnetic properties of normal metals and inclusion of hole pockets with volume density n_h for $\mathcal{R}_\infty^{-1} = e(n_e - n_h)$ see the textbook by Lifshitz and Pitaevskii [325] or the monograph by Lifshitz, Azbel and Kaganov [326]. The Bernoulli effect can be easily observed in almost compensated superconductors for which $n_e \approx n_h$ and the Hall constant is bigger.

In the superconducting phase the Hall constant $\mathcal{R}_\infty$ can be determined by the Bernoulli potential

$$\Delta\varphi = -\mathcal{R}_\infty\frac{1}{2\varepsilon_0 c^2}\lambda^2(T)j^2; \tag{3.102}$$

generalization for the anisotropic case can be obtained by the obvious replacement $\lambda^2 j^2 \to j_\alpha\lambda^2_{\alpha\beta}j_\beta$. Here we suppose that $j \ll j_c(T)$, j_c being the critical current. If the magnetic field $\mathbf{B}$ is parallel to the surface of a bulk superconductor this formula gives

$$\Delta\varphi = -\mathcal{R}_\infty\frac{B^2}{2\mu_0}. \tag{3.103}$$

All charge carriers interact with the electric potential φ, but only the superfluid part $\propto r_d(\Delta_p/2T)$ creates kinetic energy. The constancy of the electrochemical potential in the superconductor gives the change of the electric potential, i.e., the Bernoulli effect. For the temperature dependent condensation energy $\Delta G = -B_c^2(T)/2\mu_0$ the corresponding change of the electric potential is given by

$$\Delta\varphi = \mathcal{R}_\infty \frac{B_c^2(T)}{2\mu_0}. \tag{3.104}$$

For complete determination of the Hall constant $\mathcal{R}_\infty$, the penetration depth $\lambda(T)$ and the optical mass of conduction electrons in a clean superconductor [cf. Ref. [321], Eq. (20)],

$$m = \frac{e\lambda^2(0)}{\varepsilon_0 c^2 \mathcal{R}_\infty}, \tag{3.105}$$

we have to investigate the Bernoulli effect for thin, $d_{\mathrm{film}} \ll \lambda(T)$, and thick, $d_{\mathrm{film}} \gg \lambda(T)$, superconducting films of the same material. $M_{\mathrm{cp}} \equiv 2m$ can be called effective mass of the Cooper pairs; this parameter can be significantly increased by disorder.

For the temperature dependence of the electrochemical potential of the normal phase we have [Ref. [326], Eq. (12.16)]

$$e\Delta\varphi = \frac{\pi^2}{6}\frac{\nu'(E_{\mathrm{F}})}{\nu(E_{\mathrm{F}})}T^2. \tag{3.106}$$

Close to a VHS the influence of the energy derivative of the density of states can be significant and measurable.

The entropy and specific heat related to the volume density of the free energy of superconducting condensation $B_c^2(T)/2\mu_0$ can be determined by electric capacitor measurements, applying surface temperature oscillations. For discussions of possible experimental setups see Ref. [321] and references therein.

It is a matter of technical calculations to verify the identity

$$(y/\pi)^2 \sum_{n=0}^{\infty}\left[(y/\pi)^2 + \left(n+\frac{1}{2}\right)^2\right]^{-3/2} + \int_{-\infty}^{+\infty} \frac{dx}{2\cosh^2\sqrt{x^2+y^2}} = 1, \tag{3.107}$$

which transcribes into the form

$$r_a(y) + r_d(y) = 1. \tag{3.108}$$

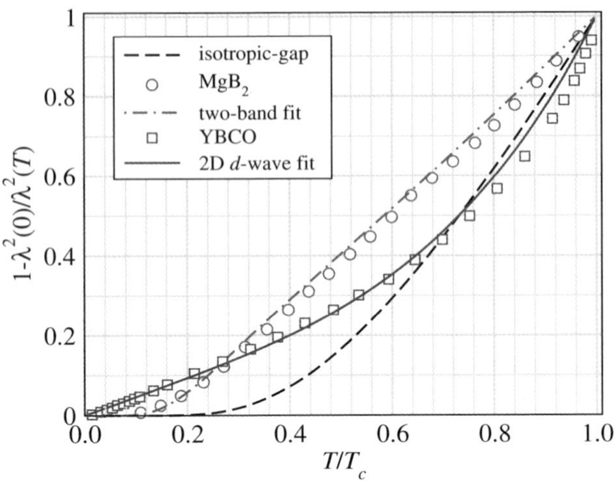

Fig. 3.7 In-plane normal fluid density $1 - \lambda^2(0)/\lambda^2(T)$ versus reduced temperature $t = T/T_{\rm c}$ computed for 3 cases: (i) isotropic-gap BCS superconductor (dashed line), (ii) two-band superconductor $\mathrm{MgB_2}$ with parameters $c_1 = 0.59$, $\delta = 7.1/2.8$ (dash-doted line) and (iii) 2D d-wave superconductor (solid line). The experimental points for $\mathrm{YBa_2Cu_3O_{7-\delta}}$ (squares) are digitized from Ref. [311] and the corresponding theoretical 2D d-wave curve is calculated according to Eq. (3.111) with renormalization factor $Z = 1.4$. Some experimental points for $\mathrm{MgB_2}$ (circles) are digitized from Ref. [327], Fig. 9; for details see the original work.

In such a way the *electrodynamic behavior* of a superconductor can be expressed in terms of the functions, defined for description of its *thermodynamic behavior*. Using Eqs. (3.108) and (3.94) we obtain

$$\rho_{\rm N}(T) = 1 - \frac{(\lambda^{-2}(T))_{\alpha\beta}}{(\lambda^{-2}(0))_{\alpha\beta}} = \frac{\langle r_a(\Delta_p/2T) v_\alpha v_\beta\rangle}{\langle v_\alpha v_\beta\rangle}. \tag{3.109}$$

Within the framework of London electrodynamics $\rho_{\rm N}(T) = 1-\lambda^2(0)/\lambda^2(T)$ is the normal fluid density, and $\rho_{\rm S}(T) = \lambda^2(0)/\lambda^2(T)$ is the superfluid one, having total charge density $\rho_{\rm S}(T)/\mathcal{R}_\infty$. For a two-band superconductor, Eqs. (3.99) and (3.109) give for the penetration depth along the principal crystal axes

$$\rho_{\rm S}(T) = \frac{\lambda_\alpha^2(0)}{\lambda_\alpha^2(T)} = \sum_{b=1,2} w_{\alpha,b}\, r_d(\Delta_b/2T), \qquad w_{\alpha,b} = c_b \frac{\langle v_\alpha^2\rangle_b}{\langle v_\alpha^2\rangle},$$
$$w_{\alpha,1} + w_{\alpha,2} = 1, \quad \langle v_\alpha^2\rangle = c_1\langle v_\alpha^2\rangle_1 + c_2\langle v_\alpha^2\rangle_2. \tag{3.110}$$

For a set of parameters see the review by Kogan and Bud'ko [295]. We take $\delta = 7.1/2.8$ according to the spectroscopic data [307,319]; see also the

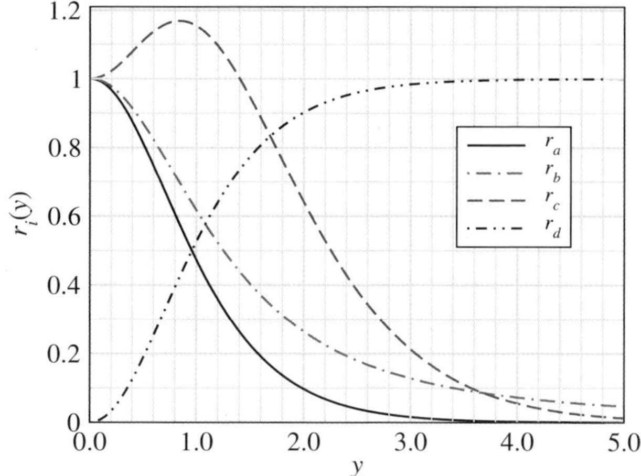

Fig. 3.8 Plot of the $r_i(y)$ functions.

point contact spectroscopy data in Ref. [328]. In Fig. 3.7 we compare our theoretical calculation with the experimental data for $\lambda(T)$ by Carrington and Manzano [327]. Here we take $c_1 = 0.59$ which gives $w_{a,1} \approx w_{a,2} \approx 0.5$.

The graphs of $r_i(y)$ and the corresponding $g_i(z)$ functions are given in Figs. 3.8 and 3.9. The temperature dependence of the penetration depth $\lambda(T)$ is also programmed for isotropic-gap, two-band and model 2D d-wave superconductor. In the 2D d-wave case the theoretical result is compared with the experimental data [311] for $YBa_2Cu_3O_{7-\delta}$, which is also depicted at the figure. The linear dependence of $1-\lambda^2(0)/\lambda^2(T)$ at low temperatures for $YBa_2Cu_3O_{7-\delta}$ is discussed in Ref. [311], Eq. (2.10). For a 2D d-wave superconductor the general formula Eq. (3.94) gives

$$\rho_S(T) = \frac{\lambda^2(0)}{\lambda^2(T)} = \int_0^{2\pi} r_d(Z\frac{\Delta_{\max}(T)}{2T}\cos 2\varphi)\frac{d\varphi}{2\pi}, \tag{3.111}$$

where the temperature dependence of the order parameter is described in Appendix 3.2. We are using an oversimplified model for cuprate superconductivity for which are neglected (i) the anisotropy of the Fermi velocity $v_F(\mathbf{p})$ along the Fermi contour; (ii) higher harmonics of the gap function Δ_p along the Fermi contour and (iii) the influence of VHS of the density of states slightly below the Fermi level. For comparison between ARPES data and a lattice model for high-T_c spectrum see Ref. [60], Fig. 3.

Let us assume now that the order parameter for $YBa_2Cu_3O_{7-\delta}$ is Z-times higher than the BCS prediction. This could be due to the influence

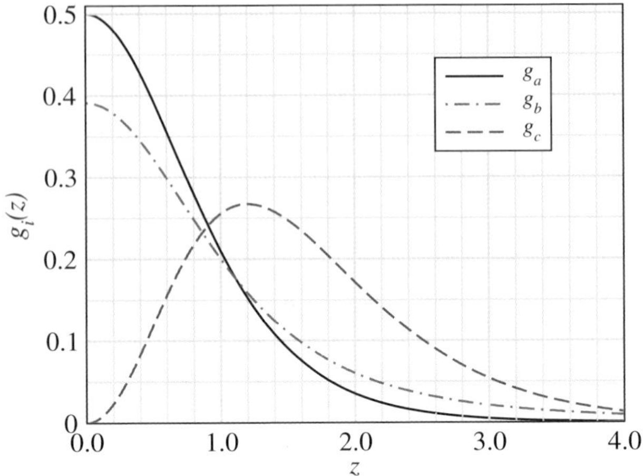

Fig. 3.9 Plot of the $g_i(z)$ functions ($i = a, b, c$).

of VHS or, which is more important, strong coupling effects. Inserting here $Z = 1.4$ we can see that such a renormalization well describes the temperature dependence of the penetration depth in the whole temperature interval. Finally, we have a good working BCS-like formula. In fact, significantly higher $\Delta_{\max}(0)/T_c$ than BCS prediction is in agreement with the ARPES data.

3.4 The case for Sr_2RuO_4

Our approach is aslo applicable to the triplet superconductor Sr_2RuO_4; for a review see Ref. [329]. We adopt the promising gap anisotropy model by Zhitomirsky and Rice, [330] which gives $E_p = \sqrt{\xi_p^2 + |\Delta_p|^2}$, with

$$|\Delta_p|^2 \propto \left[\sin^2 \frac{p_x a}{2\hbar} \cos^2 \frac{p_y a}{2\hbar} + \cos^2 \frac{p_x a}{2\hbar} \sin^2 \frac{p_y a}{2\hbar}\right] \cos^2 \frac{p_z c}{2\hbar}, \tag{3.112}$$

where $p_x a/\hbar$, $p_y a/\hbar$, $p_z c/\hbar \in (0, 2\pi)$. For the Fermi surface we take a simple cylinder $\varepsilon_p \approx \varepsilon(\sqrt{p_x^2 + p_y^2})$ with radius $p_F a/\hbar \approx 0.93\,\pi$. Our calculations are depicted in Fig. 3.10. In this model calculation we have taken into account only one band responsible for superconductivity. Although it is not *a priori* clear how "good" is this assumption, our curve reproduces the theoretical curve by Zhitomirsky and Rice [330] and passes close to the experimental points by NishiZaki *et al.* [331] This promising success encouraged us to present the theoretical prediction for the penetration depth

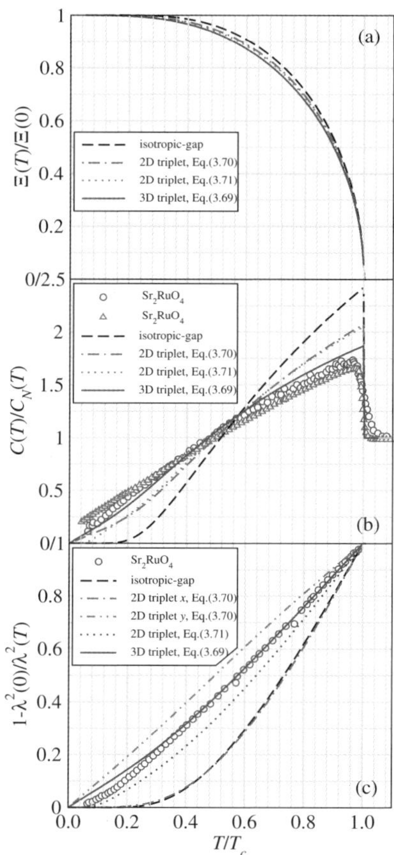

Fig. 3.10 Sr_2RuO_4. (a) Reduced order parameter for the Zhitomirsky and Rice model Eq. (3.112) (solid line), the 2D vertical line nodes model Eq. (3.113) (dot-dashed line), and for the 2D model by Deguchi *et al.* Eq. (3.114) (dotted line). (b) Specific heat ratio $C(T)/C_N(T)$ for the Zhitomirsky and Rice model (solid line), the 2D vertical line nodes model (dash-dotted line), and for the Deguchi *et al.* model (dotted line). The experimental points (circles) from Ref. [331] are digitized from Ref. [330], Fig. 1. (c) Normal fluid density $1 - \lambda^2(0)/\lambda^2(T)$ corresponding to the gap anisotropy models (3.112)–(3.114). The experimental points (circles) from Ref. [333] are digitized from Ref. [334], Fig. 2. Note that the model with vertical line nodes predicts spontaneous breaking of the symmetry of the penetration depth in the *ab*-plane.

calculated from Eq. (3.109). According to the conclusions by Zhitomirsky and Rice [330] their model with horizontal line nodes (see also Ref. [332]) describes the experimental data better than a model with vertical line nodes. For illustration, in Fig. 3.10 we present also our calculations for a simple

2D vertical line nodes model with gap anisotropy function

$$\chi_p \propto \sin\left(\frac{p_x a}{2\hbar}\right). \tag{3.113}$$

Similar model was studied by NishiZaki *et al.* [331]; see also Fig. 26 in the review by Mackenzie and Maeno [329].

From aesthetic point of view our preferences are for the recent model for the gap anisotropy by Deguchi *et al.* [335]

$$|\Delta_p|^2 \propto \sin^2(p_x a/\hbar) + \sin^2(p_y a/\hbar). \tag{3.114}$$

Such type of anisotropy can be derived in the framework of quasi-two-dimensional exchange models for perovskite superconductivity of the type of the considered for CuO_2 plane in Ref. [60]. The theoretical prediction corresponding to Eq. (3.114) is also illustrated in Fig. 3.10 together with the experimental data by Deguchi *et al.* [335]

3.5 Discussion

Let us discuss now the specific heat. We have shown that for factorizable kernels [296] the specific heat can be represented as a sum of a "normal" component $C_\nu(T)$ and a term dependent on the order parameter $C_\Delta(T)$, which has the same form as in the GL theory. There is one detail that is worth focusing on: for the *s*-*d* model for high-T_c superconductivity [60] the kernel is indeed separable because the contact interaction is localized in a single atom in the lattice unit cell. One should only substitute the spectrum of the superconductor at $T < T_c$ in the known expression for the GL coefficients from classical work of Gor'kov and Melik-Barkhudarov [294]. The final expression for the specific heat is a generalization of the result of Pokrovskii [293]. For the jump of the specific heat at the critical temperature $\Delta C|_{T_c} = C_\Delta(T_c^-)$ general consideration has already been given in Refs. [297–299]. The derived formula is not exact, but interpolates between the correct low-temperature behavior and the result by Pokrovskii [293] for the specific heat jump at T_c. That is why we believe that our interpolation formula Eq. (3.29) can be useful for preliminary analysis of the experimental data for the specific heat in superconductors; for experimental data processing the accuracy could be comparable, e.g., with the accuracy of the Debye formula for the phonon heat capacity.

We illustrated our formulas for $C(T)$ and $\lambda(T)$ for isotropic-gap BCS model and three of the best investigated anisotropic-gap superconductors

$YBa_2Cu_3O_{7-\delta}$, Sr_2RuO_4 and MgB_2. The nature of superconductivity for those superconductors is completely different: high-T_c and low-T_c, phonon- and exchange-mediated, singlet and triplet Cooper pairs. In all those cases the derived formulas work with an acceptable accuracy; in some cases we have even quantitative agreement and for high-T_c cuprates we have shown what the BCS analysis can give. We conclude that the statistical properties of the superconductors [thermodynamic $C(T)$ and kinetic $\lambda(T)$] are determined mainly by the gap anisotropy, irrespective of the underlying pairing mechanism, and the approximative weak coupling separation of variables [293] $\Delta_p(T) = \Xi(T)\chi_p$ is an adequate approach. It is worth applying the derived formulas for $C(T)$ and $\lambda(T)$ for every new superconductor. Often after the synthesis of a new superconductor single crystals are not available and only the data for heat capacity $C(T)$ can help the theory to distinguish between different models for the gap anisotropy even before detailed spectroscopic investigation is performed.

Chapter 4

Plasmons and the Cooper pair mass

4.1 Plasmons: prediction

In 1989, D. Farrel and collaborators [336] using torque magnetometry measured a giant superconducting effective mass anisotropy $m_c^*/m_a^* \simeq 3000$ in $Bi_2Sr_2CaCu_2O_8$. This leads to anisotropy in the penetration depth $\lambda_c/\lambda_a = \sqrt{m_c^*/m_a^*} \simeq 55$, where λ_c and λ_a are the London penetration depths corresponding to screening currents flowing in $\hat{c}$ direction and in the ab-plane, respectively.

Using a typical value $\lambda_a \simeq 250$ nm [337], we get the giant value $\lambda_c = \sqrt{m_c^*/m_a^*}\lambda_a \simeq 14$ μm. This, in turn, leads to an extremely low plasma frequency, $\omega_{\rm pl} = c(\lambda_c\sqrt{\epsilon}) \simeq 4$ meV, lower than the superconducting gap $2\Delta(0)$ at low temperatures $T \ll T_{\rm c}$. Here c is the speed of light, and $\epsilon \simeq 4.5$ [338] is the background dielectric constant. The electric field of these plasma oscillations is polarized in the $\hat{c}$ direction. The BCS theory gives $2\Delta(0) = 3.52T_{\rm c} \simeq 30$ meV and the experimental data—even more [339].

In conventional superconductors, $2\Delta(0) \ll \omega_{\rm pl}$ and plasmons destroy the Cooper pairs (see Chapter 16.4 in Ref. [272]). But, for the layered and extremely anisotropic Bi-based 2:2:1:2 with weak coupling between the conducting CuO_2 planes, the attenuation of the plasmons will be negligible. The predicted slow decay may encourage experimentalists to look for plasmons in their samples.

It is well known that the London electrodynamics describes the electromagnetic response for frequencies ω up to the superconducting gap $\omega < 2\Delta$. No special treatment connected with the mechanism of superconductivity is necessary for considerations of the consequences of London electrodynamics for all superconductors. The existence of plasmons connected with fluctuations of the "superfluid" electron density n is a simple consequence of the London electrodynamics if the plasma frequency is low enough:

$$\omega_{\rm pl} = (4\pi n e^{*2}/m_c^* \epsilon)^{1/2} = \frac{c}{\lambda_c \sqrt{\epsilon}} < 2\Delta(0), \tag{4.1}$$

$$1/\lambda_c^2 = 4\pi n e^{*2}/m_c^* c^2, |e^*| = 2|e|,$$

where e^* and m_c^* are the charge and mass of Cooper pairs. Plasma waves connected with the Cooper-pair motion were predicted for thin wire [340] (one-dimensional superconductors) and for thin films [5] (two-dimensional superconductors) without any connection to the mechanisms of superconductivity. Moreover, for small spheres (zero-dimensional superconductors), plasma resonances were observed experimentally [341] for many high-T_c superconductors; let us mention that the existence of such plasma oscillations is not directly connected with the mechanism of high-T_c superconductivity. The suggestion here is that similar plasma oscillations can exist in bulk (three-dimensional) superconductors, and it is worthwhile understanding the plasmons of Cooper pairs in $Bi_2Sr_2CaCu_2O_8$ and $Tl_2Ba_2CaCu_2O_8$. The experimental observation can be performed, for example, using high-resolution electron-energy-loss spectroscopy for electrons reflected by a clean *ab* surface of a high-T_c crystal. The electric field of these plasma oscillations is oriented in the *c* direction, perpendicular to the *ab* surface of high-T_c superconductor crystal, and these resonances cannot be excited by electromagnetic waves falling perpendicularly to the *ab* surface. Namely, this is the reason why the considered plasma oscillations are not discovered by chance during the far-infrared (FIR) investigations of high-T_c superconductors.

This high-T_c layers with thickness $d \ll \lambda_c$ are also intensively investigated. Let us mention that the plasma frequency for a thin layer is just the same as that for a bulk material. For thin high-T_c layers, the polarization, geometry, and plasma frequency are of the same order as intersubband plasmons in a two-dimensional electron gas in semiconductors and can be observed by the same FIR techniques [342] (for an introduction to the physics of the two-dimensional electron gas see the review by Ando, Fowler and Stern [343]). Another possibility is to use a grating coupling [344] of the FIR electromagnetic field with the plasma oscillations and the standard FIR transmission spectroscopy.

Thin *c*-oriented $Bi_2Sr_2CaCu_2O_8$ films represent an appropriate systems for observation of *c*-polarized plasma oscillations as the attainable quality of bulk crystals is not as high as that of thin films. The first experimental observation of the predicted [4] FIR transparency of and plasma oscillations in bulk superconductors was reported by Tamasaku, Nakamura and

Uchida [255] for $La_{2-x}Sr_xCuO_4$. Due to the strong anisotropy of high-T_c superconductors, the bulk plasmons are known as Josephson plasma resonances.

4.2 In search for the vortex charge and the Cooper pair mass

4.2.1 *Introduction*

The sign change of the Hall effect observed in the superconducting state of many high-T_c superconductors is one of the most puzzling problems in the electrodynamics of these materials [345, 346]. One may ask then what is the doping dependence [345] of this Hall anomaly and how the vortex-lattice melting [346] affects the Hall behavior? Alas, due to the complexity of the vortex matter many related problems are still not answered satisfactorily if at all. It is quite possible that the sign reversal of the temperature dependence of the Hall effect could be closely related to charging of the vortices [347–350]. There is no doubt [349] that the experimental solution of this enigma would provide the key towards understanding the various electromagnetic phenomena. On the other hand, the currently existing theoretical models often lead to conflicting results, thus making it difficult to discriminate between all those competing explanations. In such a situation we feel it is appealing to accelerate the selection by looking for simplicity in experiments with artificial structures where many of the complications typical for the real systems are avoided.

The aim of the present chapter is to propose an experiment for determination of the vortex charge employing transport measurement in a layered metal-insulator-superconductor (MIS) system. We shall require that the quality of the insulator-superconductor interface be extremely high and the insulator layer be very thin. Such a layered MIS structure incorporating a high-T_c film can be manufactured by the contemporary technology of atomic-level engineering of superconducting oxide multilayers and superlattices [289, 351–353]. In fact, structures of the kind are now being in use for purposes of the fundamental research [354–357] in the physics of high-T_c superconductors, therefore the vortex charge problem can find its solution thanks to the technological progress. The simplest possible idea behind the search for the vortex charge is to study the electrostatic effect related to charged vortices [347, 348], which is analogous to electrostatic effects originating, in turn, in the Bernoulli effect [358] due to circulating currents in a thin superconducting film in a vortex-free state; the numerical value of the angular momentum $\oint (m^*\mathbf{v} + e^*\mathbf{A}) \cdot d\mathbf{r}$ is irrelevant for the current-induced contact-potential difference. The chapter is organized as follow: in Sec. 4.2.2 we derive the formula for the vortex charge q_v expressed via the effective mass of Cooper pairs m^* as well as the expression for the interface Hall conductivity σ_{xy}. In Sec. 4.2.4 we will analyze our proposed experimental set-up for determination of the vortex charge by measuring the Hall resistance of the vortex charge currents. An overview is made in Sec. 4.2.5 of different experimental methods for determination of the Cooper pair mass: the surface Hall current [359], subsection 4.2.5.1; the Bernoulli effect [358], subsection 4.2.5.2, and the electrostatic charge modulation [6], subsection 4.2.5.3. It is finally concluded in Sec. 4.2.6 that the vortex charge q_v and the effective mass m^* of fluctuation Cooper pairs fall into the class of the last unresolved problems in the physics of superconductivity. These important parameters enter the theories of a number of phenomena related to electrodynamics of superconductors and can be simultaneously determined by standard electronic measurements. Contemporary layer-by-layer growth of layered oxide structures gives the unique chance for finding q_v and m^* for high-T_c materials but some of the proposed experiments can be realized also in MIS structures with conventional superconductors.

4.2.2 *Model*

4.2.3 *Type-II superconductors*

This section gives an account of the vortex charging due to the Bernoulli effect within the framework of London electrodynamics. For a superconductor in thermodynamic equilibrium the electrochemical potential ζ is constant and the space distribution of the electric potential $\varphi(\mathbf{r})$ is determined by the Bernoulli-Torricelli theorem

$$\frac{1}{2}m^*v^2(\mathbf{r})n(T) + \rho_{\text{tot}}\varphi(\mathbf{r}) = \rho_{\text{tot}}\zeta. \tag{4.2}$$

Formally, this equation can be derived within the framework of the BCS theory using the statistical mechanics methods, but its physical meaning is very simple—it is a consequence of the energy conservation. We shall further stick to the standard notations for the effective mass of Cooper pairs in the ab-plane m^*, the superfluid velocity $\mathbf{v}$ related to the current density $\mathbf{j} = e^*n(T)\mathbf{v}$, the mass density of the superfluid $m^*n(T)$, and the total charge of the conduction band $\rho_{\text{tot}} = e^* n(T = 0)$; at zero temperature all charge carriers are superfluid and according to the BCS theory $|e^*| = 2|e|$. The temperature dependence of $n(T)$ can be extracted from that of the London penetration depth for screening currents flowing in the CuO_2 plane,

$$\frac{1}{\lambda^2(T)} = \frac{\mu_0 n(T) e^{*2}}{m^*}, \tag{4.3}$$

where the use of SI units is implied, $\mu_0 = 4\pi \times 10^{-7}$. Although the temperature dependence of the superfluid ratio

$$\frac{n(T)}{n(0)} = \frac{\lambda^2(0)}{\lambda^2(T)} \tag{4.4}$$

is related to the gap anisotropy, the hydrodynamic relation Eq. (4.2) remains invariant.

Consider now a thin cuprate film thread by a perpendicular magnetic field $\mathbf{B} = B_z \hat{\mathbf{z}}$. As a first step we determine the distribution of the electric potential as a function of the distance to the vortex line $r = \sqrt{x^2 + y^2}$. For r larger than the Ginzburg–Landau (GL) coherence length in the ab-plane but smaller than the penetration depth, $\xi_{ab}(T) \ll r \ll \lambda_{ab}(T)$ one can use the Bohr-Zommerfeld relation

$$r m^* v = \hbar. \tag{4.5}$$

Substituting $v(r) = \hbar/m^* r$ from the above equation into the Bernoulli theorem Eq. (4.2) we derive the current-induced change of the electric potential

$$\varphi(r) = -\frac{\hbar^2}{2e^* m^*} \frac{n(T)}{n(0)} \frac{1}{r^2}. \tag{4.6}$$

This equation is applicable not only to the volume of the superconductor $z < 0$ but even to the superconducting surface $z = 0$ which is supposed to be clean enough as well as to expose the properties of the bulk material. The superconductor is capped by a thin insulating layer of thickness much smaller than the penetration depth, $d_{\text{ins}} \ll \lambda_{ab}(0)$. On top of the latter a thin-normal-metal layer is evaporated, hence a plane capacitor configuration is achieved, being in fact realized as a metal-insulator-superconductor (MIS) layered structure. For definiteness the electric potential of the normal plate is set to zero. Far from the vortex core, for $r > d_{\text{ins}}$, the electric field E_z of such a plane capacitor can be considered as being homogeneous,

$$E_z = \frac{\varphi}{d_{\text{ins}}} = -\frac{q^{(2D)}}{\epsilon_0 \epsilon_{\text{ins}}}, \tag{4.7}$$

which is employed to express the induced on the normal plate surface charge density $q^{(2D)}(r)$ via the Bernoulli potential $\varphi(r)$,

$$q^{(2D)} = \frac{\hbar^2}{2e^*m^*}\frac{\epsilon_0\epsilon_{\rm ins}}{d_{\rm ins}}\frac{n(T)}{n(0)}\frac{1}{r^2}, \tag{4.8}$$

where $\epsilon_0 = 1/\mu_0c^2$, c being the speed of light, and $\epsilon_{\rm ins}$ is the relative dielectric constant of the insulator. We notice that $q^{(2D)}(r)$ has the same sign as the charge of the Cooper pairs in the superconductor. On the other hand, the Bernoulli potential keeps the Cooper pairs on a circular orbit inside the vortex. The radial electric force is then equal to the centrifugal force

$$e^*\frac{\partial\varphi}{\partial r} = m^*\frac{v^2}{r}. \tag{4.9}$$

The electric potential attracts the Cooper pairs and the charges with the same sign on the normal plate of the plane capacitor. In order to derive the total charge related to the vortex we have to integrate the charge density up to some maximum radius,

$$r_{\rm max} = \min\left(\lambda(T),\ \sqrt{\frac{\Phi_0}{B}}\right), \tag{4.10}$$

corresponding to the screening length $\lambda(T)$ or the typical intervortex distance in case of high area density of vortex lines $n_v = B/\Phi_0$, where $\Phi_0 = 2\pi\hbar/|e^*| = 2.07\ \mathrm{fTm}^2$ is the flux quantum. Supposing that the insulator layer is thin enough, $d_{\rm ins} \ll r_{\rm max}$, the integration of the surface density gives for the total vortex charge

$$\begin{aligned} q_v &= \int_{d_{\rm ins}}^{r_{\rm max}} q^{(2D)}(r)\,\mathrm{d}(\pi r^2) \approx \frac{\pi\hbar^2}{e^*m^*}\frac{\epsilon_0\epsilon_{\rm ins}}{d_{\rm ins}}\frac{n(T)}{n(0)}\ln\frac{r_{\rm max}}{d_{\rm ins}} \\ &= \frac{1}{8}\mathrm{sgn}(e^*)|e|\frac{a_0\epsilon_{\rm ins}}{d_{\rm ins}}\frac{m_0}{m^*}\frac{\lambda_{ab}^2(0)}{\lambda_{ab}^2(T)}\ln\kappa_{\rm eff}, \end{aligned} \tag{4.11}$$

where $a_0 = 4\pi\epsilon_0\hbar^2/e^2m_0 = 53$ pm is the Bohr radius, $m_0 = 9.11\times10^{-31}$ kg is the mass of a free electron, and $\kappa_{\rm eff} = r_{\rm max}/d_{\rm ins}$ is a quantity analogous to the Ginzburg-Landau parameter $\kappa = \lambda_{ab}(0)/\xi_{ab}(0)$. According to our model, the charge related to vortices is localized not in the vortex core but in the adjacent conducting layers: superconducting CuO_2 planes in a real high-T_c crystal or the normal layer in the model MIS system. With this we close the electrostatic consideration of the vortex charge, but rhe reader is reffered to a number of ingenious experiments related to electrostatics of vortices which are suggested in Ref. [348]. We believe, however, that the standard transport measurement have some advantage even if they are related to observations of pA-range and below.

The next important step is to address the vortex flow regime of the superconducting film when a strong enough dc current density j_y is applied through the superconducting film. This condition will create small dissipation and give rise to an electric field E_y parallel to the current density. The electric field, in turn, creates a drift of the vortices with mean drift velocity in x-direction $v_v = E_y/B_z$. In a coordinate system moving with the vortex drift velocity $\mathbf{v}_v$ the electric field is zero. We suppose that v_v is much smaller than the critical depairing velocity $v_c = \hbar/m^*\xi_{ab}(T)$ and the Bernoulli potential is nearly the same as in the dissipation-free static regime. Along this line let us recall the fact that airplanes fly thanks to the Bernoulli theorem that holds true for a unviscous dissipationless fluid, but the significant part of the ticket price covers the dissipated energy. By the same token, for $v_v \ll v_c$ the vortex-induced charge has nearly the static value q_v. Since the charge images will follow the vortices as shadows, the vortex flow will create a two dimensional (2D) current density on the surface of the normal metal

$$j_x^{(2D)} = q_v n_v v_v = \frac{q_v}{\Phi_0} E_y = \sigma_{xy} E_y. \tag{4.12}$$

The electric field E_y resides the superconducting layer, whereas the current $j_x^{(2D)}$ exists in the normal slab. The 2D Hall conductivity directly gives the vortex charge

$$\sigma_{xy} = \frac{q_v}{\Phi_0} = \frac{q_v |e^*|}{2\pi\hbar}. \tag{4.13}$$

For $L_x \times L_y$ rectangular shape of the MIS structure the voltage drop in the superconducting layer is $V_y = E_y L_y$, the total current in the normal layer is $I_x = L_x j_x^{(2D)}$, and the interface Hall resistivity is size-independent,

$$R_{xy} \equiv \frac{V_y}{I_x} = \frac{1}{\sigma_{xy}} = \frac{2\pi\hbar}{q_v|e^*|} = \frac{e^2}{q_v|e^*|} R_{\mathrm{QHE}} = \frac{1}{2} R_{\mathrm{QHE}} \frac{|e|}{q_v}, \tag{4.14}$$

where $R_{\mathrm{QHE}} = 2\pi\hbar/e^2 = 25.813\ \mathrm{k}\Omega$ is the fundamental resistance determined by the quantum Hall effect (QHE). Since the vortex charge $q_v \ll |e|$, the experiment would face the problem of measuring huge Hall resistances. This sets the first technological requirement regarding to the quality of the insulating layer—in order to avoid the leakage currents the resistance R_{MS} of the plane capacitor should satisfy the relation $R_{\mathrm{MS}} = \rho_{\mathrm{ins}} d_{\mathrm{ins}}/(L_x L_y) \gg R_{xy}$. In the present model we used the hydrodynamic approach applicable for extreme type-II superconductors and completely neglected the influence of the geometrically small vortex core. However the states in vortex core can have some influence in the total charge of vortex core [347, 348]. In order qualitatively to "interpolate" a real situation with moderate Ginzburg-Landau parameter κ let us analyze the interface Hall current for a type I superconductor. In this case the normal "cores" are domains of normal metal surrounded by circulating superconducting currents. This problem, certainly, is only of an academic interest and is irrelevant for the oxide superconductors.

4.2.3.1 *Interface Hall current for type-I MIS structure*

If the superconducting layer of a MIS structure is of type-I superconductor, in a perpendicular magnetic field B_z the magnetic field in the normal domains is equal to the thermodynamic one $B_c(T)$ and is zero in the superconducting domains. The relative area of the normal regions is $c_N = B_z/B_c(T)$, correspondingly the part of the superconducting area is $c_S = 1 - B_z/B_c(T)$, thus $c_N + c_S = 1$ and the external field is equal to the mean field $B_z = c_N B_c(T) + c_S \times 0$. The contact potential difference between the normal and the superconducting phase (see Eq. (4.39) below) is

$$\varphi_N - \varphi_S = -\frac{1}{e^* n(0)} \frac{B_c^2(T)}{2\mu_0}. \tag{4.15}$$

This contact potential difference creates, in turn, a difference in the charge density at the surface of the normal layer in front of the normal domain

$$q^{(2D)} = \frac{\epsilon_0 \epsilon_{\rm ins} c_N}{d_{\rm ins}} (\varphi_S - \varphi_N) = \frac{\epsilon_0 \epsilon_{\rm ins} B_c(T)}{2\mu_0 d_{\rm ins} e^* n(0)} B_z, \tag{4.16}$$

where a plane capacitor configuration is implied.

When an electric field E_y is applied in the superconducting layer the normal domains acquire a drift velocity in x-direction $v_v = E_y/B_z$. Again, in the mobile coordinate system the domain structure is static and the mean electric field is zero. The extra charges induced in the normal layer follow the moving normal domains and for the 2D current $j_x^{(2D)} = q^{(2D)} v_v = \sigma_{xy} E_y$ at the surface of the normal plate we obtain

$$\sigma_{xy} = \frac{\epsilon_0}{2\mu_0} \frac{B_c(T) \epsilon_{\rm ins}}{e^* n(0) d_{\rm ins}} = \frac{1}{R_{xy}}. \tag{4.17}$$

This very small interface Hall conductivity vanishes at T_c and its detection requires fA sensitivity. For comparison with Eq. (4.11) here we also give the expression for the induced charges per flux quantum

$$q_{v,I} \equiv \frac{q^{(2D)}}{n_v} = \frac{\epsilon_0 \Phi_0}{2\mu_0} \frac{B_c(T) \epsilon_{\rm ins}}{e^* n(0) d_{\rm ins}}. \tag{4.18}$$

Of course, around every normal domain in a type-I superconductor $|e^* \oint \mathbf{A} \cdot d\mathbf{r}/\hbar| \gg 1$.

Having derived the formulae, Eq. (4.13) and Eq. (4.17), concerning the new predicted effect we proceed with more detailed discussion and description of the proposed new experiment in the next section.

4.2.4 *Experimental set-up for measuring the vortex charge*

To begin with, we have sketched a "*gedanken*" experimental setup in Fig. 4.1. The contemporary technology of layer-by-layer growth of oxide superconductors opens the possibility for realization of such a layered structure—a superconducting film protected by an insulating plate. Moreover, we consider that a MIS plane capacitor is one of the simplest possible systems employed in the fundamental research towards further technical applications. Therefore we believe that the suggested experiment

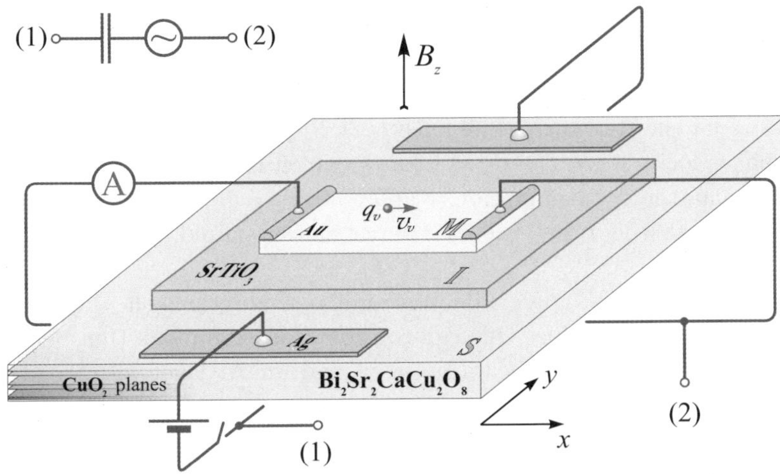

Fig. 4.1 *Gedanken* set-up proposed to determine the vortex charge and the Cooper pair mass. Thin $Bi_2Sr_2CaCu_2O_8$ layer is thread by perpendicular magnetic field B_z. The voltage V_y applied through the Ag electrodes in circuit (1) creates a drift of the vortices with mean velocity v_v. Due to the Bernoulli effect the superfluid currents around every vortex create a change in the electric potential on the superconducting surface. The Bernoulli potential of the vortex leads to an electric polarization on the normal Au surface. The charge q_v, related to the vortex, has the same drift velocity v_v. The corresponding current I_x in circuit (2) can be read by a sensitive ammeter. The quality of the $SrTiO_3$ plate should be high enough so as to allow detection of the interface Hall current without being significant perturbed by the leakage currents between circuits (1) and (2).

could become a standard tool in studying the quality of the insulator-superconductor interface. In order to check whether this idea is another case of a science fiction or, vice versa, is a smoking gun we provide below a numerical example involving an acceptable set of parameters which have been collected from various references: $m^* = 11m_0$ [6], $\xi_{ab}(0) = 1.1$ nm, $d_{\rm ins} = 15$ nm, $d_{\rm ins}/\epsilon_{\rm ins} = 1$ nm [360], $\lambda_{ab}(0) = 150$ nm [361]. For an illustration we take as well: $B_z = 100$ mT, $E_y = 1$ V/cm, and $L_x = L_y = 1$ mm.

The value of B_z we chose imply for the following parameters: $n_v = B_z/\Phi_0 = 4.83 \times 10^{13}\mathrm{m}^{-2}$, $L_x L_y n_v = 48 \times 10^6$, and $1/\sqrt{n_v} = 144$ nm $\simeq \lambda_{ab}(0)$. For a model estimate we also take $r_{\rm min} \approx 150$ nm. It is now straightforward to work out the vortex charge at liquid-helium temperature, i.e., in the temperature range far below $T_{\rm c}$. In this case the substitution of the above mentioned set of parameters in Eq. (4.11) gives

$$\frac{q_v}{|e|} = \frac{1}{8} \cdot \frac{53}{1000} \cdot \frac{1}{11} \cdot \ln(10) = 1.386 \times 10^{-3} \simeq \frac{1}{1000}. \tag{4.19}$$

The so estimated $q_v \simeq 10^{-3}|e|$ is in agreement with another model evaluation due to Khomskii and Freimuth [347]. Further, Eq. (4.14) gives $R_{xy} = 9.35\mathrm{M}\Omega$ and the electric field chosen gives for the voltage $V_y = 100$ mV, therefore for the Hall current we have $I_x = R_{xy}V_y = 11$ pA. Lastly, the vortex drift velocity $v_v = E_y/B_z = 1$ km/s, which is one order of magnitude smaller than the depairing velocity at $T = 0$, $v_c = \hbar/m^*\xi_{ab}(0) = 9.6$ km/s. We note that the resistance of the capacitor should be thus at least $R_{\rm MS} = 100$ MΩ.

For conventional superconductors similar evaluations show that effect is less but still observable. One can consider, for example, a thin Nb metal film grown by molecular beam epitaxy, and an Al layer after oxidation in natural condition could give a good insulator layer. All technologies for planar Josephson junctions provide as a rule metal-insulator interface of sufficient quality. Only the insulator layer should be thick enough to prevent leakage tunneling.

The example analyzed above shows that the proposed experiment is in principle possible to be carried out but we find it difficult to anticipate all problems that could arise in the course of it. For instance, due to a good capacitance cross-talk the noise created by the vortex motion in the superconducting layer will be transmitted to the normal layer thus disturbing the detection of the small Hall current. We believe, however, that similar problems could be surmounted, given the challenge of the novel physics underlying the vortex charge. Furthermore, it is quite possible that the charge, concentrated in the vortex core, is comparable to the charge outside, so only a detailed analysis within the microscopic theory can shine a light on the latter point. In order to verify whether a hydrodynamic approach based upon the Bernoulli effect suffices to quantitatively describe the predicted vortex-charge interface current one needs independent methods to determine the effective mass of the Cooper pairs. In the next section we will analyze similar experiments employing artificial MIS structures.

4.2.5 *How to measure the Cooper pair mass*

Before addressing the problem of measuring the Cooper pair effective mass m^* let us analyze a parallel between the latter issue and the civil engineering, where in a static approximation only the weight $W = mg$ is essential for a construction. In this approximation the masses could reach colossal values if we renormalize the earth acceleration $g \to 0$. The uncertainty, however, immediately disappears during the first earthquake when a dynamical problem should be solved. Just the same is the situation with the superconducting order parameter Ψ—in the static GL theory the superfluid density $n = |\Psi|^2$ and the effective mass m^* are inaccessible separately. They are contained in the experimental parameters, such as the penetration depth Eq. (4.3), only via the ratio n/m^*. In order to determine the effective mass one has to investigate some dynamic phenomenon, which is time-dependent. Due to phase invariance, however, the time t could participate only in the gauge invariant derivative $(i\hbar\partial/\partial t - e^*\varphi)\,\Psi$, that is why electric field effects in superconductors are to be studied. The subtle point is that the latter are already dynamic effects even if the electric fields are static. One therefore needs to perturb the thermodynamic equilibrium of the superconductor as slightly as possible and all methods for determination of the effective mass m^* of Cooper pairs thus become effectively ac methods, based on the electrostatic effects in superconductors. The set-up proposed to determine the vortex charge, Fig. 4.1, is a MIS device having four terminals. Probably the most simple method to accomplish the task would be to use the same MIS structure without making any contacts on the superconducting layer and to investigate the surface Hall current [359] as described in the next subsection.

4.2.5.1 *Surface Hall current*

This physical effect reffers to the 2D surface currents $\mathbf{j}^{(2D)}$ at the surfaces of a thin ($d_{\mathrm{film}} \ll \lambda_{ab}(0)$) superconducting film induced by a normal-to-the-layer electric induction $\mathbf{D}_n$ and parallel-to-the-layer magnetic field $\mathbf{B}_t$

$$\mathbf{j}^{(2D)} = \frac{e^*}{m^*} d_{\rm film} \frac{\lambda_{ab}^2(0)}{\lambda_{ab}^2(T)} \mathbf{D}_n \times \mathbf{B}_t, \tag{4.20}$$

where the Cooper pair mass m^* is the material constant of the effect. This is an electrostatic effect and the superconducting film is in vortex-free state. The dissipation is zero and the superconductor is in thermodynamic equilibrium. A symmetric layered structure is grown by capping of the superconducting film with an insulator layer. Two normal metal layers are evaporated on the protecting insulator layer and on the back side of the substrate thus achieving a plane capacitor configuration. The normal-metal electrodes are circles with radius R and a cartoon of the experimental set-up in Corbino geometry is shown in Fig. 4.2. Exploiting the axial symmetry of the geometry Eq. (4.20) reads as $j_\varphi^{(2D)} \propto D_z B_r$ and for the total magnetic moment of the circulating currents we have

$$M(t) = \int_0^R (\pi r^2) j_\varphi^{(2D)}(r) dr = \frac{e^*}{m^*} d_{\rm film} \frac{\lambda_{ab}^2(0)}{\lambda_{ab}^2(T)} D_z(t) \int_0^R (\pi r^2) B_r(r) dr \tag{4.21}$$

This small magnetic moment could be difficult to detect against the large background due to the dc magnets creating B_r. We derive a static magnetic moment and the next natural step is to consider in a quasistatic approximation the electric induction D_z as being time-dependent, $D_z = D_z(t)$. The ac magnetic moment can be detected by the electromotive voltage

$$\mathcal{E}(t) = -\mu_0 \nu \frac{dM(t)}{dt}, \tag{4.22}$$

induced in the solenoid having ν turns per unite length. The total charge of the capacitor is $(\pi R^2) D_z$ and the time derivative of the electric induction,

$$\frac{dD_z(t)}{dt} = \frac{I(t)}{\pi R^2} \tag{4.23}$$

can be expressed by the current $I(t)$ charging the capacitor. For the electromotive voltage we finally obtain the equation

$$\mathcal{E}(t) = R_{\rm eff} I(t) - M_{12} \frac{dI(t)}{dt}, \tag{4.24}$$

where

$$R_{\rm eff} = -\mu_0 \frac{e^*}{m^*} \frac{\nu d_{\rm film}}{\pi R^2} \frac{\lambda_{ab}^2(0)}{\lambda_{ab}^2(T)} \int_0^R (\pi r^2) B_r(r) dr \tag{4.25}$$

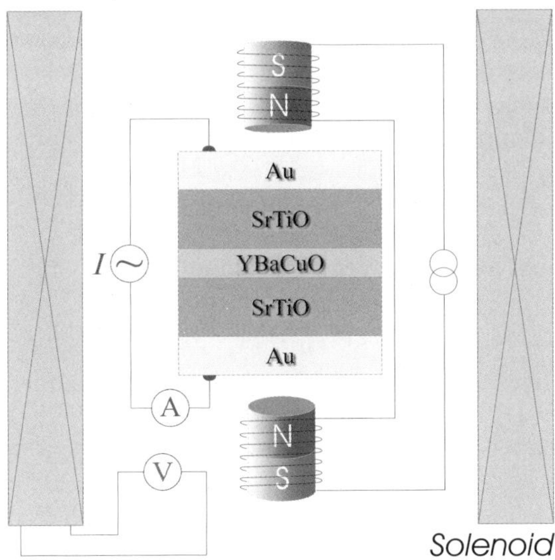

Fig. 4.2 Set-up for observation of surface Hall current induced by a normal to the superconducting film electric induction D_z and nearly homogeneous parallel-to-the-film magnetic field B_r. The core ingredient is a layered MIS structure (see text) in the field of a plane capacitor (Corbino geometry; schematically, not to be scaled). The ac voltage generator creates current I through the plane capacitor, and the dc current source generates opposite oriented magnetic poles in the drive coils and a radial magnetic field B_r in the plane of the superconducting film. A many-turn solenoid is used to detect the ac magnetic moment M_z of the circulating surface Hall currents $j_\varphi^{(2D)}$.

is the effective resistance describing this new electrodynamic effect created by the effective mass m^*. The experimental difficulties might be related with the careful compensation of the mutual inductance M_{12} between the solenoid and the ac generator charging the MIS plane capacitor. The rigorous analysis of the experiment requires the knowledge of the break-through voltages of the MIS structure and the noise induced in the detecting coil, but in any case this auxiliary experiment would be easier to perform than the detection of vortex charge currents.

In the following we will also provide an elementary derivation of the formula for the surface Hall current Eq. (4.20) using the London electrodynamics. Let us trace the trajectory of a London *superconducting electron* (i.e., a Cooper pair) crossing the circular superconducting film during the charging of the MIS plane capacitor, Fig. 4.2. The *superconducting electron* leaves the inital surface of the film with zero velocity $v_\varphi(t_i) = 0$, experiences the Lorentz force while traveling across the film

$$m^* \frac{dv_\varphi(t)}{dt} = e^*_z B_r, \tag{4.26}$$

and arrives at the opposite surface of the film at the t_f, *i.e.*,

$$\int_{t_i}^{t_f} v_z(t)dt = d_{\mathrm{film}} \tag{4.27}$$

with an additional azimuthal velocity component

$$v_\varphi = \frac{e^*}{m^*} d_{\mathrm{film}} B_r. \tag{4.28}$$

For $T = 0$ all charges are superfluid and the electric induction determines the surface (or 2D) excess charge density $D_z = e^* n^{(\mathrm{2D})}$. For the surface current density of these polarization charges we therefore have

$$j_\varphi^{(\mathrm{2D})} = e^* n^{(\mathrm{2D})} v_\varphi = D_z v_\varphi = \frac{e^*}{m^*} d_{\mathrm{film}} D_z B_r. \tag{4.29}$$

For non-zero temperatures one has to take into account the thermal dissociation of the *superconducting electrons*, $e^* \to e + e$, and the appearance of a normal fluid. Thus, taking into account the superfluid part,

$$j_\varphi^{(\mathrm{2D})}(T > 0) = \frac{n(T)}{n(T = 0)} j_\varphi^{(\mathrm{2D})}(T = 0) \tag{4.30}$$

we recover the basic equation Eq. (4.20). The BCS treatment certainly gives the same result because the London electrodynamics is not a mere, naive phenomenological alternative to the microscopic BCS theory, instead it should be viewed as an efficient tool to apply the BCS theory to low frequencies $\omega \ll \Delta/\hbar$ and small wave-vectors $k\xi_{ab}(0)$.

Analogous experiment could be performed with a bulk crystal or thick film $d_{\mathrm{film}} \gg \lambda_{ab}(0)$. In this case in the initial Eq. (4.20) and the final result, Eq. (4.25), the thickness of the film d_{film} should be replaced with the penetration depth $\lambda_{ab}(T)$ and the formula for the surface current then reads as

$$\mathbf{j}^{(\mathrm{2D})} = \frac{e^*}{m^*} \frac{\lambda_{ab}^2(0)}{\lambda_{ab}(T)} \mathbf{D}_n \times \mathbf{B}_t. \tag{4.31}$$

The investigation of the temperature dependence of this effect can give a new method for determination of the temperature dependence of the penetration depth $\lambda_{ab}(T)$. A $SrTiO_3$ layer should be grown on the fresh cleaved surface of $Bi_2Sr_2CaCu_2O_8$ crystal and a circular Au electrode needs to be overgrown on the protecting layer. One plate of the capacitor is the bulk high-T_c crystal and the other one is the Au layer. In order to avoid frozen vortices the constant magnetic field of the dc drive coil must be applied after cooling down to low temperatures. An ac voltage should be applied to the plane capacitor, a lock-in ammeter will measure the polarization current, and the induced due to the effect ac magnetic moment can be detected by a lock-in voltmeter connected to the detector coil. For derivation of the above formula Eq. (4.31) we have to use: (i) the distribution of the vector-potential at depth $|z|$ in the superconductor and some fixed radius r,

$$A_\varphi(z) = B_r \lambda_{ab}(T) \exp\left(-\frac{|z|}{\lambda_{ab}(T)}\right), \tag{4.32}$$

where $B_r(r)$ is the tangential magnetic field at the superconducting surface; (ii) the London-BCS formula for the current response of the superconductor (the polarization operator),

$$j_\varphi = -\frac{A_\varphi}{\mu_0 \lambda_{ab}^2(T)}, \tag{4.33}$$

and (iii) the formula for the bulk (3D) density of the superfluid polarization charges

$$e^* n(z) = D_z \frac{\lambda_{ab}^2(0)}{\lambda_{ab}^2(T)} \delta(z), \tag{4.34}$$

where δ stands for the Dirac δ-function. The effective mass m^* can be determined not only by the surface Hall effect but also from the Bernoulli effect for which the BCS theory was developed by Omel'yanchuk and Beloborod'ko [362] as well as from all other predictions of the London theory. The existence of Bernoulli effect for conventional superconductors is experimentally confirmed; some references can be found, for example, in Ref. [358]. In the next subsection we give a brief account of the suggested here Cooper pair mass spectroscopy.

4.2.5.2 *Bernoulli effect in thin superconducting film*

The experimental set-up for a current-induced Cooper pair mass spectroscopy is presented in Fig. 4.3. The Bernoulli effect is related to a current-induced contact-potential difference that can be measured by the electrostatic polarization of a normal metal electrode which covers the surface of the superconductor, forming a plane capacitor. For the averaged change of the electric potential beneath the electrode the Bernoulli theorem Eq. (4.2) gives

$$e^* \langle\varphi\rangle = -\frac{n(T)}{n(T=0)} \left\langle \frac{1}{2} m^* v^2 \right\rangle, \tag{4.35}$$

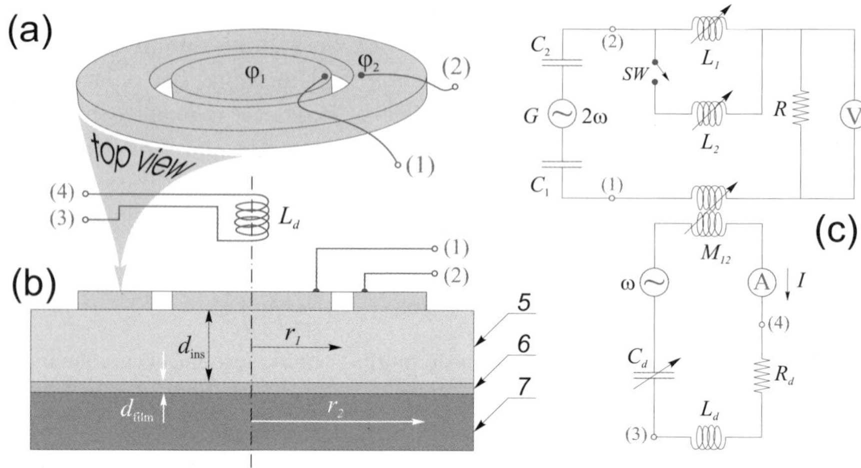

Fig. 4.3 Cooper pair mass spectroscopy based on the Bernoulli potential (after Ref. [358]). (a) top view (b) cross section, (c) equivalent electric scheme. Two electrodes, circle- (1) and ring-shaped electrode (2), should be produced on the insulating layer capping the superconducting film. (3) and (4) denote the contacts of the drive coil with inductance L_d and resistance R_d; (5)—insulator layer with thickness $d_{\rm ins}$; (6)—superconducting film with thickness $d_{\rm film} < \lambda_{ab}(0)$; (7)—substrate; M_{12}—mutual inductance; L_1, L_2—variable inductances; R—load resistor; V—voltmeter; A—ammeter; SW—switch; C_d—capacitor of the drive resonance contour with resonance frequency ω; G—Bernoulli voltage generator with doubled frequency 2ω; C_1, C_2—capacitances between the superconducting film and metal electrodes (1) and (2). This figure and the underlying author's idea have been used in the discussions in Refs. [347,348] on the vortex charge problem; for distribution of the electric force lines of the circulating currents see Fig. 1 of Ref. [348].

i.e., the Bernoulli potential is proportional to the averaged kinetic energy of Cooper pairs beneath the electrode. For thin films, $d_{\rm film} < \lambda_{ab}(0)$, the current across the layer is more or less homogeneous $j^{(\rm 2D)} = d_{\rm film} j$ and we have to substitute in this equation $v \approx j^{(\rm 2D)}/(d_{\rm film} e^* n(T))$. Then the formula for the Bernoulli potential takes the form

$$\begin{aligned}\langle\varphi\rangle &= -\frac{m^*\langle(j^{(\rm 2D)})^2\rangle}{2e^{*3}d_{\rm film}^2 n(0)n(T)} = -\frac{e^*\mu_0^2\lambda_{ab}^2(0)\lambda_{ab}^2(T)}{2m^* d_{\rm film}^2}\langle(j^{(\rm 2D)})^2\rangle \\ &= -\frac{L_\Box(T)}{2e^*n(0)d_{\rm film}}\langle(j^{(\rm 2D)})^2\rangle,\end{aligned} \tag{4.36}$$

where

$$L_\Box(T) \equiv \frac{m^*}{e^{*2}n(T)^{(\rm 2D)}} = \mu_0\frac{\lambda_{ab}^2(T)}{d_{\rm film}} \tag{4.37}$$

is the kinetic inductance which can be measured directly by means of the mutual inductance method [361, 363], $n^{(2D)}(T) = n(T)d_{\rm film}$ is the area density of Cooper pairs and $\langle (j^{(2D)})^2 \rangle$ is the averaged square of the 2D supercurrent beneath the electrode whose distribution has to be found by solving a magnetostatic problem. If two electrodes were grown on the superconductor surface, the Bernoulli voltage

$$V_{\rm Bernoulli} = \langle \varphi \rangle_2 - \langle \varphi \rangle_1 \tag{4.38}$$

can be considered as a voltage generator sequentially connected to two capacitors C_1 and C_2 as depicted in Fig. 4.3 (c). The currents induced in the superconductor film are proportional to the current through the drive coil L_d, $j^{(2D)} \propto I_{\rm drive}$, Fig. 4.3 (b,c). The coefficient A_a of this proportion $\langle (j^{(2D)}) \rangle = I^2_{\rm drive}/A_a$ has dimension of area [358]. According to Eq. (4.36) an ac current $I_d \propto \cos(\omega t)$ will create an ac Bernoulli voltage of doubled frequency $V_{\rm Bernoulli} \propto \cos(2\omega t)$. Initially, in the switched-off regime, when the detector contour resonates at frequency $\omega = 1/(L_1 C)^{1/2}$, where $C = C_1 C_2/(C_1 + C_2)$ the parasite mutual inductance between the drive coil contour and the detector contour must be carefully annulled by a small tunable mutual inductance M_{12}. After that taking $L_2 \approx L_1/3$ in switched-on regime the detecting contour will resonate at doubled frequency $2\omega = 1/(LC)^{1/2}$, $L = L_1 L_2/(L_1 + L_2)$. In resonance conditions the Bernoulli voltage can be directly detected by a lock-in voltmeter with a low noise preamplifier. If we know the penetration depth $\lambda_{ab}(T)$ the measured Bernoulli voltage, according to the Eq. (4.36), gives the effective mass of Cooper pairs m^*.

If thick films, $d_{\rm film} \gg \lambda_{ab}(0)$, or bulk single crystals are to be used for such experiment we have to substitute in Eq. (4.35) the London formula for the velocity $m^*\mathbf{v} = -e^*\mathbf{A}$, which is a trivial consequence of the Newton equation $m^* \mathrm{d}\mathbf{v}/\mathrm{d}t = e^*\mathbf{E}$ for a nearly homogeneous electric field $\mathbf{E}(t) = -\partial \mathbf{A}/\partial t$. Combining with Eq. (4.32) we obtain

$$\langle \varphi \rangle = -R_{\rm LH} \langle p_B \rangle, \qquad p_B = \frac{B_r^2}{2\mu_0}, \qquad R_{\rm LH} \equiv \frac{1}{e^* n(T=0)}, \tag{4.39}$$

where p_B is the pressure of the tangential to the superconducting surface magnetic field, and the $R_{\rm LH}$ is the temperature independent [358] London–Hall constant expressed via the total volume density of conduction band $\rho_{\rm tot} = e^* n(0) = 1/R_{\rm LH}$. We consider the Greiter, Wilczek and Witten's [364] prediction for a temperature dependence of the London–Hall constant as being erroneous and the problem still waits for its experimental solution. For type-I superconductors the Eq. (4.39) can be applied up to

$B_c(T)$ obtaining in this way the contact potential difference Eq. (4.15). It is still questionable whether the thermal-induced contact-potential difference

$$\varphi(T_2) - \varphi(T_1) = -\frac{1}{e^* n(0)} \frac{B_c^2(T_2) - B_c^2(T_1)}{2\mu_0} \tag{4.40}$$

may be measured, but if the answer is positive this effect can be used to determine the thermodynamic critical field $B_c(T)$ even for type-II superconductors. In any case the fluctuation of the temperature should be taken into account in the experiments aiming to observe the Bernoulli effect.

The realistic experiment proposed in Ref. [358] can be substantially simplified (cf. Ref. [365]): the ring electrode capacitor can be substituted by a short circuit, and the central one could cover the whole facet. We stress that at least one capacitive connection is indispensable. The voltmeters do not measure any voltage difference but just the *difference in the electrochemical potential* (even nowadays almost 99% of the experimentalists are unaware of what a voltmeter really measures)! An error of the kind has prevented Lewis [366, 367] during his pioneer investigations in the period 1953–1955 from observing the Bernoulli effect in superconductors soon after it has been predicted by London [368]. Lewis did not use the capacitive connection but he introduced all other necessary ingredients: lock-in voltmeter with nV sensitivity, ac magnetic field and doubling of the frequency. Now it is worthwhile measuring both the Bernoulli effect and the surface Hall effect in the same sample. At known total charge density ρ_{tot}, Eq. (4.39), and penetration dept $\lambda_{ab}(0)$, Eq. (4.31), the Cooper pair mass can be determined as $m^* = \mu_0 e^* \rho_{\text{tot}} \lambda_{ab}^2(0)$. Despite the $\sim 10^5$ papers published on high-T_c superconductivity (cf. Fig. 0.1 on page vii), without the Cooper pair mass the physics of superconductivity remains Hamlet without the Prince, with only the role of Ophelia performed by *onnagata*.[1] In the next subsection we briefly describe the only, to the best of our knowledge, reliable experiment for determination of effective mass m^*.

4.2.5.3 *Electric charge modulation of the kinetic inductance*

When an electric voltage is applied to a MIS plane capacitor the charging of the superconducting surface will create a change of the 2D superfluid charge density

$$e^* n^{(2\text{D})} = e^* d_{\text{film}}\, n(T) + D_z \frac{n(T)}{n(0)} = (d_{\text{film}}\rho_{\text{tot}} + D_z)\,\frac{\lambda_{ab}^2(0)}{\lambda_{ab}^2(T)}. \tag{4.41}$$

[1]Female impersonator in *kabuki* theater.

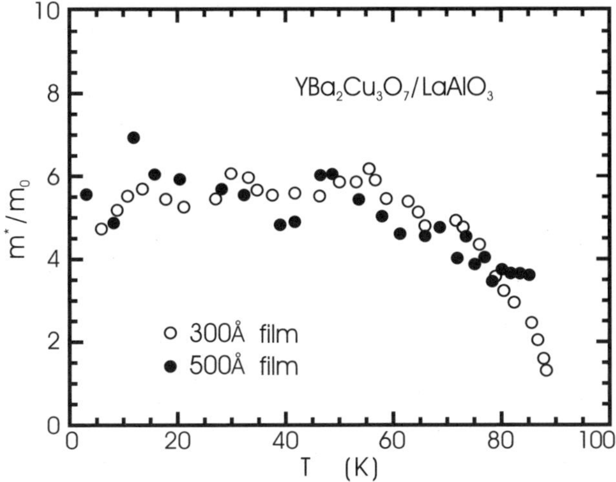

Fig. 4.4 First experimental determination of effective mass of Cooper pairs m^* relative to the doubled free electron mass m_0 (reprinted with permission from Ref. [369]; Copyright © 1991 by the American Physical Society).

It is then easily worked out from Eq. (4.37) that this creates a modulation of the kinetic inductance and the derivative determines [6] the effective mass

$$m^* = -e^* L_\square(0) L_\square(T) \frac{\delta D_z}{\delta L_\square(T)}. \tag{4.42}$$

This simple picture gets complicated due to T_c-changing upon electrostatic doping of the material, but below the critical region this experiment confirms [6, 370] a temperature independent effective mass m^*, Fig. 4.4. When m^* and all other parameters of the superconductor are already determined we can turn to the vortex charge problem. Note also that the same temperature-dependent kinetic inductance from the above equation appears in the wave-vector dependence of the frequency $\omega_{\rm 2D}$ of the 2D plasmons predicted by Mishonov and Groshev [5],

$$\omega_{\rm 2D}^2(k) = 2\pi L_\square^{-1}(T) k / \epsilon_{\rm ins}, \tag{4.43}$$

where $\epsilon_{\rm ins}$ is the mean dielectric constant of insulators on both sides of the thin superconducting layer. Shortly after the theoretical prediction [5] the 2D plasma resonances were experimentally confirmed by Buisson *et al.* [257], Fig. 4.5.

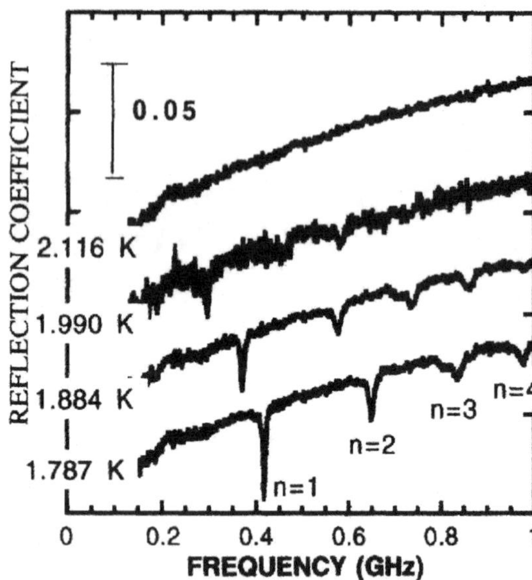

Fig. 4.5 Plasma resonances for thin granular Al film deposited on $SrTiO_3$ substrate (reprinted with permission from Ref. [257]; Copyright © 1994 by the American Physical Society). Frequency dependence of the reflection coefficient of a coaxial line terminated by the film, measured for different T. The plasmon mode index is labeled by n.

4.2.6 *Discussion*

The preceding analysis demonstrates that the proposed electronic measurements are feasible and the suggested experimental programme could be soon realized. The appearance of the first good samples would immediately lead to the solution of the problem concerning the vortex charge and Cooper pair mass. These two parameters, q_v and m^*, might fall in the lime-light of the physics of superconductivity in the nearest future. As a byproduct the Cooper pair mass spectroscopy could become a standard tool for testing the quality of the superconducting films for future superconductor electronics. Even in the present chapter we have suggested a few new effects thus there is no doubt that new physics will emerge from the development of the layer-by-layer oxide technology. Let us also list some of the main results of this study: the formulae for vortex charge Eqs. (4.11) and (4.19), vortex conductivity Eqs. (4.13) and (4.14), surface Hall current for bulk crystals Eq. (4.31), interface Hall conductivity for type-I superconductors Eq. (4.17), thermal-induced contact-potential difference Eq. (4.40), etc.

Finding a solution to the vortex charge problem by employing a model system, where the superconducting and the polarized layers are separated, will immediately trigger the answer to the question about what is the charge induced in the adjacent CuO_2 layers by a pancake vortex. One may further ask about the fate of the charge cloud when the pancake vortices "polymerize" in a vortex line, and what is the influence of the vortex charge in the vortex-vortex interaction and correlation. According to our analysis of the Bernoulli effect the charge will concentrate at the end of vortex lines, at kinks and sharp turns of stacks of pancake vortices. Needless to say, the clear solution of some model problems is always useful in the search for solution to the complex problems in material science.

The problem of determining the vortex charge by a transport measurement brings us back to one of the first ideas of the electron physics. Only two months after the discovery of the electron [371] Francis Mott made the first attempt to observe the influence of the electric fields and surface charges on the conductivity of Pt. Likewise, the vortex charge current has led us to another immortal idea of the XIX century—the Kelvin vortex model of the "atom". Starting from a hydrodynamic approach, we were able to realize that the hydrodynamic excitations can propagate as particles and that the charge related to vortex "atoms" gives a measurable electric current.

Chapter 5

Thermodynamics of Gaussian fluctuations and paraconductivity

5.1 Introduction

The writing of this chapter was provoked by the progress in the studies of fluctuation phenomena in high-T_c superconductors [372]. The small coherence lengths in the layered cuprates $\xi_c \ll \xi_a \simeq \xi_b$ give rise to a very high density of fluctuation degrees of freedom $\propto 1/(\xi_{ab}^2(0)\xi_c(0))$ which makes the fluctuation effects easier to be observe in the high-T_c- rather than in the conventional superconductors. An intriguing feature of the fluctuation effects to be pointed out is that they can be observed even in the case when the interaction between fluctuations is vanishing or can be treated in a self-consistent manner. In such a case, for high quality crystals, the fluctuations are of Gaussian nature and their theory is very simple. A number of good experimental studies have already been performed in the Gaussian regime thus initiating the Gaussian fluctuation spectroscopy for high-T_c materials.

By spectroscopy here we imply only those experiments with trivial theory where every measurement provides an immediate information for some parameter(s) important for the material science or fundamental physics of these interesting materials. Half a century ago Landau used to speak about himself as being the greatest trivializator in the theoretical physics. At present, the Ginzburg–Landau (GL) theory (called by Ginzburg also Ψ-theory) is the adequate tool to describe the fluctuation phenomena in the superconductors. The parameters of the Ψ-theory, such as coherence lengths, relaxation time $\tau_{0,\Psi}$ of the GL order parameter Ψ, the GL parameter $\kappa_{\rm GL} = \lambda_{ab}(0)/\xi_{ab}(0)$ are also "meeting point" between the theory and the experiment.

From one hand, these parameters are necessary for the description of the experimental data and from another hand they can be derived from the microscopic theory using the methods of the statistical mechanics. That is why the determination of the GL parameters is an important part of the investigations of every superconductor and the Gaussian fluctuation spectroscopy is an indispensable tool in these comprehensive investigations.

The purpose of this chapter is to systematize the known classical results for the GL Gaussian fluctuations, to derive new ones when needed, and to finally give suitable for coding formulae necessary for the further development of the Gaussian spectroscopy. The derivation of all results is described in detail and trivialized to the level of the Landau–Lifshitz encyclopedia on theoretical physics [373], the textbooks by Abrikosov [272] and Tinkham [374] or the well-known reviews by Cyrot [375] on the GL theory, by Bulaevskii [376] concerning the layered superconductors with Josephson coupling, and by Skocpol and Tinkham [377] on the fluctuation phenomena in superconductors. The present chapter is intended as a review on the theoretical results which can be used by the Gaussian spectroscopy of fluctuations but no historical survey of the experimental research is attempted. Therefore we do explicitly refer to only a limited number of experimental studies in this field. Instead, the reader is referred to the citations-reach conference proceedings [372], but even therein a number of good works are probably not included. We do not cite directly even the epoch-creating paper by Bednorz and Müller but its spirit can be traced to every contemporary paper on high-$T_c$superconductivity. Even to focus on the theoretical results related to fluctuation phenomena is a very difficult problem by itself and therefore, when referring to any result one should imply "to the best of our knowledge...". One of our goals was also to fill the gap between the textbooks and experimentalists' needs for a compilation of theoretical formulae written in common notations, appropriate for direct use.

Of course, there is a great number of interesting physical situations especially related to vortices where the fluctuations are definitely non-Gaussian. Those problems fall beyond the scope of the present review and we include only some references from this broad field in the physics of superconductivity [378–387].

The review is organized as follows: in Sec. 5.2 the case of weak magnetic fields is considered and the thermodynamic variables are expanded in power series in the dimensionless magnetic field $h = B_z/B_{c2}(0)$. The standard notations for the thermodynamic variables in a layered superconductor are then introduced in Sec. 5.2.1, and Sec. 5.2.2 is dedicated to the Euler–MacLaurin summation formula in the form appropriate for the analysis of the GL results for the free energy and its ultraviolet (UV) regularization. A systematic procedure to derive the results for a layered superconductor from the results for a two-dimensional (2D) superconductor is developed in Sec. 5.2.3 and the action of the introduced "layering" operator $\hat{\mathsf{L}}$ is illustrated on the example of the formulae for the paraconductivity. Further we consider the static paraconductivity in case of perpendicular magnetic field as well as the high-frequency conductivity in zero magnetic field. The power series for the nonlinear magnetic susceptibility and the magnetic moment in the Lawrence–Doniach (LD) model are derived in Sec. 5.2.4 and the ε-method for summation of such divergent series is described in Sec. 5.2.5.

Further in Sec. 5.2.6 we present the power series for the differential susceptibility and general weak-magnetic-field expansion formulae for the magnetization. Section 5.3 is dedicated to the study of the strong magnetic fields limit. Firstly, in Sec. 5.3.1 the general formula for the Gibbs free energy in perpendicular to the layers magnetic field is analyzed. The fluctuation part of the thermodynamic variables is found then by differentiation in Sec. 5.3.2. Section 5.3.3 is devoted to the self-consistent mean-field treatment of the fluctuation interactions in the LD model. The important limit case of an anisotropic 3D GL model is considered in Sec. 5.3.4 where we derive the Gibbs free energy and the fluctuation magnetic moment. In Sec. 5.4 an account is given of the fitting procedure for the GL parameters which rests on theoretical results and some recommendations for the most appropriate formulae are also given for determination of the cutoff energy $\varepsilon_{✄}$ in Sec. 5.4.1, the in-plane coherence length $\xi_{ab}(0)$ in Sec. 5.4.2, the Cooper pair life-time constant τ_0 in Sec. 5.4.3, and the 2D Ginzburg number in Sec. 5.4.4. All new results derived throughout this review are summarized in Sec. 5.5 and some perspectives for the Gaussian spectroscopy are discussed as well.

5.2 Weak magnetic fields

5.2.1 *Formalism*

Before embarking on a detailed analysis we shall briefly introduce all entities entering the basic for our further considerations quantity—the GL functional G for the Gibbs free energy in external magnetic field $\mathbf{H}^{\mathrm{ext}}$. For compliance with the previous works we follow the standard notations in which G reads

$$G[\Psi_{j,n}(x,y),\mathbf{A}(\mathbf{r})] = \sum_{n=-\infty}^{+\infty}\sum_{j=1}^{N}\int dx\,dy\Bigg\{\sum_{l=x,y}\frac{1}{2m_{ab}}\left|\left(\frac{\hbar}{\mathrm{i}}\frac{\partial}{\partial x_l}-e^*A_l\right)\Psi_{jn}\right|^2$$

$$+\,a_0\epsilon\left|\Psi_{j,n}\right|^2+\frac{1}{2}\tilde{b}\left|\Psi_{j,n}\right|^4+a_0\gamma_j\left|\Psi_{j+1,n}-\Psi_{j,n}\exp\left(\frac{\mathrm{i}e^*}{\hbar}\int_{z_{j,n}}^{z_{j+1,n}}A_z dz\right)\right|^2\Bigg\}$$

$$+\int\frac{1}{2\mu_0}\left(\nabla\times\mathbf{A}-\mu_0\mathbf{H}^{\mathrm{ext}}\right)^2 dx\,dy\,dz, \qquad (5.1)$$

with $\mathbf{A}$ being the vector potential of the magnetic field $B=\nabla\times\mathbf{A}$.

The material parameters in this sizable expression are illustrated in Fig. 5.1, thus we only need to note that the GL potential $a(\epsilon)=a_0\epsilon$ is parameterized by $a_0=\hbar^2/2m_{ab}\xi_{ab}^2(0)$, and $\epsilon\equiv\ln(T/T_{\mathrm{c}})\approx(T-T_{\mathrm{c}})/T_{\mathrm{c}}$ is the reduced (dimensionless) temperature. If not otherwise stated we shall make use of the SI units, thus the magnetic permeability of vacuum $\mu_0=4\pi\times10^{-7}$.

Here we will restrict ourselves to the study of fluctuations in the Gaussian regime in the normal phase not too close to the critical line $H_{\mathrm{c}2}(T)$. In this case the nonlinear term in $G[\Psi,\mathbf{A}]$ is negligible, $\tilde{b}|\Psi|^4/2\to0$. For the normal phase the magnetization is also very small and with high accuracy $\mu_0\mathbf{H}^{\mathrm{ext}}\approx\mathbf{B}=\mu_0(\mathbf{H}+\mathbf{M})\approx\mu_0\mathbf{H}$.

To begin with, consider the simplest case of zero external magnetic field $\mathbf{H}^{\mathrm{ext}}=0$. Given the above assumption for Ψ, the GL functional is a quadratic form and one needs to sum over all eigenvalues of the energy spectrum

$$\varepsilon_j(\mathbf{p},p_z)=\frac{\mathbf{p}^2}{2m_{ab}}+\varepsilon_{cj}(p_z),\quad \varepsilon_{cj}(p_z)=a_0\omega_j^{(N)}(\theta), \qquad (5.2)$$

where $\varepsilon_{cj}(p_z)$ are the tight-binding energy bands describing the motion of Cooper pair in z (c) direction, $\mathbf{p}=(p_x,p_y)$ is the in-plane (ab-plane) momentum of the fluctuating Cooper pairs and $\theta=p_z s/\hbar\in(0,2\pi)$ is the Josephson phase. For a single layered material, $N=1$, this corresponds to the well known Lawrence–Doniach model [388],

$$\omega_1^{(\mathrm{LD})}(\theta)=2\gamma_1(1-\cos\theta) \qquad (5.3)$$

while the case $N=2$ is the Maki-Thompson (MT) model, [389] proposed independently by Hikami and Larkin [390] as well,

$$\omega_j^{(\mathrm{MT})}(\theta)=\gamma_1+\gamma_2+(-1)^j\sqrt{\gamma_1^2+\gamma_2^2+2\gamma_1\gamma_2\cos\theta},\qquad j=1,2. \qquad (5.4)$$

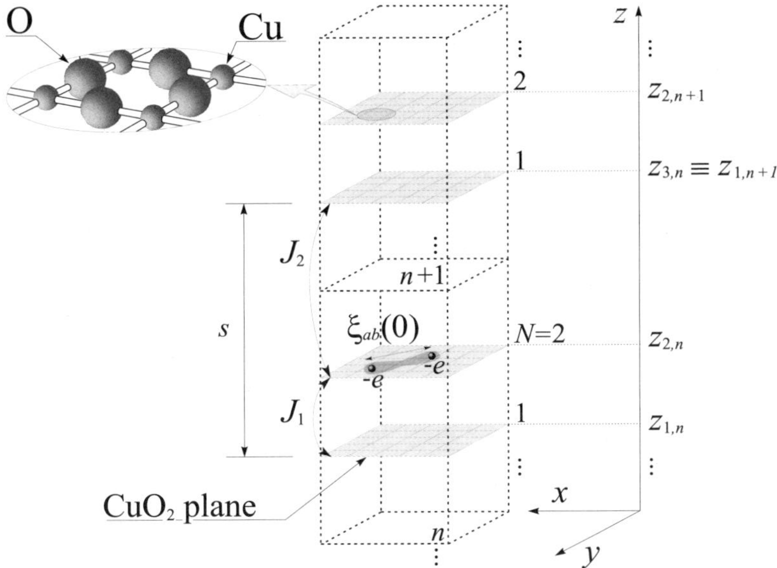

Fig. 5.1 Pictorial representation of material parameters that enter Eq. (5.1). The z-coordinates of the CuO_2 planes, satisfying periodic boundary conditions, are $z_{N+1,n} = z_{1,n+1}$, where n labels the unit cell and $j = 1, \ldots, N$ is the index of the CuO_2 plane within each unit cell. Periodicity in c-direction is designated by s, i.e., $0 \leqslant z_{j,0} \leqslant s$. $\xi_{ab}(0)$ stands for the in-plane coherence length extrapolated to $T = 0$. The Josephson coupling energies between the neighboring planes $J_j = a_0\gamma_j$ are parameterized via dimensionless quantities γ_j. Lastly, the effective charge and the in-plane effective mass of the Cooper pairs are, respectively, $|e^*| = 2|e|$ and m_{ab}.

Thus, the sum over the energy spectrum gives

$$G[\Psi] = \sum_{\mathbf{p},p_z,j} (\varepsilon_j(\mathbf{p},p_z) + a) \left|\Psi_{\mathbf{p},p_z,j}\right|^2, \tag{5.5}$$

where $\Psi_{\mathbf{p},p_z,j}$ is the wave function of the superconducting condensate in momentum space. We use standard periodic boundary conditions for a bulk domain of volume $V = L_x \times L_y \times L_z$ which give

$$\sum_{\mathbf{p}} = L_x L_y \int \frac{dp_x\, dp_y}{(2\pi\hbar)^2}, \qquad \sum_{p_z} = L_z \oint \frac{d\theta}{2\pi s}. \tag{5.6}$$

In order to calculate the fluctuation part of the Gibbs free energy $G(T)$ at zero magnetic field, one usually solves for every point in the momentum space $\mathbf{p}, p_z, j$ the Gaussian integral

$$\exp\left[-\frac{G}{k_{\mathrm{B}}T}\right] = \iint \frac{d\Psi'\, d\Psi''}{2\pi} \exp\left\{-\frac{\varepsilon + a}{k_{\mathrm{B}}T}\left[(\Psi')^2 + (\Psi'')^2\right]\right\}$$

$$= \frac{1}{2}\int_0^\infty \exp\left[-\frac{\varepsilon + a}{k_{\mathrm{B}}T}\rho\right] d\rho = \frac{k_{\mathrm{B}}T/2}{\varepsilon + a}, \tag{5.7}$$

where $\Psi = \sqrt{\rho}\,\mathrm{e}^{\mathrm{i}\varphi} \equiv \Psi' + \mathrm{i}\Psi''$, and $\varphi \in (0, 2\pi)$. Making use of this auxiliary result the calculation of the fluctuation part of the Gibbs free energy reduces to summation over the spectrum of an effective Hamiltonian, i.e.,

$$G = -k_{\mathrm{B}}T \sum_{\mathbf{p}, p_z, j} \ln \frac{k_{\mathrm{B}}T/2}{\varepsilon_j(\mathbf{p}, p_z) + a}, \tag{5.8}$$

or, taking into account Eqs. (5.2) and (5.6),

$$\frac{G}{V} = k_{\mathrm{B}}T \int \frac{d\left(\pi \mathbf{p}^2\right)}{(2\pi\hbar)^2} \frac{1}{N}\sum_{j=1}^{N} \oint \frac{d\theta}{2\pi s} \ln\left[\frac{(\mathbf{p}^2/2m_{ab}) + a_0\omega_j^{(N)}(\theta) + a_0\epsilon}{a_0}\,\frac{a_0}{\frac{1}{2}k_{\mathrm{B}}T}\right]. \tag{5.9}$$

In view of the further calculations it is also useful to introduce a dimensionless in-plane kinetic energy

$$\tilde{x} = \frac{\mathbf{p}^2}{2m_{ab}a_0} = \left(\frac{\xi_{ab}(0)\mathbf{p}}{\hbar}\right)^2 \in (0, c), \tag{5.10}$$

bound by a dimensionless cutoff parameter c which we consider to be an important parameter of the GL theory when applied to copper oxide superconductors. Later in Sec. 5.4.1 we demonstrate how the value of the dimensional cutoff energy $\varepsilon_{\text{ж}}$,

$$\frac{\mathbf{p}^2}{2m_{ab}} < \varepsilon_{\text{ж}} = c a_0 = \frac{p_c^2}{2m_{ab}}, \tag{5.11}$$

can be determined by fitting to the experimental data. An immediate simplification to Eq. (5.9) can be achieved by dropping the $\frac{1}{2}k_{\mathrm{B}}T/a_0 \to \mathrm{const}$ multiplier in the argument of the logarithm as it is irrelevant for the critical behavior of the material. Furthermore, since fluctuational observables are related to non analytical dependence of the Gibbs free energy on the reduced temperature, we can substitute $T = T_{\mathrm{c}}(1 + \epsilon) \approx T_{\mathrm{c}}$ and the free energy per unit volume $F(\epsilon)$ is cast in more elegant form,

$$F(\epsilon) \approx \frac{G}{L_x L_y L_z} = \frac{k_{\mathrm{B}}T_{\mathrm{c}}}{4\pi\xi_{ab}^2(0)}\frac{N}{s}\int_0^c d\tilde{x}\,\frac{1}{N}\sum_{j=1}^{N}\oint \frac{d\theta}{2\pi}\ln\left(\tilde{x} + \omega_j^{(N)}(\theta) + \epsilon\right), \tag{5.12}$$

that could easily include the $(1+\epsilon)$-factor in all cases when necessity appears. The physical meaning of this important for our further considerations expression is fairly transparent: one has to integrate with respect to the Josephson phase θ, which describes the motion of Cooper pairs in c-direction, and to take into account as many different Cooper pair energy bands as are there the different superconducting layers per unit cell. Finally, integration with respect to the in-plane Cooper pair kinetic energy is to be carried out.

Consider now the important case of an external magnetic field applied parallel to the c-direction, i.e., perpendicular to the CuO_2 planes, $\mathbf{B} = (0, 0, B)$. In this case, the in-plane kinetic energy of the Cooper pairs acquires oscillator spectrum [391], corresponding to the quantum mechanical problem of an electron in an external magnetic field [373],

$$\frac{\mathbf{p}^2}{2m_{ab}} \to \hbar\omega_c\left(n + \frac{1}{2}\right), \tag{5.13}$$

where $n = 0, 1, 2, 3, \ldots$ is a non-negative integer and $\omega_c = |e^*|\,B/m_{ab}$ is the cyclotron frequency. The integration over the momentum space is thus reduced to summation over oscillator energy levels

$$\int_{|\mathbf{p}|<p_c} \frac{d^2\mathbf{p}}{(2\pi\hbar)^2} \to \frac{B}{\Phi_0}\sum_{n=0}^{n_c-1}, \tag{5.14}$$

where $n_c \equiv c/2h$ and $\Phi_0 = 2\pi\hbar/\,|e^*| = 2.07$ fT m^2 is the flux quantum. The energy cutoff is to be applied now to the oscillator levels [373], $\hbar\omega_c(n_c+\frac{1}{2}) = ca_0$. Let us recall that the equation for the upper critical field $H_{c2}(T)$ within the GL theory is nothing but the equation for annulment of the lowest energy level, $\frac{1}{2}\hbar\omega_c + a(\epsilon) = 0$. Thereby introducing the upper critical field linearly extrapolated to zero temperature,

$$\mu_0 H_{c2}(0) = B_{c2}(0) \equiv -T_c \left.\frac{dB_{c2}(T)}{dT}\right|_{T_c} = \frac{\Phi_0}{2\pi\xi_{ab}^2(0)}, \tag{5.15}$$

and the dimensionless reduced magnetic field,

$$h \equiv \frac{B}{B_{c2}(0)} = \frac{H}{H_{c2}(0)}, \tag{5.16}$$

we obtain a linear approximation for the critical line about T_c,

$$h_{c2}(\epsilon) = \frac{H_{c2}(T)}{H_{c2}(0)} \approx -\epsilon \ll 1. \tag{5.17}$$

With the help of the dimensionless variables introduced so far it is easily worked out that the influence of the external magnetic field is reduced to discretization of the dimensionless in-plane kinetic energy,

$$\tilde{x} \to h(2n+1) \tag{5.18}$$

and the integrals of an arbitrary function f with respect to $\tilde{x}$ are converted to sums,

$$\int_0^c f(\tilde{x})d\tilde{x} \to 2h\sum_{n=0}^{n_c-1} f(h(2n+1)). \tag{5.19}$$

In fact, Max Planck discovered the quantum statistics of the black-body radiation using the same replacement. Applying this procedure to the previously derived free energy at zero magnetic field, Eq. (5.12), we obtain

$$F(\epsilon) \to F(\epsilon,h) = \frac{\Delta G}{L_x L_y L_z}, \tag{5.20}$$

$$F(\epsilon,h) = \frac{k_{\mathrm{B}}T_{\mathrm{c}}}{4\pi\xi_{ab}(0)^2}\frac{N}{s}2h\sum_{n=0}^{n_c-1}\frac{1}{N}\sum_{j=1}^{N}\oint\frac{d\theta}{2\pi}\ln\left[h(2n+1)+\omega_j^{(N)}(\theta)+\epsilon\right]. \tag{5.21}$$

This expression represents the starting point for all further considerations. As a first step we address in the next section the Euler–MacLaurin method and its application to the sum over the Landau levels which appears in Eq. (5.21).

5.2.2 *Euler–MacLaurin summation for the free energy*

Near the critical temperature, when $\epsilon \ll c$, one can consider formally $c \to \infty$, and $n_c(h) \approx c/2h \to \infty$. Within such a local approximation the previous finite sums are transformed into infinite ones,

$$2h\sum_{n=0}^{\infty} f(\epsilon + h(2n+1)) = \hat{\Sigma}_{\mathsf{EM}}\int_{\epsilon}^{\infty} f(\tilde{x})d\tilde{x}, \tag{5.22}$$

where

$$\begin{aligned}\hat{\Sigma}_{\mathsf{EM}} \equiv \frac{h\frac{\partial}{\partial\epsilon}}{\sinh\left(h\frac{\partial}{\partial\epsilon}\right)} &= \sum_{n=0}^{\infty}(-1)^n\frac{2}{\pi^{2n}}\left(1-\frac{1}{2^{2n-1}}\right)\zeta(2n)\left(h\frac{\partial}{\partial\epsilon}\right)^{2n}\\ &= 1-\frac{1}{6}h^2\frac{\partial^2}{\partial\epsilon^2}+\frac{7}{360}h^4\frac{\partial^4}{\partial\epsilon^4}-\frac{31}{15120}h^6\frac{\partial^6}{\partial\epsilon^6}+\cdots\end{aligned} \tag{5.23}$$

is the Euler–MacLaurin operator for summation of series, in which we employ the Riemann and Hurwitz zeta functions, respectively,

$$\zeta(\nu) = 1 + \frac{1}{2^\nu} + \frac{1}{3^\nu} + \cdots = \zeta(\nu, 1), \qquad \zeta(\nu, z) = \sum_{n=0}^{\infty} \frac{1}{(n+z)^\nu}, \tag{5.24}$$

instead of Bernoulli numbers. The $\hat{\Sigma}_{\mathsf{EM}}$ operator can be easily obtained exploiting the exponential representation of the standard translation operator $\hat{\mathsf{T}}$, whose action is defined as follows

$$f(b+\epsilon) = \hat{\mathsf{T}}_z(b) f(z)\Big|_{z=\epsilon} = \exp\left(b\frac{\partial}{\partial z}\right) f(z)\Big|_{z=\epsilon}. \tag{5.25}$$

If summed up from zero to infinity the above expression would give an infinite geometric progression,

$$\sum_{n=0}^{\infty} \left[\hat{\mathsf{T}}_z(b)\right]^n = \sum_{n=0}^{\infty} \left[\exp\left(b\frac{\partial}{\partial z}\right)\right]^n = \frac{1}{1-\exp\left(b\frac{\partial}{\partial z}\right)}. \tag{5.26}$$

Let us introduce now the fluctuational part of the heat capacity,

$$C(\epsilon) = -\frac{1}{T_c}\frac{\partial^2}{\partial \epsilon^2} F(\epsilon). \tag{5.27}$$

Using this physical observable and Eq. (5.21), one can extract the magnetic field dependent part of the free energy,

$$\begin{aligned} F(\epsilon, h) - F(\epsilon) &= \sum_{n=1}^{\infty} (-1)^{n-1} \frac{2}{\pi^{2n}} \left(1 - \frac{1}{2^{2n-1}}\right) \zeta(2n) h^{2n} \frac{\partial^{2(n-1)}}{\partial \epsilon^{2(n-1)}} T_c C(\epsilon) \\ &= \left[\frac{1}{6}h^2 - \frac{7}{360}h^4 \frac{\partial^2}{\partial \epsilon^2} + \frac{31}{15120} h^6 \frac{\partial^4}{\partial \epsilon^4} - \cdots \right] T_c C(\epsilon). \end{aligned} \tag{5.28}$$

It is then straightforward to calculate the magnetization M and the nonlinear susceptibility defined as $\chi(\epsilon, h) \equiv M/H$, i.e.,

$$M = -\frac{\partial[F(\epsilon,h) - F(\epsilon)]}{\partial B} = \chi(\epsilon, h) H. \tag{5.29}$$

For the Meissner–Ochsenfeld (MO) phase, for example, $\chi^{(\mathrm{MO})} = -1$. Importantly, Eq. (5.29) incorporates the *regularized* free energy which, by virtue of Eq. (5.28), does not contain zero magnetic field part,

$$F_{\mathrm{reg}}(\epsilon, h) \equiv F(\epsilon, h) - F(\epsilon) := \widehat{\mathsf{Reg}}_{\mathsf{EM}} F(\epsilon) \tag{5.30}$$

where

$$\widehat{\mathsf{Reg}}_{\mathsf{EM}} = \hat{\Sigma}_{\mathsf{EM}} - \hat{1} = 2h \sum_{n=0}^{n_c} - \int_0^c d\tilde{x} \tag{5.31}$$

is the Euler–MacLaurin regularization operator. This method was applied [392–394] for calculation of zero-field limit of the magnetic susceptibility, cf. Ref. [373]. Hence, inserting the h^2-term of Eq. (5.28) into Eq. (5.29) one finds [395] for $H \to 0$

$$\chi(\epsilon) = -\frac{\mu_0 T_c C(\epsilon)}{3B_{c2}^2(0)} = -\frac{4\pi^2\mu_0}{3\Phi_0^2}\xi_{ab}^4(0)T_c C(\epsilon). \tag{5.32}$$

For illustration, let us analyze how this relation between susceptibility and heat capacity can be applied to the LD model. For arbitrary multilayered structure we can calculate the curvature of the lowest dimensionless energy band in c-direction $\omega_1(\theta) = \epsilon_{c1}(p_z)/a_0$,

$$r \equiv 2\frac{\partial^2}{\partial\theta^2}\omega_1(\theta)|_{\theta=0}. \tag{5.33}$$

According to this definition, r parameterizes the effective mass in c-direction m_c for an anisotropic GL model, $|p_z| \ll \pi\hbar/s$,

$$\varepsilon_{c1} \approx a_0\frac{r}{4}\theta^2 = \frac{p_z^2}{2m_c}. \tag{5.34}$$

It is now easily realized that the identity holds true,

$$r = \left(\frac{2N\xi_c(0)}{s}\right)^2, \tag{5.35}$$

which for $N = 1$ is the LD parameter r that determines the effective dimensionality of the superconductor, cf. the review by Varlamov *et al.* [396] In many other studies, e.g., Ref. [377], the wave vector $\mathbf{k} = \mathbf{p}/\hbar$ has been used as well. In terms of the latter, for the dimensionless kinetic energy in the long-wavelength approximation we have

$$\frac{[\varepsilon_{ab}(\hbar\mathbf{k}) + \varepsilon_{c1}(\hbar k_z)]}{a_0} \approx \xi_{ab}^2(0)\mathbf{k}^2 + \xi_c^2(0)k_z^2 = \xi_{ab}^2(0)\mathbf{k}^2 + r\theta^2, \tag{5.36}$$

where $\epsilon_{ab}(\hbar\mathbf{k})$ represents the in-plane part of the kinetic energy. Let us mention that the LD model is not only applicable to single layered cuprates with $N = 1$, but is it also to bi-layered cuprates ($N = 2$) in the limit cases $\gamma_1 \simeq \gamma_2$ as well as in the case $\gamma_1 \ll \gamma_2$ when formally $N = 1$. That is why we use in our formulae an effective periodicity of the LD-model

$$s_{\text{eff}} = \frac{s}{N}. \tag{5.37}$$

For completeness we list below without deriving some of the well-known results within the LD-model. The single energy band has the form

$$\varepsilon_c(p_z) = a_0\omega_1\left(\frac{p_z s}{2\pi\hbar}\right) = \frac{\hbar^2}{m_c(s/N)^2}(1-\cos\theta), \text{ where} \tag{5.38}$$

$$\omega_1(\theta) = \frac{1}{2}r(1-\cos\theta) = r\sin^2\frac{\theta}{2}, \tag{5.39}$$

being parameterized by the Josephson coupling energy

$$J_1 = a_0\gamma_1 = \frac{\hbar^2}{m_c(s/N)^2}, \qquad r = 4\gamma_1 = (2N\xi_c(0)/s)^2 . \tag{5.40}$$

For the heat capacity one has

$$C^{(\mathrm{LD})}(\epsilon) = \frac{k_{\mathrm{B}}}{4\pi\xi_{ab}^2(0)}\frac{N}{s}\frac{1}{\sqrt{\epsilon\,(\epsilon+r)}}, \tag{5.41}$$

and the magnetic susceptibility for a weak magnetic field applied in c-direction, according to Tsuzuki [397] and Yamayi [398], reads

$$-\chi^{(\mathrm{LD})}(\epsilon) = \frac{\pi}{3}\mu_0\frac{k_{\mathrm{B}}T_{\mathrm{c}}}{\Phi_0^2}\xi_{ab}^2(0)\frac{N}{s}\,\frac{1}{\sqrt{\epsilon}}\,\frac{1}{\sqrt{\epsilon+r}} = \frac{1}{6}\frac{M_0}{H_{c2}(0)}\frac{1}{\sqrt{\epsilon\,(\epsilon+r)}}, \tag{5.42}$$

where

$$M_0 \equiv \frac{k_{\mathrm{B}}T_{\mathrm{c}}}{\Phi_0}\frac{N}{s}. \tag{5.43}$$

Before proceeding we feel it, appealing to make some technical remarks concerning the representation of the general formulae for fluctuations in arbitrary layered superconductor. To be specific, we shall demonstrate how the expressions for the magnetic susceptibility, Eq. (5.42), and heat capacity, Eq. (5.41), within the LD model can be obtained as special cases of a general procedure described in the next subsection.

5.2.3 *Layering operator* $\hat{\mathsf{L}}$ *illustrated on the example of paraconductivity*

In the general formula for the density of the free energy, Eq. (5.12), the energies related to motion in c-direction $\epsilon_{cj}(p_z)$ enter the final result solely via the fragment $\epsilon + \omega_j(\theta)$. Thus, in all such cases one can first solve the corresponding 2D problem and then for a layered superconductor the result can be derived by merely averaging the 2D result with respect to the motion of the fluctuation Cooper pairs in perpendicular to the layers direction. Formally, this method reduces to introducing a *layering* operator $\hat{\mathsf{L}}$ acting on functions of ϵ; e.g., for the conductivity one would have the relation

$$\sigma(\epsilon) = \hat{\mathsf{L}}\sigma^{(2\mathrm{D})}(\epsilon) \equiv \frac{1}{N}\sum_{j=1}^{N}\oint\frac{d\theta}{2\pi}\sigma^{(2\mathrm{D})}\left(\epsilon+\omega_j^{(N)}(\theta)\right). \tag{5.44}$$

In terms of the so introduced operator $\hat{\mathsf{L}}$ the expression for the free energy, Eq. (5.21), takes the form

$$F(\epsilon) = F_0 \int_0^c d\tilde{x} \sum_{j=1}^{N} \hat{\mathsf{L}} \ln\left(\tilde{x} + \omega_j^{(N)}(\theta) + \epsilon\right), \qquad F_0 \equiv \frac{k_{\mathrm{B}} T_{\mathrm{c}}}{4\pi\xi_{ab}^2(0)} \frac{N}{s}. \tag{5.45}$$

Besides thermodynamic variables, this operator works also for the fluctuation in-plane conductivity. Conforming with the work of Hikami and Larkin [390] for the conductivity within the LD model we have to integrate the 2D conductivity with respect to the Josephson phase,

$$\sigma(\epsilon) = \hat{\mathsf{L}}^{(\mathrm{LD})} \sigma^{(2\mathrm{D})}(\epsilon) \equiv \oint \frac{d\theta}{2\pi} \sigma^{(2\mathrm{D})}\left(\epsilon + \frac{1}{2} r(1 - \cos\theta)\right). \tag{5.46}$$

Given a system with independent 2D layers, having density in c-direction N/s, for zero magnetic field we have to average the well-known Aslamazov–Larkin expression for the static (zero-frequency) conductivity,

$$\sigma_{\mathrm{AL}}(\epsilon) = \frac{e^2}{16\hbar} \frac{N}{s} \frac{1}{\epsilon} = \frac{\pi}{8} R_{\mathrm{QHE}}^{-1} \frac{N}{s} \frac{1}{\epsilon}, \tag{5.47}$$

where $R_{\mathrm{QHE}} \equiv 2\pi\hbar/e^2 = 25.813\ \mathrm{k\Omega}$. A simple integration gives

$$f_{\mathrm{LD}}(\epsilon; r) \equiv \hat{\mathsf{L}}^{(\mathrm{LD})} \frac{1}{\epsilon} = \oint \frac{d\theta}{2\pi} \frac{1}{\epsilon + \frac{1}{2} r\,(1 - \cos\theta)} = \frac{1}{\sqrt{\epsilon(\epsilon + r)}}. \tag{5.48}$$

We note that this integral determines both the heat capacity and magnetic susceptibility for the LD model and is widely used for fitting to experimental data. Another important integral is

$$\begin{aligned} \hat{\mathsf{L}}^{(\mathrm{LD})} \int_\epsilon^c \ln\tilde{\epsilon}\, d\tilde{\epsilon} &= \int_\epsilon^c 2\ln\left(\frac{\sqrt{\tilde{\epsilon}} + \sqrt{\tilde{\epsilon} + r}}{2}\right) d\tilde{\epsilon} \\ &= \left[(2\tilde{\epsilon} + r)\ln\left(\sqrt{\tilde{\epsilon}} + \sqrt{\tilde{\epsilon} + r}\right) - \sqrt{\tilde{\epsilon}(\tilde{\epsilon} + r)} - \ln(4)\,\tilde{\epsilon}\right]\Big|_\epsilon^c, \end{aligned} \tag{5.49}$$

which is used in representing the free energy at zero magnetic field, Eq. (5.12), cf. also Eqs. (5.131) and (5.132) below. Further, the ϵ-derivative of this equation,

$$\hat{\mathsf{L}}^{(\mathrm{LD})} \ln\epsilon = \oint \frac{d\theta}{2\pi} \ln\left(\epsilon + \frac{1}{2} r\,(1 - \cos\theta)\right) = 2\ln\left(\frac{\sqrt{\epsilon} + \sqrt{\epsilon + r}}{2}\right), \tag{5.50}$$

is important for the calculation of the fluctuation part of the entropy and the density of fluctuation Cooper pairs. We provide also two other integrals employed in calculating the magnetoconductivity [396]

$$\hat{\mathsf{L}}^{(\mathrm{LD})} \frac{1}{\epsilon^2} = -\hat{\mathsf{L}}^{(\mathrm{LD})} \frac{\partial}{\partial\epsilon} \frac{1}{\epsilon} = \frac{\epsilon + \frac{1}{2} r}{[\epsilon(\epsilon + r)]^{3/2}}, \tag{5.51}$$

$$\hat{\mathsf{L}}^{(\mathrm{LD})} \frac{1}{\epsilon^3} = \frac{1}{2} \frac{\partial^2}{\partial\epsilon^2} \hat{\mathsf{L}}^{(\mathrm{LD})} \frac{1}{\epsilon} = \frac{(\epsilon + r)\epsilon + \frac{3}{8} r^2}{[\epsilon(\epsilon + r)]^{5/2}}. \tag{5.52}$$

Analogously, in the Maki-Thompson bi-layered model [389] ($N = 2$) one summation should precede the integration [395, 399],

$$\sigma(\epsilon) = \hat{\mathsf{L}}^{(\mathrm{MT})} \sigma^{(2\mathrm{D})}(\epsilon) = \frac{1}{2} \oint \frac{d\theta}{2\pi} \left[\sigma^{(2\mathrm{D})}(\epsilon + \omega_1(\theta)) + \sigma^{(2\mathrm{D})}(\epsilon + \omega_2(\theta)) \right], \tag{5.53}$$

i.e., in order to calculate the conductivity [389] and susceptibility [395] we have to add the terms

$$\frac{1}{\epsilon + \omega_1} + \frac{1}{\epsilon + \omega_2} = \frac{2\epsilon + (\omega_1 + \omega_2)}{\epsilon^2 + (\omega_1 + \omega_2)\epsilon + \omega_1 \omega_2}. \tag{5.54}$$

In this expression both $\omega_1 + \omega_2$ and $\omega_1 \omega_2$ are rational, cf. Eq. (5.4), and the integral (5.53) is reduced to the integral (5.48),

$$\begin{aligned} \hat{\mathsf{L}}^{(\mathrm{MT})} \frac{1}{\epsilon} &= \frac{\epsilon + \gamma_1 + \gamma_2}{\sqrt{\epsilon \left[\epsilon + 2(\gamma_1 + \gamma_2)\right] (\epsilon + 2\gamma_1)(\epsilon + 2\gamma_2)}} \\ &= \frac{\epsilon + \frac{1}{2} r w}{\sqrt{(\epsilon^2 + r w \epsilon)\left(\epsilon^2 + r w \epsilon + \frac{1}{4} r^2 w\right)}} \equiv f_{\mathrm{MT}}(\epsilon, h; r, w), \end{aligned} \tag{5.55}$$

where

$$\begin{aligned} r &\equiv 4 \frac{2}{\frac{1}{\gamma_1} + \frac{1}{\gamma_2}}, \qquad u \equiv \frac{J_{\max}}{J_{\min}} = \frac{\gamma_{\max}}{\gamma_{\min}}, \\ w &\equiv \frac{1}{4} \left(2 + \frac{\gamma_1}{\gamma_2} + \frac{\gamma_2}{\gamma_1} \right) = \frac{1}{4} \left(2 + u + \frac{1}{u} \right). \end{aligned} \tag{5.56}$$

Such a form, involving the w parameter, is convenient for fitting the experimental data since for both $w = 1$ and $w \gg 1$ cases the LD approximation holds true, which is often found to give a satisfactory explanation of the experimental observations. For more detailed discussion the reader is referred to Refs. [400–404]. The inverse relations for the above introduced parameters read as

$$u = (2w - 1) + 2\sqrt{w(w-1)}, \quad \gamma_{\min} = \frac{1}{2}\left(1 + \frac{1}{u}\right)\frac{r}{4}, \quad \gamma_{\max} = \frac{1}{2}(1 + u)\frac{r}{4}, \tag{5.57}$$

and

$$J_{\max} = a_0 \gamma_{\max} = \frac{\hbar^2}{2 m_{ab} \xi_{ab}^2(0)} \frac{1}{2} (1 + u) \frac{r}{4}. \tag{5.58}$$

Importantly, at known effective mass m_{ab} the last equation gives one the possibility to determine the Josephson coupling energy between double CuO_2 planes.

Let us now illustrate in more detail the action of the $\hat{\mathsf{L}}$ operator. Towards this end we consider the famous Aslamazov–Larkin [405] formula for the 2D conductivity $\sigma_{ab}^{(2D)}(\epsilon)$, the results for the susceptibility $-\chi_{ab}^{(2D)}(\epsilon)$ due to V. Schmidt [394], A. Schmid [393] and H. Schmidt [392], and Ferrell [406] and Thouless [407] fluctuation part of the heat capacity $C^{(2D)}(\epsilon)$. With the help of the $\hat{\mathsf{L}}$ operator for *arbitrary layered superconductor*, cf. Refs. [408] and [409], these three quantities can be generally written as

$$\sigma_{ab}(\epsilon) = \frac{1}{R_{\mathrm{QHE}}}\left(\frac{2\tau_0 k_{\mathrm{B}} T_{\mathrm{c}}}{\hbar}\right)\frac{N}{s}\,\hat{\mathsf{L}}\frac{1}{\epsilon}, \tag{5.59}$$

$$-\chi_{ab}(\epsilon) = \frac{\pi}{3}\mu_0\frac{k_{\mathrm{B}} T}{\Phi_0^2}\xi_{ab}^2(0)\frac{N}{s}\,\hat{\mathsf{L}}\frac{1}{\epsilon}, \tag{5.60}$$

$$C(\epsilon) = \frac{k_{\mathrm{B}}}{4\pi\xi_{ab}^2(0)}\frac{N}{s}\,\hat{\mathsf{L}}\frac{1}{\epsilon}, \tag{5.61}$$

where the "ab"-subscript in Eq. (5.59) indicates that the conductivity is in the ab-planes, while in Eq. (5.60) it indicates that the vanishing magnetic field is perpendicular to the same planes. It is immediately apparent that the common for all these expressions function $\hat{\mathsf{L}}\epsilon^{-1}$ cancels when calculating the χ/C, σ/C, and σ/χ quotients [400–404]. In particular, the temperature independent ratio

$$\tau_0 = \frac{\mu_0}{3}\xi_{ab}^2(0)\frac{\sigma_{ab}(\epsilon)}{-\chi_{ab}(\epsilon)} = \mathrm{const} \tag{5.62}$$

provides the best method for probing the time constant τ_0 parameterizing the life time of the fluctuation Cooper pairs with zero momentum,

$$\tau(\epsilon) = \frac{\tau_0}{\epsilon}, \qquad |\psi_{\mathbf{p}=0}(t)|^2 \propto \exp\left(-\frac{t}{\tau(\epsilon)}\right). \tag{5.63}$$

The τ_0 constant participates in time-dependent GL (TDGL) theory; see for example the reviews by Cyrot [375], Skocpol and Tinkham, [377] and the textbooks by Abrikosov [272] and Tinkham [374]. Within the weak coupling BCS theory in the case of negligible depairing mechanisms [272, 375, 377, 410–417] τ_0 satisfies the relation

$$\frac{\tau_{0,\Psi}^{(\mathrm{BCS})} k_{\mathrm{B}} T_{\mathrm{c}}}{\hbar} = \frac{\pi}{8}, \quad \tau_0^{(\mathrm{BCS})} = \frac{\pi}{16}\frac{\hbar}{k_{\mathrm{B}} T_{\mathrm{c}}}, \quad \tau^{(\mathrm{BCS})}(\epsilon) = \frac{\tau_0^{(\mathrm{BCS})}}{\epsilon} = \frac{\pi}{16}\frac{\hbar}{k_{\mathrm{B}} T_{\mathrm{c}}}\frac{1}{\epsilon}, \tag{5.64}$$

where $\tau_{0,\Psi}^{(\mathrm{BCS})} \equiv 2\tau_0^{(\mathrm{BCS})}$ is the relaxation time constant for the order parameter being two times larger [374]. At the present experimental accuracy, this

BCS value agrees well with the experimental data for the layered cuprates. Thus, the above observation led us to propose the dimensionless ratio [418]

$$\tilde{\tau}_{\mathrm{rel}} \equiv \frac{\tau_0}{\tau_0^{(\mathrm{BCS})}} = 32\frac{k_{\mathrm{B}}T_{\mathrm{c}}\tau_0}{(2\pi\hbar)} = \frac{8k_{\mathrm{B}}T_{\mathrm{c}}\tau_{0,\Psi}}{\pi\hbar} = \frac{16\mu_0}{3\pi\hbar}\xi_{ab}^2(0)\frac{k_{\mathrm{B}}T\sigma_{ab}(\epsilon)}{-\chi_{ab}(\epsilon)} = \mathrm{const}, \tag{5.65}$$

to be used for more reliable experimental data processing; any deviation of $\tilde{\tau}_{\mathrm{rel}}$ from unity should be interpreted as a hint towards unconventional behavior and presence of depairing mechanisms. Notice also that the BCS value $\pi/8 = 0.393$ in Eqs. (5.47), (5.64), and (5.65) is extremely robust, being originally derived for dirty 3D superconductors, and the $\tau_0 T_{\mathrm{c}}$ product remains the same [396, 415, 416] for clean 2D superconductors and is not affected by the multilaminarity. The general formula for the fluctuation conductivity of a layered superconductor in perpendicular magnetic field can be also rewritten via the layering operator and relative life-time employing the 2D results by Redi [419], and Abrahams, Prange and Stefen [420] (APS), cf. also Ref. [395, 399],

$$\sigma_{ab}(\epsilon, h) = \tilde{\tau}_{\mathrm{rel}}\frac{e^2}{16\hbar}\frac{N}{s}\,\hat{\mathsf{L}} f_{\mathrm{APS}}(\epsilon, h), \tag{5.66}$$

where, for $\epsilon + h > 0$,

$$f_{\mathrm{APS}}(\epsilon, h) \equiv \frac{1}{\epsilon}2\left(\frac{\epsilon}{h}\right)^2\left[\psi\left(\frac{1}{2}+\frac{\epsilon}{2h}\right) - \psi\left(1+\frac{\epsilon}{2h}\right) + \frac{h}{\epsilon}\right], \tag{5.67}$$

is an universal dimensionless function of dimensionless reduced temperature ϵ and dimensionless magnetic field h. The functions

$$\Gamma(z) \equiv \int_0^\infty e^{-t}t^{z-1}dt, \quad \psi(z) \equiv \frac{d}{dz}\ln\Gamma(z), \quad \psi^{(1)}(z) \equiv \frac{d}{dz}\psi(z) = \zeta(2, z) \tag{5.68}$$

are respectively the Euler gamma, digamma, and trigamma functions. This general formula is often utilized to process the experimental data for the paraconductivity. We provide also several useful asymptotics of $f_{\mathrm{APS}}(\epsilon, h)$ in different physical conditions,

$$f_{\mathrm{APS}}(\epsilon, h) \approx \begin{cases} \dfrac{2}{h}\left[1 - \dfrac{\epsilon}{2h}\ln 2\right], & h \gg |\epsilon| \\ \dfrac{4}{\epsilon + h} = 4\,\dfrac{T_{\mathrm{c}}}{T - T_{\mathrm{c2}}(H)}, & \epsilon + h \ll h \\ \left[1 - \dfrac{1}{2}\left(\dfrac{h}{\epsilon}\right)^2\right]\dfrac{1}{\epsilon} = \left[1 - \dfrac{h^2}{4}\dfrac{\partial^2}{\partial\epsilon^2}\right]\dfrac{1}{\epsilon}, & h \ll \epsilon \end{cases} \tag{5.69}$$

For the LD model, for example, the $(h \ll \epsilon)$-asymptotics gives [395, 396, 399, 421] according to Eq. (5.52),

$$\sigma_{ab}(\epsilon, h) \approx \tilde{\tau}_{\rm rel} \frac{e^2}{16\hbar} \frac{N}{s} \left[\frac{1}{\sqrt{\epsilon(\epsilon + r)}} - \frac{h^2}{2} \frac{\epsilon(\epsilon + r) + \frac{3}{8}r^2}{[\epsilon(\epsilon + r)]^{5/2}} \right]. \tag{5.70}$$

Note that the classical Aslamazov–Larkin result, Eq. (5.47), is recovered for $r = 0$, $h = 0$, and $\tilde{\tau}_{\rm rel} = 1$. In the practical application to layered cuprates, however, we need to take into account the nonlocality effects. In $\varepsilon_{\text{\ae}}$-approximation to the GL theory we have to subtract the part of the corresponding cutoff area in the 2D momentum space. Thereby, the fluctuation conductivity is given by the difference

$$\sigma_{ab}(\epsilon, h; c) = \hat{\mathsf{C}}\sigma_{ab}(\epsilon, h) \equiv \sigma_{ab}(\epsilon, h) - \sigma_{ab}(c + \epsilon, h) \approx \sigma_{ab}(\epsilon, h) - \sigma_{ab}(c, h), \tag{5.71}$$

where a cutoff operator $\hat{\mathsf{C}}$ is introduced, and the approximation is valid for $\epsilon \ll c$. Similarly, for the magnetization we have the same "cutoff" expression which appears when calculating the truncated sums over the Landau levels, $\sum_0^{n_c-1} = \sum_0^\infty - \sum_{n_c}^\infty$, or integral with respect to the dimensionless in-plane kinetic energy,

$$\int_0^c d\tilde{x} = \int_0^\infty d\tilde{x} - \int_c^\infty d\tilde{x}. \tag{5.72}$$

As a rule the GL theory allows for ultraviolet (UV) regularization—every expression can be easily regularized in the local $(c \to \infty)$-approximation. Therefore the energy cutoff parameter c is not viewed as a tool for UV regularization, it is simply an important and immanent parameter of the GL theory, being of the order $c \simeq 1$. The cutoff procedure has been essentially introduced *from the beginning* in the GL theory (see Sec. 147, problem 1 in Ref. [66]). Unfortunately, for many superconductors systematic studies for determination of the energy cutoff parameter are still missing. Here we suggest only the simplest possible interpolation formula within the LD model for $\epsilon \ll c$,

$$\sigma_{ab}(\epsilon, h) = \frac{\pi}{8} \frac{\tilde{\tau}_{\rm rel}}{R_{\rm QHE}} \frac{N}{s} \hat{\mathsf{C}} f_{\rm LD}(\epsilon, r) \approx \tilde{\tau}_{\rm rel} \frac{e^2}{16\hbar} \frac{N}{s} \left[\frac{1}{\sqrt{\epsilon(\epsilon + r)}} - \frac{1}{\sqrt{c(c + r)}} \right], \tag{5.73}$$

which takes into account only the fist nonlocal correction. This simple expression fits very well [422] the experimental data for $YBa_2Cu_3O_{7-\delta}$ and $Bi_2Sr_2CaCu_2O_8$.

As a last example of the action of the layering operator $\hat{\mathsf{L}}$ we consider the 2D frequency-dependent paraconductivity at zero magnetic field. Taking the general expression for D-dimensional GL model [423] and performing carefully the limit $D \to 2$ (note, that there is an omitted term in the expression for the 2D conductivity in Ref. [423]) we get for the in-plane complex conductivity

$$\begin{aligned}\sigma^*_{ab}(\omega\tau(\epsilon)) &= \sigma'_{ab}\left(\frac{\omega\tau_0}{\epsilon}\right) + \mathrm{i}\sigma''_{ab}\left(\frac{\omega\tau_0}{\epsilon}\right) \\ &= \tilde{\tau}_{\mathrm{rel}}\frac{e^2}{16\hbar}\frac{N}{s}\,\hat{\mathsf{L}}\left[\frac{1}{\epsilon}\,\varsigma_1\left(\frac{\omega\tau_0}{\epsilon}\right) + \frac{\mathrm{i}}{\epsilon}\,\varsigma_2\left(\frac{\omega\tau_0}{\epsilon}\right)\right],\end{aligned} \tag{5.74}$$

or in expanded notation for singe layered superconductor,

$$\sigma^*_{ab}(\omega) = \frac{2\tau_0 k_{\mathrm{B}} T_{\mathrm{c}}/\hbar}{s_{\mathrm{eff}} R_{\mathrm{QHE}}} \int\limits_0^{\pi/2} \frac{d\phi}{\pi/2}\, \frac{\varsigma_1\left(\frac{\omega\tau_0}{\epsilon + r\sin^2\phi}\right) + \mathrm{i}\varsigma_2\left(\frac{\omega\tau_0}{\epsilon + r\sin^2\phi}\right)}{\epsilon + r\sin^2\phi}, \tag{5.75}$$

where we have for the dimensionless real and imaginary conductivity, $\varsigma_1(0) = 1$,

$$\varsigma_1(z) \equiv \frac{2}{z^2}\left[z\arctan(z) - \frac{1}{2}\ln\big(1+z^2\big)\right] = \frac{2}{\pi}\mathcal{P}\int\limits_0^\infty \frac{y\,\varsigma_2(y)}{y^2 - z^2}dy, \tag{5.76}$$

$$\varsigma_2(z) \equiv \frac{2}{z^2}\left[\arctan(z) - z + z\frac{1}{2}\ln\big(1+z^2\big)\right] = -\frac{2z}{\pi}\mathcal{P}\int\limits_0^\infty \frac{\varsigma_1(y)}{y^2 - z^2}dy. \tag{5.77}$$

As usual in the above Kramers-Kronig integrations $\mathcal{P}$ indicates that the principal value of the integral is taken. For computer implementation of the $\hat{\mathsf{L}}$ operator we have to verify that, for $r \gg 1$, $\hat{\mathsf{L}}$ is simply equivalent to an incremental operator for the spatial dimensionality,

$$\sigma^{(\mathrm{D}+1)}(\epsilon) \approx \hat{\mathsf{L}}\sigma^{(\mathrm{D})}(\epsilon). \tag{5.78}$$

In many cases the GL results for integer dimensionality are well-known and we can derive a generalization for a layered system. For both the MT and LD models the integration in Eq. (5.74) can be easily programmed, so we have a useful formula for fitting of the ultra high frequency measurements of $\sigma^*_{ab}(\omega)$. The original explicit expressions derived from retarded electromagnetic operator by Aslamazov and Varlamov [424] are too cumbersome to be used by experimentalists. Hence, one may realize that the GL theory is not some phenomenological alternative to the microscopic BCS theory (this scorn, dating back to the beginning of fifties, is still living even nowadays among students). The GL theory is a tool for applying the theory of superconductivity for the important for applications, let us say "hydrodynamic", case of low frequencies and small wave-vectors. For $\epsilon \ll r$

the frequency dependent conductivity $\sigma^*_{ab}(\omega)$, having dimension $(\Omega\ \mathrm{cm})^{-1}$, from Eq. (5.75) displays 3D behavior, while in the opposite case of $\epsilon \gg r$ it shows 2D character. For thin films of layered superconductors with thickness d_{film} we have to calculate the 2D conductivity $\sigma^{(2\mathrm{D})} = d_{\mathrm{film}}\sigma$, while for single layered films of conventional superconductors, for example, we have to substitute in Eq. (5.75) formally $s_{\mathrm{eff}} = d_{\mathrm{film}}$, and certainly $r = 0$.

Having analyzed in detail the action of the $\hat{\mathsf{L}}$ operator, we developed practically all technical tools necessary to proceed our investigation of the thermodynamics of Gaussian fluctuations and fluctuation magnetization.

5.2.4 *Power series for the magnetic moment within the LD model*

We will calculate in this subsection the nonlinear susceptibility by substituting first into the free energy, Eq. (5.28), the heat capacity, expressed via the susceptibility from Eq. (5.32). Then, the formula for the magnetization, Eq. (5.29), gives

$$\chi(\epsilon,h) = 6\sum_{n=1}^{\infty}(-1)^{n-1}\frac{2n}{\pi^{2n}}\left(1-\frac{1}{2^{2n-1}}\right)\zeta(2n)h^{2(n-1)}\frac{\partial^{2(n-1)}}{\partial\epsilon^{2(n-1)}}\chi(\epsilon). \tag{5.79}$$

Taking the LD expression for the susceptibility at zero field Eq. (5.42), calculating the derivatives with respect to ϵ by means of the relation

$$\frac{\partial^m}{\partial\epsilon^m}\frac{1}{\sqrt{\epsilon}} = \frac{(2m-1)!!}{2^m\epsilon^m}\frac{1}{\sqrt{\epsilon}}, \tag{5.80}$$

and defining the relative susceptibility as

$$\tilde{\chi}_{\mathrm{rel}}(\epsilon,h) \equiv \frac{\chi(\epsilon,h)}{\chi(\epsilon)} \tag{5.81}$$

we obtain

$$\begin{aligned}\tilde{\chi}_{\mathrm{rel}}(\epsilon,h;r) &= 12\sum_{n=0}^{\infty}\frac{(-1)^n}{2n+1}\left(1-\frac{1}{2^{2n+1}}\right)\frac{(2n+2)!}{2^{2n+1}}\frac{\zeta(2n+2)}{\pi^{2n+2}}\\ &\quad\times\left(\frac{h^2}{\epsilon^2}\right)^n\sum_{m=0}^{2n}\frac{(2m-1)!!(4n-2m-1)!!}{m!(2n-m)!(1+r/\epsilon)^{2n-m}}\\ &= 1-\frac{7}{15}\frac{\epsilon^2+r\epsilon+3r^2/8}{(\epsilon+r)^2}\left(\frac{h}{\epsilon}\right)^2+\cdots.\end{aligned} \tag{5.82}$$

Although these series is found to be a solution to the problem of calculating the fluctuational magnetization,

$$M(\epsilon, h) = -\frac{k_{\rm B} T_{\rm c}}{\Phi_0 s_{\rm eff}} \left\{ \frac{\tilde{\chi}_{\rm rel}(\epsilon, h; r)}{6\sqrt{\epsilon(r+\epsilon)}} - \frac{\tilde{\chi}_{\rm rel}(c+\epsilon, h; r)}{6\sqrt{(c+\epsilon)(c+r+\epsilon)}} \right\}, \tag{5.83}$$

for the physical conditions of interest, i.e., an observable effect of magnetic field on the susceptibility, one needs to extend the series summation onto arguments h^2/ϵ^2 beyond the radius of convergence. Analogous series has been already reported for the 3D paraconductivity [425]. One of the best devices for extending the convergence of series and also for calculating slowly convergent series is the ε-algorithm [426, 427] based on Padé approximants [428]. In the next section we describe a simplified version of this algorithm suitable for computer implementation.

5.2.5 *The epsilon algorithm*

The epsilon algorithm is a method for finding the limit L of infinite series

$$L = \lim_{N\to\infty} S_N \equiv \lim_{N\to\infty} \sum_{i=0}^{N} a_i, \tag{5.84}$$

in case where only the first $N+1$ terms a_i, $i = 0, \ldots N$, are known. The algorithm operates by employing two rows. The first one, called here auxiliary A-row, is initially set to zero, i.e.,

$$A_0^{[0]} = 0, \qquad A_1^{[0]} = 0, \qquad A_2^{[0]} = 0, \ldots \qquad A_N^{[0]} = 0. \tag{5.85}$$

The second one is sequential S-row loaded in zero-order approximation with the partial sums of the series

$$S_0^{[0]} = a_0, \qquad S_1^{[0]} = a_0 + a_1, \quad \ldots \qquad S_N^{[0]} = a_0 + a_1 + a_2 + \cdots + a_N. \tag{5.86}$$

The above assignments, as indicated by Eqs. (5.85) and (5.86), constitute the initialization phase of the ϵ-algorithm. The essence of the latter consists of filling in the so called ϵ-table

$$\begin{pmatrix} A_0^{[0]} & A_1^{[0]} & A_2^{[0]} & A_3^{[0]} & \ldots \\ S_0^{[0]} & S_1^{[0]} & S_2^{[0]} & S_3^{[0]} & \ldots \\ A_0^{[1]} & A_1^{[1]} & A_2^{[1]} & A_3^{[1]} & \ldots \\ S_0^{[1]} & S_1^{[1]} & S_2^{[1]} & S_3^{[1]} & \ldots \\ \ldots & \ldots & \ldots & \ldots & \ldots \end{pmatrix} = \begin{pmatrix} \epsilon_0^{[0]} & \epsilon_1^{[0]} & \epsilon_2^{[0]} & \ldots \\ \epsilon_0^{[1]} & \epsilon_1^{[1]} & \epsilon_2^{[1]} & \ldots \\ \epsilon_0^{[2]} & \epsilon_1^{[2]} & \epsilon_2^{[2]} & \ldots \\ \epsilon_0^{[3]} & \epsilon_1^{[3]} & \epsilon_2^{[3]} & \ldots \\ \ldots & \ldots & \ldots & \ldots \end{pmatrix} \equiv \begin{pmatrix} \ldots & \ldots & \ldots & \ldots \\ [0/0] & [1/0] & [2/0] & \ldots \\ \ldots & \ldots & \ldots & \ldots \\ [1/1] & [2/1] & [3/1] & \ldots \\ \ldots & \ldots & \ldots & \ldots \end{pmatrix}, \tag{5.87}$$

where according to the standard notations $[j/k] = P_j(z)/P_k(z)|_{z=1}$ designates a Padé approximant of power j in the numerator and, respectively, k in the denominator [428].

Starting from the $A^{[0]}$- and $S^{[0]}$ rows every subsequent row is derived by applying the *cross rule* (known also as the missing identity of Frobenius). To be specific, for calculation of the kth A-row we have to solve the cross rule equation

$$\begin{array}{c|c|c|c} \ddots & \vdots & \vdots & \cdots \\ \hline \cdots & \cdots & A_{i+1}^{(k-1)} & \cdots \\ \hline \cdots & S_i^{(k-1)} & S_{i+1}^{(k-1)} & \cdots \\ \hline \cdots & A_i^{(k)} & \cdots & \cdots \\ \hline \cdots & \vdots & \vdots & \ddots \end{array} \to \begin{array}{c|c} & \text{North} \\ \hline \textbf{West} & \textbf{East} \\ \hline \text{South} & \end{array} \to (\mathbf{S}-\mathrm{N})(\mathbf{E}-\mathbf{W}) = 1 \tag{5.88}$$

Likewise, for calculating the k-th S row we have to apply the same cross rule

$$\begin{pmatrix} \cdots\cdots\cdots\cdots & \\ \cdots & S_{i+1}^{[k-1]} \\ A_i^{[k]} & A_{i+1}^{[k]} \\ S_i^{[k]} & \cdots \end{pmatrix} = \begin{pmatrix} \cdots\cdots\cdots\cdots & \\ \cdots & \text{North} \\ \text{West} & \text{East} \\ \text{South} & \cdots \end{pmatrix} \to \text{South} = \text{North} + (\text{East} - \text{West})^{-1} . \tag{5.89}$$

Having applied the algorithm we get in the S-rows of the ϵ-table, Eq. (5.87), a set of different Padé approximants to the limit L. The ith term of the kth A-row can be easily obtained by

$$A_i^{[k]} = A_{i+1}^{[k-1]} + \left(S_{i+1}^{[k-1]} - S_i^{[k-1]}\right)^{-1}, \qquad \text{for } i = 0, 1, \ldots, N-2k+1, \tag{5.90}$$

but for practical implementation of the algorithm, we can omit the index of the approximation and to use only one auxiliary row, updating it each time,

$$A_i := A_{i+1} + (S_{i+1} - S_i)^{-1}, \qquad \text{for } i = 0, 1, \ldots, N-2k+1. \tag{5.91}$$

For the k-th S-row, the i-th term reads as

$$S_i^{[k]} = S_{i+1}^{[k-1]} + \left(A_{i+1}^{[k]} - A_i^{[k]}\right)^{-1}, \qquad \text{for } i = 0, 1, \ldots, N-2k, \tag{5.92}$$

and can be updated in the same manner as described for the A-row,

$$S_i := S_{i+1} + (A_{i+1} - A_i)^{-1} \qquad \text{for } i = 0, 1, \ldots, N-2k. \tag{5.93}$$

In order to find an estimate for the limit L of the infinite series, two different empirical criteria can be implemented. In the first one, the ϵ-table is scanned for a minimal difference $|S_{i+1}^{[k-1]} - S_i^{[k-1]}|$. The limit L is then given by

$$\min_{i,k} \left| S_{i+1}^{[k-1]} - S_i^{[k-1]} \right| \Longrightarrow L \approx S_i^{[k-1]}. \tag{5.94}$$

This minimal difference gives also an estimate for the empirical error of the method. In the second criterion the ϵ-table is scanned for the maximum of the East A-row element, cf Eqs. (5.89) and (5.92),

$$\max_{i,k} \left| A_{i+1}^{[k]} \right| \Longrightarrow L \approx S_i^{[k]}. \tag{5.95}$$

The reciprocal of the maximum auxiliary value gives in this case the estimate for the empirical error of the method. It is the second criterion that we have used in the FORTRAN90 implementation of the ϵ-algorithm in Ref. [429]. Therein we have also made use of pseudo-inverse numbers in order to ensure provisions against division by zero in Eqs. (5.91) and (5.93),

$$z^{-1} := \begin{cases} 0, & \text{for } z = 0 \\ 1/z, & \text{for } z \neq 0 \end{cases}. \tag{5.96}$$

For an illustration, consider the first approximation. In the beginning we have for the first A-row according to Eq. (5.90)

$$A_0^{[1]} = [(a_0 + a_1) - (a_0)]^{-1} = \frac{1}{a_1}, \qquad A_1^{[1]} = \frac{1}{a_2}, \quad \ldots, \qquad A_{N-1}^{[1]} = \frac{1}{a_N}. \tag{5.97}$$

The first S-row then reads

$$S_0^{[1]} = a_0 + a_1 + \frac{1}{1/a_2 - 1/a_1}, \qquad S_1^{[1]} = a_0 + a_1 + a_2 + \frac{1}{1/a_3 - 1/a_2}, \ldots, \tag{5.98}$$

and for the last element of the $S^{[1]}$-row we have

$$S_{N-2}^{[1]} = a_0 + a_1 + a_2 + \cdots + a_{N-2} + a_{N-1} + (1/a_N - 1/a_{N-1}). \tag{5.99}$$

The above approximation $S_{N-2}^{[1]}$ to the limit L is nothing but the well-known Aitken's Δ^2-method, which gives an *exact result* for the geometric progression

$$S_N^{[0]} = 1 + q + q^2 + \cdots + q^N, \qquad S_0^{[1]} = S_1^{[1]} = S_2^{[1]} = \cdots = S_{N-2}^{[1]} = \frac{1}{1-q}, \tag{5.100}$$

for an arbitrary $q \neq 1$. This fact can rationalize the success of the ϵ-algorithm when applied to weak magnetic field series expansion of susceptibility. In the Euler–MacLaurin summation, Eqs. (5.22) and (5.26), we have a hidden geometric progression of translation operators.

As a rule divergent series do not exist in physics; 99% of the divergent series born by real physical problems can be summed up by some combination of the Euler–MacLaurin method and the ϵ-algorithm and the reason lies in the analytical dependence of the coefficients on the index. In the Gaussian spectroscopy of superconductors, for example, it is necessary series related to asymptotic expansion of Euler polygamma and Hurwitz zeta functions to be summed up, but the same methods could be applied to many other physical problems. The solution often can be derived by less efforts than required to verify that a series is divergent accordingly some strict mathematical criterion. Nowadays the mathematical education in the physical departments is conquered by scholastic mathematicians. Alas, none of the students of physics knows what really happens when we press the sin key of a calculator. On the other hand this is a commercial secret of the manufacturer. Physicists do not even lightly touch the brilliant achievements of mathematics indispensable not only for the theoretical physics but for experimentalist to fit their data as well. This is the motivation why we, following the spirit of the century of enlightenment, have provided in Ref. [429] a simple FORTRAN90 program illustrating the operation of the ϵ-algorithm. Certainly, *fysics is phun*[1], being in part art *cosa mentale*[2] and every new software cannot be foolproof, but there are methods that must be taken into account in every complicated calculation.

5.2.6 *Power series for differential susceptibility*

Having calculated the relative dimensionless susceptibility by employing the ϵ-algorithm we can recover the usual susceptibility from the dimensionless one,

$$\chi(\epsilon, h) = \tilde{\chi}_{\mathrm{rel}}(\epsilon, h)\, \chi(\epsilon). \tag{5.101}$$

In order to take into account the effects of nonlocality the cutoff area in the momentum space should subtracted out from the susceptibility

[1]As R. Feynman used to say, "Fysics is Phun" (this was the title of a book he intended to write) [430, 431]

[2]During the Renaissance, Leonardo da Vinci wrote a treatise in which he explained that art is *cosa mentale* and something more than other handicrafts as is now physics among the other sciences.

$$\chi_{✂}(\epsilon, h) = \hat{\mathsf{C}}\chi(\epsilon, h) = \tilde{\chi}_{\mathrm{rel}}(\epsilon, h)\ \chi(\epsilon) - \tilde{\chi}_{\mathrm{rel}}(c+\epsilon, h)\ \chi(c+\epsilon). \tag{5.102}$$

Then we can easily find the magnetization

$$M(H,T) = \chi_{✂}(\epsilon, h)H. \tag{5.103}$$

The calculation of the differential susceptibility

$$\chi^{(\mathrm{dif})}(\epsilon, h) = \left(\frac{\partial M}{\partial H}\right)_T, \tag{5.104}$$

where $H = H_{\mathrm{c2}}(0)h$, gives an alternative method to determine the magnetization. Next we define a dimensionless relative differential susceptibility

$$\tilde{\kappa}\,(\epsilon, h, r) \equiv \chi^{(\mathrm{dif})}(\epsilon, h)/\chi(\epsilon). \tag{5.105}$$

For this variable, using Eq. (5.82), we have the series

$$\begin{aligned}\tilde{\kappa} &= 12\sum_{n=0}^{\infty}\left(1-\frac{1}{2^{2n+1}}\right)\frac{(2n+2)!}{2^{2n+1}}\frac{\zeta(2n+2)}{\pi^{2n+2}}\left(-\frac{h^2}{\epsilon^2}\right)^n \\ &\qquad\times\sum_{m=0}^{2n}\frac{(2m-1)!!\,(4n-2m-1)!!}{m!\,(2n-m)!\,(1+r/\epsilon)^{2n-m}} \\ &= 1-\frac{7}{5}\frac{\epsilon^2+r\epsilon+3r^2/8}{(\epsilon+r)^2}\left(\frac{h}{\epsilon}\right)^2+\dots,\end{aligned} \tag{5.106}$$

which, just as done in deriving Eq. (5.82), can be summed up by means of the ϵ-algorithm. For instance, in the local GL limit we have for the magnetization

$$\begin{aligned}M = \int_0^H \chi^{(\mathrm{dif})}(T,H)dH &= \chi(\epsilon)H_{\mathrm{c2}}(0)\int_0^h \tilde{\kappa}(\epsilon, h')dh' \\ &= \chi(\epsilon)H_{\mathrm{c2}}(0)\ \tilde{\chi}_{\mathrm{rel}}(\epsilon, h)h.\end{aligned} \tag{5.107}$$

For the further analysis, however, it is more suitable to introduce a dimensionless magnetization

$$\tilde{m} \equiv -\frac{M}{M_0} = -\frac{\Phi_0}{k_{\mathrm{B}}T_{\mathrm{c}}}\frac{s}{N}M, \qquad M_0 \equiv \frac{k_{\mathrm{B}}T_{\mathrm{c}}}{\Phi_0}\frac{N}{s}. \tag{5.108}$$

Then, using the relation

$$\tilde{\chi}_{\mathrm{rel}}(\epsilon, h) = \frac{1}{h}\int_0^h \tilde{\kappa}(\epsilon, h')dh' \tag{5.109}$$

the result for the dimensionless fluctuation magnetization takes the form

$$\tilde{m}(\epsilon, h) = \frac{1}{6}\frac{1}{\sqrt{\epsilon(\epsilon+r)}}\tilde{\chi}_{\mathrm{rel}}(\epsilon, h)h = \frac{1}{6}\frac{1}{\sqrt{\epsilon(\epsilon+r)}}\int_0^h \tilde{\kappa}(\epsilon, h')dh'. \tag{5.110}$$

In this section we have calculated the magnetization by means of power series in the magnetic field assuming in the beginning $h/\epsilon \ll 1$. In the next section we develop another method for calculating the fluctuation magnetic moment which is appropriate for strong magnetic fields and allows for studying the high magnetic field asymptotics for large enough values of the reduced magnetic field, $h/\epsilon \gg 1$. The overlap between these expansions about $h/\epsilon \simeq 1$ would be a test for the accuracy of the calculations.

5.3 Strong magnetic fields

5.3.1 *General formula for the free energy*

In order to derive a general formula for the Gibbs free energy for arbitrary non-vanishing magnetic field we will start again by representing the free energy density as a sum over the energy spectrum, Eq. (5.21),

$$F(\epsilon, h) = F_0\, 2h\, \hat{\mathsf{L}} \sum_{n=0}^{n_c-1} \left[\ln\left(n + \frac{1}{2} + \frac{\epsilon}{2h}\right) + \ln(2h)\right], \tag{5.111}$$

where

$$F_0 \equiv \frac{k_{\mathrm{B}} T_{\mathrm{c}}}{4\pi\xi_{ab}^2(0)} \frac{N}{s} = \frac{1}{2} M_0 B_{\mathrm{c}2}(0). \tag{5.112}$$

The first way to go in deriving convenient for programming formula is to calculate the action of the $\hat{\mathsf{L}}$ operator on the integrand, cf. Ref. [432]. In this case we write down the free energy as a finite sum over the Landau levels

$$F(\epsilon, h) = \frac{k_{\mathrm{B}} T_{\mathrm{c}}}{4\pi\xi_{ab}^2(0)} \frac{N}{s}\, 2h \sum_{n=0}^{n_c-1} \hat{\mathsf{L}} \ln[\epsilon + h(2n+1)], \tag{5.113}$$

where, according to Eq. (5.50),

$$\hat{\mathsf{L}}^{(\mathrm{LD})} \ln[\epsilon + h(2n+1)] = 2 \ln \frac{\sqrt{\epsilon + h(2n+1)} + \sqrt{\epsilon + h(2n+1) + r}}{2}. \tag{5.114}$$

This formula is useful especially in the case of strong magnetic fields when the finite series are not too long. However, in order to have a good working expression, applicable to all cases, it is much better to solve the problem analytically. Towards this end consider the last term in the integrand of Eq. (5.111). The summation of this constant term and simply yields the cutoff parameter c

$$2h \sum_{n=0}^{n_c-1} 1 = (2h) n_c = c. \tag{5.115}$$

Next we introduce a dimensionless function

$$x(\epsilon, h) \equiv \frac{1}{2} + \frac{\epsilon}{2h} = \frac{\epsilon + h}{2h} \\ = \frac{1}{2H}\left(T - T_{c2}(H)\right)\left(-\frac{\partial H_{c2}(T)}{\partial T}\right)\bigg|_{T=T_c-0}, \tag{5.116}$$

which is the argument of some of the analytical functions we use in the following. Further, we have to represent the sum in Eq. (5.111) as a difference of two appropriately regularized infinite series

$$\sum_{n=0}^{n_c-1} \ln(n+x) = \widehat{\mathsf{Reg}}_\zeta \sum_{n=0}^{\infty} \ln(n+x) - \widehat{\mathsf{Reg}}_\zeta \sum_{n_c}^{\infty} \ln(n+x). \tag{5.117}$$

In fact, one does not have any other possibility except the ζ-regularization

$$-\widehat{\mathsf{Reg}}_\zeta \sum_{n=0}^{\infty} \ln(n+z) = \frac{\partial}{\partial \nu}\zeta(\nu, z)\bigg|_{\nu=0} = \ln \frac{\Gamma(z)}{\sqrt{2\pi}}, \tag{5.118}$$

based on one relation between the Euler Γ-function and the Hurwitz ζ-function, and the definition of the logarithmic function

$$\ln z = \lim_{\nu \to 0} \frac{z^\nu - 1}{\nu} = \frac{\partial}{\partial \nu} z^\nu \bigg|_{\nu=0}. \tag{5.119}$$

According to the famous results by Riemann, the analytical continuations of the ζ-function and the factorial $n!$ are unique. Therefore the UV regularization of the partition function in the GL model in external magnetic field is practically included in the Gauss definition (second equality) of the Γ-function as an infinite product, see e.g., Ref. [433],

$$\int_0^\infty t^{z-1} \mathrm{e}^{-t} dt \equiv \Gamma(z) \equiv \lim_{n_c \to \infty} \frac{n_c!\, n_c^{z-1}}{z(z+1)(z+2)\ldots(z+n_c-1)}. \tag{5.120}$$

Let us recall some particular values,

$$\Gamma(n+1) = n!, \qquad \Gamma(1) = 0! = 1, \qquad \Gamma(1/2) = \pi^{1/2}, \tag{5.121}$$

and the Stirling's approximation for $n_c \gg 1$, derived by Gaussian saddle point approximation applied to the first, Euler definition of the Γ-function, left equality in Eq. (5.120),

$$n_c! \approx \left(\frac{n_c}{\mathrm{e}}\right)^{n_c} \sqrt{2\pi n_c}, \qquad \ln(n_c!) \approx \left(n_c + \frac{1}{2}\right) \ln n_c - n_c + \ln \sqrt{2\pi}. \tag{5.122}$$

For the local limit or for the case of weak magnetic fields we shall also make use of the asymptotic formulae for $z \gg 1$

$$\ln\Gamma(z) \approx \left(z-\frac{1}{2}\right)\ln z - z + \frac{1}{2}\ln(2\pi) + \frac{1}{12z}, \tag{5.123}$$

$$\psi^{(-1)}(z) \approx \left(z-\frac{1}{2}\right)\ln z - z + \frac{1}{12z}, \tag{5.124}$$

$$\psi(z) \approx \ln z - \frac{1}{2z} - \frac{1}{12z^2}, \tag{5.125}$$

$$\psi^{(1)}(z) \equiv \zeta(2,z) \approx \frac{1}{z} + \frac{1}{2z^2} + \frac{1}{6z^3}. \tag{5.126}$$

Substituting the Stirling asymptotics in the second Gauss definition, Eq. (5.120), and taking a logarithm we arrive at the function $\psi^{(-1)}(z)$, generating the polygamma functions

$$\begin{aligned}\psi^{(-1)}(z) &\equiv \lim_{n_c\to\infty}\left\{-\sum_{n=0}^{n_c-1}\ln(n+z) + \left(n_c - \frac{1}{2} + z\right)\ln(n_c) - n_c\right\}\\ &= \ln\frac{\Gamma(z)}{\sqrt{2\pi}}.\end{aligned} \tag{5.127}$$

As a result, the above Gauss definition for $\ln\Gamma(z)$ solves the problem for UV regularization of the infinite sum of logarithms, Eq. (5.118). The first derivative of this equation gives the well-known definition of the digamma function $\psi(z) \equiv \psi^{(0)}(z)$,

$$\begin{aligned}\psi^{(0)}(z) \equiv \frac{d}{dz}\psi^{(-1)}(z) &= -\widehat{\mathsf{Reg}}_\zeta \sum_{n=0}^{\infty}\frac{1}{n+z}\\ &= \lim_{n_c\to\infty}\left\{-\sum_{n=0}^{n_c-1}\frac{1}{n+z} + \ln(n_c)\right\}.\end{aligned} \tag{5.128}$$

In particular,

$$-\psi(1) = C_{\mathrm{Euler}} = \lim_{n_c\to\infty}\left\{\sum_{n=1}^{n_c-1}\frac{1}{n} - \ln(n_c)\right\} = 0.577216\ldots\,. \tag{5.129}$$

All other polygamma functions are actually Hurwitz ζ-functions with integer first argument $\geqslant 2$ and the sums are trivially convergent,

$$\psi^{(N)}(z) = \frac{d^N}{dz^N}\psi(z) = (-1)^N N!\,\zeta(N+1,z). \tag{5.130}$$

To summarize, we have applied the well known ζ-technique [434–438] for UV regularization of the partition function and revealed that the archetype of this powerful method comes from the century of enlightenment and finally we can bring the free energy, Eq. (5.111), to the form

$$F(\epsilon,h) = \frac{T}{T_{\mathrm{c}}} F_0 \left\{ 2h\, \hat{\mathsf{L}} \left[-\ln\Gamma\left(\frac{1}{2} + \frac{\epsilon}{2h}\right) + \ln\Gamma\left(\frac{1}{2} + \frac{\epsilon+c}{2h}\right)\right] + c\ln(2h) \right\}$$
$$= \frac{k_{\mathrm{B}}T}{4\pi\xi_{ab}^2(0)} \frac{N}{s}\, \hat{\mathsf{L}}\hat{\mathsf{C}} \left[-(2h)\ln\frac{\Gamma\left(\frac{\epsilon+h}{2h}\right)}{\sqrt{2\pi}} - \epsilon\ln(2h)\right]. \tag{5.131}$$

For weak magnetic field, $h \ll \epsilon$, cf. Eqs. (5.49), (5.50), (5.124), and (5.145),

$$-(2h)\ln\frac{\Gamma\left(\frac{\epsilon+h}{2h}\right)}{\sqrt{2\pi}} - \epsilon\ln(2h) \approx -\epsilon\left[\ln(\epsilon) - 1\right] + \frac{1}{6}\frac{h^2}{\epsilon}. \tag{5.132}$$

This is our main analytical result and all thermodynamic properties now can be obtained via derivatives. However, having this analytical result it is trivially to check that it can be derived by finite sums. The latter do not require UV regularization and the Euler Γ-function is commonly available in many textbooks on mathematical analysis.

5.3.2 *Fluctuation part of thermodynamic variables*

Having an analytical result for the free energy we can easily find other thermodynamic variables by differentiation. The magnetization, for example, is given by the derivative

$$M = -\left(\frac{\partial F}{\partial B}\right)_T = -\frac{1}{B_{\mathrm{c2}}(0)}\left(\frac{\partial F}{\partial h}\right)_\epsilon = -M_0\,\tilde{m}, \tag{5.133}$$

where a dimensionless diamagnetic moment is introduced

$$\tilde{m}(\epsilon,h) \equiv -\frac{M}{M_0} = \frac{c}{2h} - \hat{\mathsf{L}}\left[\ln\Gamma\left(\frac{\epsilon+h}{2h}\right) - \ln\Gamma\left(\frac{\epsilon+c+h}{2h}\right)\right]$$
$$+ \hat{\mathsf{L}}\left[\frac{\epsilon}{2h}\,\psi\left(\frac{\epsilon+h}{2h}\right) - \frac{\epsilon+c}{2h}\,\psi\left(\frac{\epsilon+h+c}{2h}\right)\right]. \tag{5.134}$$

In expanded notations within the LD model this formula, according to Eqs. (5.35), (5.37), (5.39), (5.43), and (5.46), reads as

$$M^{(\mathrm{LD})}(\epsilon,h) = -\frac{k_{\mathrm{B}}T_{\mathrm{c}}}{\Phi_0 s_{\mathrm{eff}}}\left(\frac{c}{2h} + \frac{2}{\pi}\int_0^{\frac{\pi}{2}} d\phi \left\{ -\left[\ln\Gamma\left(\frac{\epsilon + r\sin^2\phi + h}{2h}\right)\right.\right.\right.$$
$$\left. - \ln\Gamma\left(\frac{c+\epsilon+r\sin^2\phi+h}{2h}\right)\right] + \left[\frac{\epsilon + r\sin^2\phi}{2h}\,\psi\left(\frac{\epsilon+r\sin^2\phi+h}{2h}\right)\right.$$
$$\left.\left.\left. - \frac{c+\epsilon+r\sin^2\phi}{2h}\,\psi\left(\frac{c+\epsilon+r\sin^2\phi+h}{2h}\right)\right]\right\}\right), \tag{5.135}$$

where $\phi = \frac{1}{2}\theta$. For $|\epsilon|, h \ll r, c$ this general expression recovers the local 3D result, Eq. (5.182), analyzed later in Sec. 5.3.4, while in the opposite case of extremely high anisotropy $r < |\epsilon|, h \ll c$ we get the local 2D result, Eq. (5.144). Here we want to emphasize the existence to mention a universal magnetization law at $T = T_c$, or $\epsilon = 0$, which can be observed for many high-T_c materials at strong magnetic fields $h \gg r$

$$-M(T_c, B)\,\frac{\Phi_0 s_{\text{eff}}}{k_B T_c} = \tilde{m} = \frac{1}{2}\ln 2\; U_M\left(\frac{2}{c}\frac{B}{B_{c2}(0)}\right), \tag{5.136}$$

where the universal function of the nonlocal magnetization

$$U_M(y) \equiv \frac{2}{\ln 2}\left\{\ln\Gamma\left(\frac{1}{y}+\frac{1}{2}\right) - \frac{1}{2}\ln\pi + \frac{1}{y}\left[1-\psi\left(\frac{1}{y}+\frac{1}{2}\right)\right]\right\} \tag{5.137}$$

For conventional bulk superconductors the nonlocality effects on magnetization are well understood, see for example Refs. [439–445]. To the best of our knowledge, the first observation of fluctuation-induced diamagnetism for a cuprate superconductor well inside the finite-magnetic-field regime was reported by Carretta *et al.* [446] for $YBa_2Cu_3O_{6+x}$. Soon after, analogous measurement was reported for $La_{1.9}Sr_{0.1}CuO_4$ by Carballeira *et al.* [447]. Being familiar with the preliminary version of Ref. [429], Carballeira *et al.* have entirely based their interpretation and theoretical analysis on Eq. (5.135) and Eq. (5.144) below. Alas, we find it very disappointing and impolite that the authors of Ref. [447] did not give any credits (e.g., in the author list, acknowledgments, or references section) to the author of the theory (the first author of this book) they have used. Further details can be found in Ref. [448].

Returning now to the general expression for the magnetization, Eq. (5.134), we derive another expression for the relative differential susceptibility based on Eq. (5.110)

$$\begin{aligned}\kappa(\epsilon, h) &= 6\sqrt{\epsilon(\epsilon+r)}\,\left(\frac{\partial\tilde{m}}{\partial h}\right)\\ &= 6\sqrt{\epsilon(\epsilon+r)}\;\hat{\mathrm{L}}\Bigg[-\frac{c}{2h^2} - \frac{\epsilon^2}{4h^3}\,\psi^{(1)}\left(\frac{\epsilon+h}{2h}\right)\\ &\quad + \frac{(\epsilon+c)^2}{4h^3}\,\psi^{(1)}\left(\frac{\epsilon+h+c}{2h}\right)\Bigg].\end{aligned} \tag{5.138}$$

The comparison of this result with Eq. (5.106) is one of the best methods to check the accuracy of the programmed formulae. Analogously, differentiating the free energy with respect to the temperature $T = (1+\epsilon)T_c$ we derive the general formula for the most singular part of the entropy (neglecting the derivative of the T-prefactor in Eq. (5.131)),

$$S \equiv -\frac{1}{T_c}\frac{\partial F}{\partial \epsilon} = \frac{k_B}{4\pi\xi_{ab}^2(0)}\frac{N}{s}\ \hat{\mathsf{L}}\left[\psi\left(\frac{\epsilon+h}{2h}\right) - \psi\left(\frac{\epsilon+h+c}{2h}\right)\right], \qquad (5.139)$$

and the most singular part of the heat capacity

$$C(\epsilon,h) = \frac{\partial S}{\partial \epsilon} = -\frac{1}{T_c}\frac{\partial^2 F}{\partial \epsilon^2}$$
$$= \frac{k_B}{4\pi\xi_{ab}^2(0)}\frac{N}{s}\frac{1}{2h}\ \hat{\mathsf{L}}\left[\psi^{(1)}\left(\frac{\epsilon+h}{2h}\right) - \psi^{(1)}\left(\frac{\epsilon+h+c}{2h}\right)\right]. \qquad (5.140)$$

This expression for C can be directly derived from the starting formulae (5.21) and (5.111). The sums for the heat capacity are convergent, cf. Ref. [66], and do not require any regularization. The simplest way to reproduce the analytical result for the free energy density, Eq. (5.131), is to integrate two times the result for its second derivative, i.e., that for the heat capacity, cf. Ref. [449, 450]. In general finite sums from 0 to $n_c - 1$ for logarithms and powers can be found in many textbooks on mathematics and all our results can thus be easily checked even by experimentalists.

The fluctuation part of the entropy S is proportional to the mean square of the order parameter Ψ, i.e., the volume density of fluctuation Cooper pairs. The thermally averaged density deserves a special attention because it is the main ingredient of the self-consistent treatment of the interaction of order parameter fluctuations. This Hartree type approximation due to Ullah and Dorsey [451] will be briefly described in the next subsection.

In the following, for completeness, we will derive the local 2D asymptotics applicable for $|\epsilon|, h \ll c$. The substitution of the first term from Eq. (5.124) into the general formula for the free energy, Eq. (5.131), gives

$$\tilde{f}_{2D} \equiv \frac{F(\epsilon,h)}{F_0} = -(2h)\,\psi^{(-1)}\left(\frac{\epsilon+h}{2h}\right) - \epsilon\ln(2h) + A(c)\epsilon + B(c) + O(1/c), \qquad (5.141)$$

where for $\epsilon \ll c$, cf. Eq. (5.12),

$$f_c(\epsilon) \equiv A(c)\epsilon + B(c) \approx \int_0^{c+\epsilon} \ln\tilde{x}\, d\tilde{x} \approx \epsilon\ln c + c(\ln c - 1). \qquad (5.142)$$

This irrelevant for the fluctuation phenomena linear function of ϵ gives constant additions to the free energy $F_c = F_0 B(c)$, and entropy $S_c = -F_0 A(c)/T_c$ and can be omitted hereafter. The subtraction of $F_0 f_c$ from the free energy, Eq. (5.131), can be considered as a cutoff procedure for UV regularization,

$$\widehat{\mathsf{Reg}}_{\text{✂}} F(\epsilon,h) = F(\epsilon,h) - (F_c - T_c S_c \epsilon), \qquad (5.143)$$

which, when applied, allows the analysis of the local GL approximation to be carried out simply as $(c \to \infty)$-limit. Now a trivial differentiation gives for the dimensionless magnetization, being a positive quantity,

$$\tilde{m}_{2\mathrm{D}}(\epsilon, h) = \frac{-M(\epsilon, h)}{M_0} = \frac{\epsilon}{2h}\left[\psi\left(\frac{\epsilon}{2h} + \frac{1}{2}\right) - 1\right] - \psi^{(-1)}\left(\frac{\epsilon + h}{2h}\right)$$
$$= \frac{1}{2}\frac{\partial \tilde{f}_{2\mathrm{D}}}{\partial h} \tag{5.144}$$

This result is also a local $(c \to \infty)$-asymptotic of Eq. (5.134), which for $h \ll \epsilon$ yields

$$\tilde{m}_{2\mathrm{D}} \approx h/6\epsilon. \tag{5.145}$$

In the general case the local approximation gives $\tilde{m} = \hat{\mathsf{L}}\tilde{m}_{2\mathrm{D}}$, or for the LD model

$$\tilde{m}(\epsilon, h; r) = -\frac{M}{M_0} = \int_0^{\pi/2} \frac{d\phi}{\pi/2}\left\{\frac{\epsilon + r\sin^2\phi}{2h}\left[\psi\left(\frac{\epsilon + r\sin^2\phi}{2h} + \frac{1}{2}\right) - 1\right]\right.$$
$$\left. - \ln\Gamma\left(\frac{\epsilon + r\sin^2\phi}{2h} + \frac{1}{2}\right) + \frac{1}{2}\ln(2\pi)\right\}. \tag{5.146}$$

The next differentiation with respect to the magnetic field, using Eqs. (5.42) and (5.104), gives the relative dimensionless susceptibility

$$\tilde{\kappa}_{2\mathrm{D}}(\epsilon, h) = 6\epsilon\left(\frac{\partial \tilde{m}}{\partial h}\right) = 12\left(\frac{\epsilon}{2h}\right)^2\left[1 - \frac{\epsilon}{2h}\psi^{(1)}\left(\frac{\epsilon}{2h} + \frac{1}{2}\right)\right]$$
$$= \frac{\chi^{(\mathrm{dif})}(\epsilon, h)}{\chi(\epsilon)}, \tag{5.147}$$

which is also a local $c \gg h, |\epsilon|$ asymptotic of Eq. (5.138). For the LD model after averaging with respect to the Josephson phase, according to Eq. (5.39), we obtain

$$\kappa(\epsilon, h; r) = 12\int_0^{\pi/2} \frac{d\phi}{\pi/2}\left(\frac{\epsilon + r\sin^2\phi}{2h}\right)^2\left[1 - \frac{\epsilon + r\sin^2\phi}{2h}\zeta\left(2, \frac{\epsilon + r\sin^2\phi}{2h}\right)\right]. \tag{5.148}$$

This final result can be directly compared to low field series expansion Eq. (5.106). Similar differentiations of the free energy, Eq. (5.141), with respect to the temperatures gives the most singular part of the entropy

$$\tilde{s}_{2\mathrm{D}} \equiv -\frac{\partial}{\partial\epsilon}\tilde{f}_{2\mathrm{D}} = \left[\psi\left(\frac{\epsilon + h}{2h}\right) + \ln(2h)\right] = T_c S(\epsilon, h)/F_0, \tag{5.149}$$

and of the heat capacity

$$\hat{\mathsf{L}}\,\tilde{c}_{\mathrm{2D}} \equiv -\,\hat{\mathsf{L}}\frac{\partial^2}{\partial\epsilon^2}\tilde{f}_{\mathrm{2D}}(\epsilon,h) = \frac{1}{2h}\psi^{(1)}\left(\frac{\epsilon+h}{2h}\right) = T_{\mathrm{c}}C(\epsilon,h)/F_0. \qquad (5.150)$$

Restoring the T prefactor instead of T_{c} in Eq. (5.12), as was done in Eq. (5.131), we arrive at a slightly different expression for the fluctuation part of the free energy $F = F_0(1+\epsilon)\tilde{f}_{\mathrm{2D}}(\epsilon,h)$ and the heat capacity

$$\begin{aligned}\hat{\mathsf{L}}\,\tilde{c}_{\mathrm{2D}} &= -(1+\epsilon)\frac{\partial^2}{\partial\epsilon^2}(1+\epsilon)\,\hat{\mathsf{L}}\,\tilde{f}_{\mathrm{2D}}(\epsilon,h)\\ &= (1+\epsilon)\left[(1+\epsilon)\,\hat{\mathsf{L}}\,\tilde{c}_{\mathrm{2D}} + 2\,\hat{\mathsf{L}}\,\tilde{s}_{\mathrm{2D}}\right] = \frac{T_{\mathrm{c}}C(\epsilon,h)}{F_0},\end{aligned} \qquad (5.151)$$

which gives

$$C(\epsilon,h) = \frac{(1+\epsilon)k_{\mathrm{B}}}{4\pi\xi_{ab}^2(0)}\frac{N}{s}\,\hat{\mathsf{L}}\left[\frac{1+\epsilon}{2h}\psi^{(1)}\left(\frac{\epsilon+h}{2h}\right) + 2\psi^{(0)}\left(\frac{\epsilon+h}{2h}\right) + 2\ln(2h)\right]. \qquad (5.152)$$

For zero magnetic field we have

$$C(\epsilon,h=0) = \frac{(1+\epsilon)k_{\mathrm{B}}}{4\pi\xi_{ab}^2(0)}\frac{N}{s}\left[(1+\epsilon)\,\hat{\mathsf{L}}\,\frac{1}{\epsilon} - 2\epsilon\,\hat{\mathsf{L}}\,\ln\frac{1}{\epsilon}\right], \qquad (5.153)$$

which in the LD model takes the form

$$C(\epsilon,r) = \frac{(1+\epsilon)k_{\mathrm{B}}}{4\pi\xi_{ab}^2(0)}\frac{N}{s}\left[\frac{(1+\epsilon)}{\sqrt{\epsilon(r+\epsilon)}} - 2\epsilon\,2\ln\frac{2}{\sqrt{\epsilon}+\sqrt{r+\epsilon}}\right]. \qquad (5.154)$$

This expression differs from Eqs. (5.41) and (5.61). However, the $(1+\epsilon)^2 \approx 1+2\epsilon$ correction and the less singular part of the heat capacity $2(1+\epsilon)F_0\,\hat{\mathsf{L}}\tilde{s}_{\mathrm{2D}}/T_{\mathrm{c}}$, which appears due to differentiation of T in the numerator of Eq. (5.12) and Eq. (5.131), are difficult to be identified experimentally.

For the superconducting phase below the critical temperature, $0 < -\epsilon \ll 1$, one has to take into account more or less space homogeneous order parameter Ψ_ϵ which minimizes the nongradient part of the free energy density $F = a(\epsilon)n_\epsilon + \frac{1}{2}n_\epsilon^2$,

$$\Psi_\epsilon = \sqrt{a_0(-\epsilon)/b}, \qquad n_\epsilon = \Psi_\epsilon^2 = a_0(-\epsilon)/b. \qquad (5.155)$$

The fluctuations around this minimum

$$\Psi = \Psi_\epsilon + \Psi' + i\Psi'', \qquad n = \Psi^2 = n_\epsilon + 2\Psi_\epsilon\Psi' + (\Psi')^2 + (\Psi'')^2 \qquad (5.156)$$

should be considered as a small perturbation, thus only the quadratic term in the free energy is taken into account,

$$F(\epsilon<0) = a(\epsilon) + \frac{1}{2}bn^2 \approx -\frac{1}{2b}a_0^2\epsilon^2 + a_0(-2\epsilon)\left[1\,(\Psi')^2 + 0\,(\Psi'')^2\right]. \qquad (5.157)$$

The first term in this equation corresponds to the jump in the heat capacity $\Delta C = a_0^2/bT_c$ at T_c. The linear term $\propto \Psi'$ simply cancels. The phase fluctuations $\propto (\Psi'')^2$ are coupled to the plasmons and vortexes but they are irrelevant for the thermodynamic fluctuations significantly below T_c. In this way mainly fluctuations related to the modulus of the order parameter are essential for the heat capacity below T_c. Finally, the comparison of the second term $\propto \Psi'$ in Eq. (5.157) with the corresponding expression above T_c

$$F(\epsilon > 0) = a(\epsilon) + \frac{1}{2}bn^2 \approx a_0(\epsilon)\left[(\Psi')^2 + (\Psi'')^2\right], \tag{5.158}$$

provides a prescription to derive the fluctuation part below T_c from the fluctuation expression for the normal phase above T_c

$$\frac{1}{2}\hat{\mathsf{L}}\frac{1}{(-2\epsilon)} \leftarrow \hat{\mathsf{L}}\frac{1}{\epsilon}. \tag{5.159}$$

Applying this prescription to Eq. (5.154) results in the following expression

$$C(\epsilon < 0, r) = \frac{(1+\epsilon)k_B}{4\pi\xi_{ab}^2(0)}\frac{N}{s}\frac{1}{2}\left[\frac{(1+\epsilon)}{\sqrt{(-2\epsilon)(r-2\epsilon)}} - 2\epsilon\, 2\ln\frac{2}{\sqrt{(-2\epsilon)}+\sqrt{r-2\epsilon}}\right]. \tag{5.160}$$

This fluctuation part as well as the phonon heat capacity should be subtracted from the experimental data in order to extract the jump ΔC and related to it penetration depth $\lambda_{ab}(0)$. Such a procedure, in fact, gives a purely thermodynamic method to determine the latter quantity.

The dimensionless functions Eqs. (5.141), (5.144), (5.147), and (5.150)) derived with the local approximation are just as important for the thermodynamics of the layered superconductors as is the APS function for the paraconductivity, Eq. (5.67). The operator $\hat{\mathsf{L}}$ gives the possibility to extend the 2D analytical result for layered or even isotropic 3D superconductor. Additionally the $\hat{\mathsf{C}}$ operator gives the energy cutoff approximation for the nonlocality effects in the conducting CuO_2 planes. Therefore the analytical 2D result plays a key role for the fluctuation phenomena in layered superconductors.

We will finish the analysis of the local $c \gg |\epsilon|, h$ 2D approximation $h \gg r$, i.e.,

$$\mu_0 H \gg rB_{c2}(0) = \left(\frac{2\xi_c(0)N}{s}\right)^2 \frac{\Phi_0}{2\pi\xi_{ab}^2(0)}, \tag{5.161}$$

with the important case of strong magnetic field $h \gg |\epsilon|$. Under these conditions ($|\epsilon|, r \ll h \ll c$) the layered superconductors display a magnetization corresponding to the local 2D one in strong magnetic fields. The substitution of $\epsilon = 0$ in Eq. (5.144), using Eq. (5.121), recovers the result by Klemm, Beasley, and Luther [452]

$$\tilde{m}(h \gg r, \epsilon \ll h) \approx 0.3465735902799726\ldots,$$
$$-M \approx \frac{\ln 2}{2}\, \frac{k_{\mathrm{B}} T_{\mathrm{C}}}{\Phi_0}\, \frac{N}{s}. \tag{5.162}$$

In the concise review by Koshelev [453] on the properties of 2D GL model the calculation of $\frac{1}{2}\ln 2 \approx 0.346$ by infinite series with three decimal digits accuracy is described in great details.

5.3.3 *Self-consistent approximation for the LD model*

The bulk (3D) density of the fluctuation Cooper pairs $n(\epsilon, h)$ can be calculated from the general expression for the Gibbs free energy Eqs. (5.8) and (5.12). The differentiation with respect of the "chemical potential" of Cooper pairs $\mu_{\mathrm{CP}} = -a_0\epsilon$, according to the relation $N_{\mathrm{CP}} = (\partial G/\partial \mu_{\mathrm{CP}})_{T,H}$, gives

$$n(\epsilon, h) = \frac{N}{s}\left\langle |\Psi_n|^2 \right\rangle = \frac{1}{a_0}\frac{\partial}{\partial \epsilon} F(\epsilon, h) = -\frac{k_{\mathrm{B}} T_{\mathrm{C}}}{a_0} S(\epsilon, h; c). \tag{5.163}$$

This formula can be alternatively derived by summation of the Rayleigh–Jeans asymptotics of the energy distribution of the fluctuation Cooper pairs

$$n(\epsilon, h) = \frac{1}{V} \sum_{\mathbf{p}, p_z, j} \frac{k_{\mathrm{B}} T}{\epsilon_j(\mathbf{p}, p_z) + a}$$
$$= \frac{N}{s} \hat{\mathsf{L}}^{(\mathrm{LD})} \int_{|\mathbf{p}|<p_c} \frac{d(\pi p^2)}{(2\pi\hbar)^2} \frac{k_{\mathrm{B}} T_{\mathrm{C}}}{p^2/2m_{ab} + a_0\epsilon}, \tag{5.164}$$

see for example the monograph by Patashinskii and Pokrovsky [454].

Let us give an illustration for zero magnetic field. In this case for the density of fluctuation Cooper pairs, using Eq. (5.112) and Eq. (5.50), we obtain

$$n(\epsilon, 0) = \frac{F_0}{a_0}\, 2 \ln \frac{\sqrt{c+\epsilon} + \sqrt{c+r+\epsilon}}{\sqrt{\epsilon} + \sqrt{\epsilon + r}}. \tag{5.165}$$

This formula sets the stage for the self-consistent treatment of the order parameter fluctuations in the LD model in which the nonlinear term is replaced by its average. The idea has its origin in the Maxwell consideration of the ring of Saturn; probably it is the first work on collective phenomena in physics. Having no possibility to consider motion of all particles in detail we must search for some approximation. Within a self-consistent picture, the motion of every particle creates an average potential in which the others are moving. From the dust of the ring of Saturn to the Cooper pairs in cuprates the idea is the same, only the mechanics slightly changes. In the self-consistent approximation the nonlinear term in GL equations gives an addendum to the linear one

$$a_{\mathrm{ren}}(\epsilon, h) = a_0\epsilon + b\, n\left(\frac{a_{\mathrm{ren}}}{a_0}, h\right), \tag{5.166}$$

where the coefficient $b = \tilde{b}N/s$ can be expressed via the jump of the heat capacity ΔC at the phase transition or, which is more convenient for the high-T_{c} cuprates, via the extrapolated to zero temperature penetration depth $1/\lambda_{ab}^2(T) = \mu_0 n(T)e^{*2}/m_{ab},\ n(T) = -a(T)/b$,

$$b = \frac{a_0^2}{T_{\mathrm{c}}\,\Delta C} = 2\mu_0\left(\frac{\pi\hbar^2\kappa_{\mathrm{GL}}}{\Phi_0 m_{ab}}\right)^2,$$

$$T_{\mathrm{c}}\,\Delta C = \frac{1}{8\pi^2\mu_0}\left(\frac{\Phi_0}{\lambda_{ab}(0)\xi_{ab}(0)}\right)^2, \tag{5.167}$$

where $\kappa_{\mathrm{GL}} \equiv \lambda_{ab}(0)/\xi_{ab}(0)$ is the GL parameter. One can easily check that Eq. (5.167) has the same form in Gaussian units, where $\mu_0^{(\mathrm{Gauss})} = 4\pi$. Introducing the renormalized reduced temperature $\epsilon_{\mathrm{ren}} > 0$ for the normal phase we have the self-consistent equation

$$\epsilon_{\mathrm{ren}} = \ln\frac{T}{T_{\mathrm{c}}} + \frac{b}{a_0}n(\epsilon_{\mathrm{ren}}, h), \tag{5.168}$$

where $n(\epsilon, h)$ is calculated by means of Gaussian saddle point approximation [451]. For the LD model this equation, by virtue of Eq. (5.165), takes the form

$$\begin{aligned}\epsilon_{\mathrm{ren}} &= \ln\frac{T}{T_{\mathrm{c}}} + \epsilon_{\mathrm{Gi}}\,2\ln\frac{\sqrt{c+\epsilon_{\mathrm{ren}}} + \sqrt{c+\epsilon_{\mathrm{ren}}+r}}{\sqrt{\epsilon_{\mathrm{ren}}} + \sqrt{\epsilon_{\mathrm{ren}}+r}}\\ &= \epsilon + \epsilon_{\mathrm{Gi}}\ \hat{\mathsf{L}}^{(\mathrm{LD})}\ln\frac{c+\epsilon_{\mathrm{ren}}}{\epsilon_{\mathrm{ren}}},\end{aligned} \tag{5.169}$$

where the dimensionless parameter

$$\epsilon_{\mathrm{Gi}} \equiv \frac{bF_0}{a_0^2} = 2\pi\mu_0\frac{N}{s}\left(\frac{\lambda_{ab}(0)}{\Phi_0}\right)^2 k_{\mathrm{B}}T_{\mathrm{c}} = \frac{1}{4\pi\xi_{ab}^2(0)}\frac{N}{s}\frac{k_{\mathrm{B}}}{\Delta C} \tag{5.170}$$

is closely related to the Ginzburg number; cf. Eq. (5.61) which now reads

$$\frac{C(\epsilon)}{\Delta C} = \epsilon_{\mathrm{Gi}}\ \hat{\mathsf{L}}\hat{\mathsf{C}}\frac{1}{\epsilon}, \tag{5.171}$$

and the review article by Varlamov *et al.* [396]. At T_c, for $\epsilon_{\rm Gi} \ll r \ll c$, Eq. (5.169) gives

$$\epsilon_{\mathrm{ren},\, c} \approx \epsilon_{\rm Gi} \ln \frac{4c}{r} \tag{5.172}$$

and the effective heating $\Delta T = T_c \epsilon_{\mathrm{ren},\, c}$ constrains the fluctuation variables at T_c. To provide an order estimate we take for illustration $s_{\rm eff} = 1$ nm, $\lambda_{ab}(0) = 207$ nm, $T_c = 100$ K, $\xi_{ab}(0) = 2.07$ nm, $\kappa_{\rm GL} = 100$, $k_{\rm B} = 1.381 \times 10^{-23}$ J/K. The substitution of these values in Eq. (5.170) gives

$$\epsilon_{\rm Gi} = \frac{8\pi^2 \times 1.381}{1000} \approx 11\%, \qquad \frac{\epsilon_{\rm Gi}}{6\kappa_{\rm GL}^2} \approx 2 \times 10^{-6}. \tag{5.173}$$

In the case of nonzero magnetic field the self-consistent equation for the renormalized reduced temperature, Eq. (5.169), according to Eqs. (5.139), (5.163), and (5.168), takes the form

$$\epsilon_{\rm ren} = \ln \frac{T}{T_c} + \epsilon_{\rm Gi}\, \hat{\mathsf{L}} \left[-\psi\left(\frac{\epsilon_{\rm ren} + h}{2h} \right) + \psi\left(\frac{c + \epsilon_{\rm ren} + h}{2h} \right) \right], \tag{5.174}$$

or, within the LD model,

$$\begin{aligned} \epsilon_{\rm ren} = \ln \frac{T}{T_c} + \epsilon_{\rm Gi} \int_0^{\pi/2} \frac{d\phi}{\pi/2} \Bigg[& -\psi\left(\frac{\epsilon_{\rm ren} + h + r\sin^2\phi}{2h} \right) \\ & + \psi\left(\frac{c + \epsilon_{\rm ren} + h + r\sin^2\phi}{2h} \right) \Bigg], \end{aligned} \tag{5.175}$$

cf. also Ref. [455]. For weak magnetic fields, $h \ll \epsilon$, using the asymptotic formula for the digamma function, Eq. (5.126), we recover Eq. (5.169). The formulae pointed out could be easily programmed for the self-consistent LD fit to the paraconductivity near to the critical temperature T_c. With the foregoing discussion we finish the analysis of the thermodynamics of layered superconductors. We only note that all final formulae can be used to fit the experimental data. Before proceeding however, for reliability sake, it is necessary to check if the formulae implementation correctly reproduces the 3D limit case $r \to \infty$.

5.3.4 *3D test example*

Every layered superconductor near the critical point $|\epsilon|, h \ll r$ displays 3D behavior. For high-T_c cuprates, however, $r \ll 1$ and 3D behavior can be observed only in crystals of extremely high quality. Due to fluctuation of the stoichiometry and of the T_c 3D regime of Gaussian fluctuations may not occur. However there are many conventional layered compounds with moderate anisotropy, $r \lesssim 1$, to which the 3D behavior has broader applicability. The 3D case can be derived as $(r \to \infty)$-asymptotics if the parabolic band approximation $\omega_1(\theta) \approx r\theta^2/4$, Eq. (5.34), is substituted into the $\hat{\mathsf{L}}$ operator, Eq. (5.46). Using the variable $x(\epsilon, h)$ from Eq. (5.116) and a new dimensionless variable q, defined as

$$q = \sqrt{\frac{r}{8h}}\frac{sp_z}{\hbar}, \qquad q^2 \equiv \frac{r}{8h}\theta^2, \qquad d\theta = 2\sqrt{\frac{2h}{r}}dq, \tag{5.176}$$

we get for the regularized sum of logarithms in Eq. (5.113) the local approximation

$$\hat{\mathsf{L}}^{(\mathrm{LD})}\widehat{\mathsf{Reg}}_\zeta \sum_{n=0}^{\infty} \ln\left(n + \frac{1}{2} + \frac{\epsilon}{2h}\right) \approx 2\sqrt{\frac{2h}{r}}\,\widehat{\mathsf{Reg}}_\zeta \sum_{n=0}^{\infty} \int_{-\infty}^{+\infty} \frac{dq}{2\pi} \ln\left(n + x + q^2\right). \tag{5.177}$$

The UV regularization in this expression is carried out with the help of the equation

$$\widehat{\mathsf{Reg}}_\zeta \sum_{n=0}^{\infty} \int_{-\infty}^{+\infty} \frac{dq}{2\pi} \ln\left(n + x + q^2\right) = \zeta\left(-\frac{1}{2}, x\right), \tag{5.178}$$

which can be easily proved using derivatives of ζ-functions

$$\frac{d}{dx}\zeta(\nu, x) = -\nu\zeta(\nu + 1, x). \tag{5.179}$$

The second derivative of Eq. (5.178) is trivially convergent; the essence of the ζ-function regularization lies in the omission of an arbitrary linear function $A(c)x + B(c)$, being analytical with respect to ϵ and therefore irrelevant to the critical behavior, cf. Eq. (5.143). In fact $c \simeq 1$ but having dropped $A(c)$ and $B(c)$ we can obtain the local approximation, $|\epsilon|, h \ll c$, as $c \to \infty$ even if $A(\infty) = \infty$ and $B(\infty) = \infty$. The substitution of this UV regularization in Eq. (5.113), using Eq. (5.35), gives the result by Mishonov [449, 450] for the fluctuation part of the Gibbs free energy

$$\begin{aligned} G(T,H) &= VF(\epsilon, h) \\ &= \frac{\sqrt{2}}{2\pi}\,k_{\mathrm{B}}T\,\frac{V}{\xi_a(0)\xi_b(0)\xi_c(0)}\,h^{3/2}\,\zeta\left(-\frac{1}{2}, \frac{1}{2} + \frac{\epsilon}{2h}\right). \end{aligned} \tag{5.180}$$

This result was confirmed by Baraduc *et al.* [432] using the same notations, with the ζ-function presented implicitly.

In order to bridge the 3D result with the notations introduced for layered systems we can rewrite the coefficient in Eq. (5.180) as

$$\frac{\sqrt{2}}{2\pi}\,\frac{k_{\mathrm{B}}T}{\xi_a(0)\xi_b(0)\xi_c(0)} = 4\sqrt{\frac{2}{r}}\left(\frac{1}{2}M_0 B_{c2}(0)\right). \tag{5.181}$$

Now differentiation $F(\epsilon, h)$ with respect to the magnetic field we obtain for the dimensionless magnetization, in agreement with the result by Kurkijärvi, Ambegaokar and Eilenberger [443]

$$\tilde{m}(\epsilon, h) = 3\left(\frac{2}{r}\right)^{1/2}\sqrt{h}\left[\zeta\left(-\frac{1}{2}, \frac{1}{2}+\frac{\epsilon}{2h}\right) - \frac{1}{3}\zeta\left(\frac{1}{2}, \frac{1}{2}+\frac{\epsilon}{2h}\right)\frac{\epsilon}{2h}\right]. \tag{5.182}$$

The subsequent differentiation with respect to the magnetic field gives the differential susceptibility. In the particular case of strong magnetic fields, $\epsilon \ll h$, the local approximation to the GL model, Eq. (5.182), gives the well known-result by Prange [456] with an anisotropy correction multiplier [449, 450] $\xi_{ab}(0)/\xi_c(0)$

$$\tilde{m}(0, h) = 3\sqrt{2} \times 0.0608885\sqrt{\frac{h}{r}}, \tag{5.183}$$

$$-M = 3\pi^{1/2}\zeta\left(-\frac{1}{2}, \frac{1}{2}\right)\frac{k_{\mathrm{B}}T_{\mathrm{c}}}{\Phi_0^{3/2}}\frac{\xi_{ab}(0)}{\xi_c(0)}\sqrt{\mu_0 H}, \tag{5.184}$$

where for the values of the ζ-function we have

$$\zeta\left(-\frac{1}{2}, \frac{1}{2}\right) = \left[-1+\frac{1}{\sqrt{2}}\right]\zeta\left(-\frac{1}{2}\right) = \texttt{Zeta[-1/2,1/2]} = 0.0608885\ldots,$$

$$\zeta\left(-\frac{1}{2}\right) = \texttt{Zeta[-1/2]} = -0.207886\ldots. \tag{5.185}$$

The syntax `Zeta[...]` is used in the commercial software MATHEMATICA [457]. We stress, however, that these are only test mathematical asymptotics for $c \to \infty$. For the magnetization, as well as for every quantity exhibiting UV divergences in the local limit, the nonlocal effects are strongly pronounced simply because the contribution of high momenta is significant. That is why the local approximation could be quantitatively fairly good for fitting to the data for fluctuation conductivity and heat capacity. For the magnetization in strong magnetic field regime we have to take into account the effect of nonlocality by fitting the energy cutoff parameter $\varepsilon_{✂}$. A systematic procedure for determination of the parameters of the GL theory is developed in the next section.

5.4 Some remarks on the fitting of the GL parameters

5.4.1 *Determination of the cutoff energy* $\varepsilon_{✂}$

Let us start with designing a general procedure to fit some parameters of the GL theory which employs only data for the in-plane paraconductivity.

Later on we shall address the advantage of investigating several variables simultaneously. The first step is to extract the fluctuation part of the conductivity from the temperature dependence of the resistivity $R(T)$. For layered cuprates the resistivity of the normal phase is to within good accuracy a linear function of temperature, $R_N(T) = A_R + B_R T$, and we can fit the coefficients A_R and B_R far enough from the critical temperature T_c, e.g., in the temperature interval $(1.5\,T_c, 3\,T_c)$. After that we can determine the experimental data for the fluctuation conductivity

$$\sigma_i = R(T_i)^{-1} - (A_R + B_R T_i)^{-1} \tag{5.186}$$

for all experimental points $i = 1, \ldots, N_{\mathrm{exp}}$. For bi-layered cuprates, such as $YBa_2Cu_3O_{7-\delta}$ and $Bi_2Sr_2CaCu_2O_8$, one can attempt fitting the data with the formula for the bilayered model, Eq. (5.55), where an arbitrary life-time $\tilde{\tau}_{\mathrm{rel}}$ and cutoff parameter c are included in the interpolation

$$\sigma(\epsilon; \tilde{\tau}_{\mathrm{rel}}, r, w, c) = \frac{e^2}{16\hbar}\tilde{\tau}_{\mathrm{rel}}\frac{N}{s}\left[f_{\mathrm{MT}}(\epsilon; r, w) - f_{\mathrm{MT}}(c+\epsilon; r, w)\right], \tag{5.187}$$

where

$$\begin{aligned} f_{\mathrm{MT}}(\epsilon; r, w) &\equiv \frac{\epsilon + \frac{1}{2}rw}{\sqrt{\left(\epsilon^2 + rw\epsilon\right)\left(\epsilon^2 + rw\epsilon + \frac{1}{4}r^2 w\right)}} \\ &= \hat{\mathsf{L}}^{(\mathrm{MT})} f_{\mathrm{APS}}(\epsilon, h=0) = \hat{\mathsf{L}}^{(\mathrm{MT})}\frac{1}{\epsilon}. \end{aligned} \tag{5.188}$$

As a next step, if necessary, one may fit the data using logarithmic plot that generates the dimensionless deviations from the $\ln(\sigma_{ab}(T))$-values,

$$x_i(r, w, c) = \ln\left[f_{\mathrm{MT}}(\epsilon_i; r, w) - f_{\mathrm{MT}}(c+\epsilon_i; r, w)\right] - \ln\left(\frac{\sigma_i}{\frac{e^2}{16\hbar}\frac{N}{s}}\right). \tag{5.189}$$

For the x_i data we can calculate the mean value, the averaged square

$$\langle x\rangle = \frac{1}{N_{\mathrm{exp}}}\sum_{i=1}^{N_{\mathrm{exp}}} x_i, \qquad \langle x^2\rangle = \frac{1}{N_{\mathrm{exp}}}\sum_{i=1}^{N_{\mathrm{exp}}} x_i^2, \tag{5.190}$$

and the dispersion

$$S(r, w, c) = \langle x^2\rangle - \langle x\rangle^2. \tag{5.191}$$

The fitting procedure is then reduced to numerically finding the minimum of the dispersion

$$S(r_0, w_0, c_0) \leqslant S(r, w, c) \tag{5.192}$$

in the space of parameters (r, w, c). We have to start from some acceptable set of parameters, for example, $r_0 = \frac{1}{7}$, $w_0 = 1$, and $c_0 = \frac{1}{2}$, and to search for the minimal value in certain range, e.g., $r_0 =\in (0,\ 1)$, $w_0 \in (1, 30)$ and $c_0 = \frac{1}{2} \in (0.2, 2)$. It is possible that the parameters of the normal resistivity be corrected by the same procedure for minimization of the dispersion $S(r, w, c, A_R, B_R)$. The contemporary methods of the mathematical statistics, such as the bootstrap and Jack-knife, can tell us how reliable is the set of the fitted parameters; the simplest possible realization is to decrement sequentially $N_{\rm exp}$ by one and to investigate the distribution of the fitted GL parameters at every step. For example, the w parameter is almost inaccessible since for $w = 1$ and $w \rightarrow \infty$ we have LD-type temperature dependence of the paraconductivity. On the other hand if we try to fit the paraconductivity far from the critical temperature, e.g., $T \in (1.02\, T_{\rm c}, 1.15\, T_{\rm c})$ we can easily find some estimate for the cutoff parameter c. In any case a good fit would be useful because as a by-product we determine the life-time of the fluctuation Cooper pairs

$$\tilde{\tau}_{\rm rel} = \exp\left(-\langle x\rangle(r_0, w_0, c_0)\right). \tag{5.193}$$

The same procedure can be applied to the magnetic susceptibility at vanishing magnetic field which, according to Eq. (5.62), is proportional to the conductivity, or for the susceptibility in the LD model which, according to Eqs. (5.42), (5.73), and (5.170), reads

$$-\chi_{\rm LD}(\epsilon) = \frac{1}{6}\frac{\epsilon_{\rm Gi}}{\kappa^2_{\rm GL}}\left[\frac{1}{\sqrt{\epsilon(\epsilon+r)}} - \frac{1}{\sqrt{(c+\epsilon)(c+\epsilon+r)}}\right], \tag{5.194}$$

where

$$\frac{\epsilon_{\rm Gi}}{\kappa^2_{\rm GL}} = 2\pi\mu_0\frac{k_{\rm B}T_{\rm c}}{\Phi_0^2}\xi^2_{ab}(0)\frac{N}{s} = \frac{M_0}{H_{\rm c2}(0)}. \tag{5.195}$$

The general formula for the conductivity, Eqs. (5.59), (5.71),

$$\sigma_{ab}(\epsilon, h) = \frac{\pi}{8}\frac{\tau_{\rm rel}}{R_{\rm QHE}}\ \hat{\mathsf{L}}^{(\rm LD)}\hat{\mathsf{C}} f_{\rm APS}(\epsilon, h), \tag{5.196}$$

which for single layered superconductor reads, cf. Eq. (5.135),

$$\begin{aligned}\sigma_{ab}(\epsilon, h; r, C) = \tilde{\tau}_{\rm rel}\frac{e^2}{16\hbar s_{\rm eff}}\frac{2}{h^2}\int_0^{\pi/2}\frac{d\phi}{\pi/2}\Bigg\{&(\epsilon + r\sin^2\phi)\left[\psi\left(\frac{1}{2} + \frac{\epsilon + r\sin^2\phi}{2h}\right)\right.\\ &\left.-\psi\left(1 + \frac{\epsilon + r\sin^2\phi}{2h}\right) + \frac{h}{\epsilon + r\sin^2\phi}\right]\\ &-(c+\epsilon + r\sin^2\phi)\left[\psi\left(\frac{1}{2} + \frac{c+\epsilon + r\sin^2\phi}{2h}\right)\right.\\ &\left.-\psi\left(1 + \frac{c+\epsilon + r\sin^2\phi}{2h}\right) + \frac{h}{c+\epsilon + r\sin^2\phi}\right]\Bigg\}\end{aligned} \tag{5.197}$$

gives another possibility for determining the energy cutoff parameter c. As most appropriate regime we recommend that the measurements of the conductivity as a function of the magnetic field to be carried out at the critical temperature $T = T_c$. In this case for strong magnetic field, $h \gg r$, the layered superconductors with strong anisotropy $r \ll 1$ show 2D behavior. The substitution $\epsilon = 0$ in Eq. (5.197) gives another universal law derived within the GL theory with energy cutoff

$$\begin{aligned} \frac{1}{2}\frac{B}{B_{c2}(0)}\frac{\sigma_{ab}(\epsilon=0,h)}{\tilde{\tau}_{\rm rel}(e^2/16\hbar s_{\rm eff})} &= \hat{\mathsf{C}}\,\frac{h}{2}\,f_{\rm APS}(\epsilon=0,h) \\ &= \frac{\pi\hbar}{e}s_{\rm eff}\frac{\xi_{ab}^2(0)}{\tau_0}\,B\sigma_{ab}(T_c,B) = U_\sigma\left(\frac{2}{c}\frac{B}{B_{c2}(0)}\right), \end{aligned} \tag{5.198}$$

where $y = 2h/c$, cf. Eq. (5.137),

$$U_\sigma(y) = \frac{2}{y}\left[\psi\left(1+\frac{1}{y}\right) - \psi\left(\frac{1}{2}+\frac{1}{y}\right)\right], \tag{5.199}$$

$U_\sigma(0) = 1$, and $U_\sigma(\infty) = 0$. At best, the universal dimensionless conductivity $U_\sigma \propto B\sigma(B)$ and magnetization $U_M \propto M$ have to be fitted simultaneously using the data for the same crystal and common dimensionless argument $\propto B$. Similar universal scaling law for the heat capacity can be derived from Eq. (5.140)

$$\frac{2}{\zeta\left(2,\frac{1}{2}\right)}\frac{B}{B_{c2}(0)}\frac{C(T_c,B)}{\epsilon_{\rm Gi}\Delta C} = U_C\left(\frac{2}{c}\frac{B}{B_{c2}(0)}\right), \tag{5.200}$$

where

$$U_C(y) = 1 - \frac{\zeta\left(2,\frac{1}{2}+\frac{1}{y}\right)}{\zeta\left(2,\frac{1}{2}\right)}, \tag{5.201}$$

but the accuracy of thermal measurements is probably not high enough in order for this to be experimentally confirmed.

5.4.2 *Determination of the coherence length* $\xi_{ab}(0)$

The fit of every fluctuation variable as a function of the dimensional magnetic field h, the conductivity

$$\sigma(\epsilon,h) = \sigma(\epsilon) + \Delta\sigma(\epsilon,h), \tag{5.202}$$

for example, provides a method for determination of $B_{c2}(0)$ and $\xi_{ab}(0)$. At weak magnetic fields, $h \ll \epsilon$, the magnetoconductivity is proportional to the square of the magnetic field $\Delta\sigma(\epsilon,h) \equiv \sigma(\epsilon,h) - \sigma(\epsilon) \propto B^2$. For this small negative quantity, $0 < -\Delta\sigma(\epsilon, h) \ll \sigma(\epsilon)$, the APS result, Eq. (5.69), reads [418]

$$-\Delta\sigma(\epsilon, h) \approx \frac{h^2}{4}\frac{\partial^2}{\partial\epsilon^2}\sigma(\epsilon), \tag{5.203}$$

where $h = 2\pi\xi_{ab}^2(0)B_z/\Phi_0$. The common multiplier τ_0 from Eq. (5.66) is obviously canceled in this relation because, roughly speaking, the transport takes time even in the presence of magnetic field. We note that a multiplier $\tilde{\tau}_{\mathrm{rel}}$ was misintroduced in Ref. [418] in the right-hand-side of the above equation [see Eq. (4) in Ref. [418]]. Thereby the old experimental data in Ref. [418] have been apparently processed by employing erroneous expression and therefore the discussion related to Fig. 2 in Ref. [418] is physically unsound. As a consequence, the life-time constant of metastable Cooper pairs in cuprates is still waiting for its first experimental determination. Nevertheless the novel theoretical result that the life-time constant τ_0 and the diffusion coefficient of the fluctuation Cooper pairs $\xi_{ab}^2(0)/\tau_0$ can be determined from the σ/χ-ratio remains unchanged.

Returning to Eq. (5.203) we note that after two-fold integration of the relation (5.203) in some temperature interval, e.g., $(\epsilon_a,\ \epsilon_b) = (0.03, 0.09)$, the "noise" in the experimental data is already irrelevant and we can rewrite Eq. (5.203) as [418]

$$\xi_{ab}(0) = l_B \left[\frac{\displaystyle\int_{\epsilon_a}^{\epsilon_b} d\epsilon' \int_{\epsilon'}^{\epsilon_b} (-\Delta\sigma(\epsilon'', h))\, d\epsilon''}{\displaystyle\sigma(\epsilon_a) - \sigma(\epsilon_b) + (\epsilon_b - \epsilon_a)\frac{d\sigma}{d\epsilon}(\epsilon_b)}\right]^{1/4}, \tag{5.204}$$

where l_B is the magnetic length

$$l_B = \sqrt{\frac{\Phi_0}{\pi B}} = \sqrt{\frac{\hbar}{eB}} = \frac{25.6\ \mathrm{nm}}{\sqrt{B(T)}}. \tag{5.205}$$

For practical application we have to take into account that far from the critical temperature, even for $T - T_{\mathrm{c}} = 15\%\ T_{\mathrm{c}}$ the fluctuation conductivity is negligible $\sigma(0.15) \approx \sigma(\infty) = 0$. That is why in acceptable approximation we can take $\epsilon_b = 0.15 \approx$ "∞". For $\epsilon > \epsilon_b$ the temperature dependence of the magnetoconductivity in the numerator of Eq. (5.204) can be an extrapolated LD fit.

However, due to the strong critical behavior $-\Delta\sigma \propto h^2/\epsilon^3$ for $\epsilon \gg r$ the influence of the interval (ϵ_b, ∞) can be neglected. In such a way, after a partial integration, we arrive at a simpler equation for determination of the in-plane coherence length, cf. Ref. [418],

$$\xi_{ab}(0) \approx l_B \left[\frac{1}{\sigma(\epsilon)} \left(\int_{\epsilon}^{\infty} \epsilon' \left(-\Delta\sigma(\epsilon', h)\right) d\epsilon' - \epsilon \int_{\epsilon}^{\infty} \left(-\Delta\sigma(\epsilon'', h)\right) d\epsilon'' \right) \right]^{\frac{1}{4}}$$
$$= \text{const}, \tag{5.206}$$

where the integrations should be performed in the whole experimentally accessible temperature range above $(1+\epsilon)T_c$. This result of the Gaussian fluctuation theory does not depend upon the τ_0 parameter, effective mass of Cooper pairs m_{ab}, and the space dimensionality. We consider this procedure for determination of the coherence length $\xi_{ab}(0)$ as being the best one, as it is model-free and does not depend on the multilaminarity of the superconductor, i.e., on the dispersion of Cooper pairs in c-direction $\varepsilon_{c,j}(p_z)$. Equation (5.206) has the same form for both strongly anisotropic high-T_c cuprates and bulk conventional dirty alloys. Of course, methods particularly based on the proximity to the critical line $H_{c2}(T)$ can be very useful in determining $\xi_{ab}(0)$ especially in the case of strong magnetic fields. For example, Eq. (5.69) gives another appropriate formula

$$\sigma_{ab}(\epsilon, h) \approx \tilde{\tau}_{\mathrm{rel}} \frac{e^2}{16\hbar} \frac{N}{s} \frac{4}{\sqrt{(\epsilon+h)(\epsilon+h+r)}} \tag{5.207}$$

applicable for $\epsilon_{\mathrm{Gi}} \ll \epsilon + h \ll h$. Similar result, cf. also Eq. (5.144),

$$M = -M_0\, \tilde{m} \approx -\frac{k_{\mathrm{B}} T_{\mathrm{c}}}{\Phi_0} \frac{N}{s} \frac{h}{\sqrt{(\epsilon+h)(\epsilon+h+r)}} \tag{5.208}$$

can be derived under the same physical conditions from the formula for the fluctuation magnetic moment, Eq. (5.135), using the approximations for $0 < x \ll 1$,

$$\Gamma(x) \approx -\ln x - \frac{1}{2}\ln(2\pi), \qquad \psi(x) \approx -\frac{1}{x}. \tag{5.209}$$

The experimental investigation of the conductivity, Eq. (5.207), and magnetization, Eq. (5.208), is probably the best way to extract the upper critical field $H_{c2}(T)$ for high-T_c cuprates; the $H\sigma/M$ quotient near the critical line is 2/3 of the σ/χ quotient for weak magnetic fields.

5.4.3 *Determination of the Cooper pair life-time constant τ_0*

Having a reliable estimate for the coherence length, the life-time constant of the metastable Cooper pairs above T_c can be determined via the σ/χ-quotient, Eq. (5.62). We believe that this method will become a standard procedure in the physics of high-$T_{\rm c}$ materials. Certainly the most transparent method is just the fit to the phase angle of high-frequency complex fluctuation conductivity

$$\phi_\sigma(\omega\tau(\epsilon_{\rm ren})) = \arctan\frac{\sigma''(\omega)}{\sigma'(\omega)} = \arctan\frac{\hat{\mathsf{L}}\,\varsigma_2(\omega\tau_0/\epsilon_{\rm ren})}{\hat{\mathsf{L}}\,\varsigma_1(\omega\tau_0/\epsilon_{\rm ren})}. \tag{5.210}$$

The state-of-the-art electronics gives such a possibility, but unfortunately the first experiments of the type [458, 459] was not performed in the Gaussian region. For the development of Gaussian spectroscopy which will give results relevant for the microscopic mechanisms of superconductivity we recommend the use of the conventional thin films and high-quality low temperature cuprate films, such as $Bi_2Sr_2CaCu_2O_8$.

5.4.4 *Determination of the Ginzburg number and penetration depth $\lambda_{ab}(0)$*

The applicability of the self-consistent approximation in the theory of fluctuation phenomena in superconductors is strongly limited by the quality of the samples. The fluctuation of the critical temperature $\Delta T_{\rm c}$, e.g., due to the oxygen stoichiometry in cuprates, should be small enough, $\Delta T_{\rm c} \ll \epsilon_{\rm Gi} T_{\rm c}$, and this has to be verified empirically. If the σ/χ ratio remains temperature independent for $\epsilon < 3\%$ and both $\sigma(\epsilon)$ and $\chi(\epsilon)$ demonstrate weak deviation from the LD fit obtained from the range $\epsilon \in (3\%, 9\%)$, this could be considered as a hint in favor of the self-consistent approximation. In this case $\epsilon_{\rm Gi}$ can be fitted by substituting the solution $\epsilon_{\rm ren}(\epsilon)$ of Eq. (5.169) into the LD fit to $\sigma^{\rm (LD)}(\epsilon_{\rm ren})$ and $\chi^{\rm (LD)}(\epsilon_{\rm ren})$. We note that the reliability in fitting $\epsilon_{\rm Gi}$ is determined by the condition whether the self-consistent approach and the use of $\epsilon_{\rm ren}$ significantly improve the accuracy of the fit to the experimental data near $T_{\rm c}$. According to Eq. (5.170) we can parameterize $\epsilon_{\rm Gi}$ with the help of the penetration depth $\lambda_{ab}(0)$. In any case, an evaluation of such type should be a part of the complete set of GL parameters of the superconductor. Another possibility for the thermodynamic determination of the penetration depth $\lambda_{ab}(0)$ is provided by the jump in the specific heat at the critical temperature, Eq. (5.167). As a rule the accuracy of the determination of the penetration depth by the thermodynamic methods cannot be high, especially for high-$T_{\rm c}$ cuprates where the phonon part strongly dominates. An acceptable value of $\ln\kappa_{\rm GL}$ derived from the heat capacity is necessary for the establishment of a coherent understanding of the superconductivity; there is no doubt that the direct investigation of the vortex phase of the superconductors or vortex-free high-frequency measurements constitute the best methods for determination of $\lambda_{ab}(0)$.

5.5 Discussion

In the attempts to systematize the available results we had to derive in parallel new ones too. We shall summarize the most important of them starting with remarks concerning the theory. As the ultimate result we consider the representation of the fluctuation part of the Gibbs free energy by the Euler Γ-function Eq. (5.131) in Gaussian approximation. This result trivializes the derivation of all thermodynamic variables, such as fluctuation magnetization Eq. (5.135), or fluctuation heat capacity Eq. (5.152). To our knowledge this is a novel result, but we find it strange that it remained unobserved given the great attention which the fluctuations in high-T_c superconductors have attracted. The importance of fluctuations was mentioned even in the classical work by Bednorz and Müller. Fluctuations in superconductors were among the main topics in many scientific activities; the Γ-function is well-known to all physicists; the mathematical physics behind the 2D statistical mechanics is well developed, polygamma functions can be found in a number of BCS papers, and finally the solution turns out to be on a textbook level. Just the same is the situation for the 3D GL model. In this case the solution for the free energy is given in terms of the Hurwitz ζ-functions. Analogous result gave the name of one of the most powerful methods in the field theory—ζ-function method for ultraviolet regularization, but this method was never applied to the most simple problem of a 3D GL model related to numerous experiments in the physics of superconductivity.

Another simple but useful detail is the layering operator $\hat{\mathsf{L}}$, Eq. (5.44), which allows us to extend the 2D result onto layered superconductors and even to 3D superconductors. The method can be applied not only to the thermodynamic variables but to the fluctuation part of the kinetic coefficients as well. In this way we obtained useful formulae for the in-plane fluctuation conductivity in perpendicular magnetic field, Eq. (5.197), and for the high-frequency Aslamazov–Larkin conductivity in layered superconductors, Eq. (5.75). We proposed further convenient r-w parameters, Eq. (5.55), for the bi-layered model which could be utilized for experimental data processing of the fluctuation phenomena in bi-layered cuprates, such as $YBa_2Cu_3O_{7-\delta}$ for example.

The representation of the thermodynamic variables via polygamma functions is very helpful at strong magnetic fields but due to the presence of the magnetic field in denominator these results cannot be directly applied to zero-magnetic-field limit. For small magnetic fields, on the other hand, we have to use asymptotic formulae for polygamma and ζ-functions with large arguments. This is the reason why the weak-magnetic-field expansion of the magnetization and the other thermodynamic variables has so bad convergence. In order to fit the experimental data for the magnetization in weak magnetic field using the new analytical result for the LD model, Eq. (5.83), we arrive at the problem for summation of divergent asymptotic series. At least for experimentalists this is a nontrivial problem which led us to give a prescription for usage of series from the theoretical papers. There is no doubt that the ϵ-method is one of the brilliant achievements of the applied mathematics of XX century. However, it turns out that this method was not cast in a suitable form to be employed by users like experimentalists having no time to understand how the underlying mathematics can be derived. That is why we have provided in Ref. [429] a simplified FORTRAN90 version of this algorithm. The latter can also be used for calculation of the differential nonlinear susceptibility at finite magnetic field, Eq. (5.106), which was another novel result in Ref. [429].

Let us now address the simple final formulae that can be directly used for experimental data processing. First of all we advocate that the relation between fluctuation conductivity and magnetoconductivity, Eq. (5.206), provides the best method (shortly announced in Ref. [418]) for determination of the in-plane coherence length $\xi_{ab}(0)$ in layered high-T_c cuprates and conventional superconductor superlattices and thin films. Having such a reliable method for determination of $\xi_{ab}(0)$, the Cooper pair life-time spectroscopy can be created [418] on the basis of determination of the life-time constant τ_0 by the σ/χ quotient, Eq. (5.62).

Usually science starts with some simplicity, thus it is surprising that the temperature independence of the σ/χ, χ/C, and σ/C quotients has not attracted any attention in physics. The question of whether the high-T_c cuprates are BCS superconductors, or have a non BCS behavior, consumed more paper and caused more information pollution than that about the sense of life, about the smile of Mona Lisa. Now we possess a perfect tool to check whether this sacramental $\frac{\pi}{8}$ BCS ratio, Eq. (5.64), still exists in the physics of high-T_c superconductivity. A careful study of the relative life-time constant τ_0 by the σ/χ ratio, Eq. (5.65), will provide a unique information on the presence of depairing impurities in the superconducting cuprates. The doping dependence of this ratio will give important information for the limits of applicability of the self-consistent BCS approximation. In principle the same life-time spectroscopy can be applied to heavy fermions and other exotic superconductors.

The methods we proposed in this chapter can be initially tested by means of alternative methods for determination of $\xi_{ab}(0)$, e.g., from the slope of the upper critical field $H_{c2}(T)$ defined by the fluctuation magnetization of the normal phase near the critical line, Eq. (5.208), being another new result derived here, or from the fluctuation conductivity of the LD model, Eq. (5.207). Addressing the conductivity we consider that the fitting to high frequency experimental data with the help of formulae (5.74) and (5.210) will give a direct method for determination of the relaxation time of the superconducting order parameter. A good monocrystal of layered cuprate or high-quality thin film with as low as possible critical temperature could ensure the overlap of both the suggested methods for Cooper pair life-time spectroscopy. At present we only know [418] that for 93 K $YBa_2Cu_3O_{7-\delta}$ $\tau_{0,\Psi} = 2\tau_0 = 32$ fs. We hope, however, that several experimental methods for determination of $\xi_{ab}(0)$ and τ_0 will be mutually verified in the nearest future. Thus, the investigation of the Gaussian fluctuations may become a routine procedure in the materials science of superconductors.

We also believe that the development of the Gaussian spectroscopy will lead to determination of the Ginzburg number ϵ_{Gi}, the energy cutoff, i.e., the maximal kinetic energy of the Cooper pairs $\varepsilon_{\varkappa} = c\hbar^2/2m_{ab}\xi_{ab}^2(0)$. Up to now these parameters of the GL theory are inaccessible. We hope that our derived self-consistent equation for the reduced temperature, Eq. (5.175), will stimulate experimentalists to reexamine the data for high-quality crystals in the region close to $\epsilon + h \simeq 3\%$ in order to extract ϵ_{Gi}. Virtually all final results are presented by taking into account the energy cutoff parameter c. The nonlocality corrections can be extracted from almost all fluctuation variables, if $\epsilon + h > 10\%$, but we suggest special new experiments to be conducted for investigation of nonlocality effects in quasi 2D superconductors at T_c. Analogous investigations for fluctuation diamagnetism for classical bulk superconductors are already classics in physics of superconductivity; see for example Fig. 8.5 in the well-known textbook by Tinkham [374]. The universal scaling law for the heat capacity, Eq. (5.201), for the magnetization, Eq. (5.137), and conductivity, Eq. (5.199), versus the reduced magnetic field $y = 2h/c$ are depicted in Fig. 5.2.

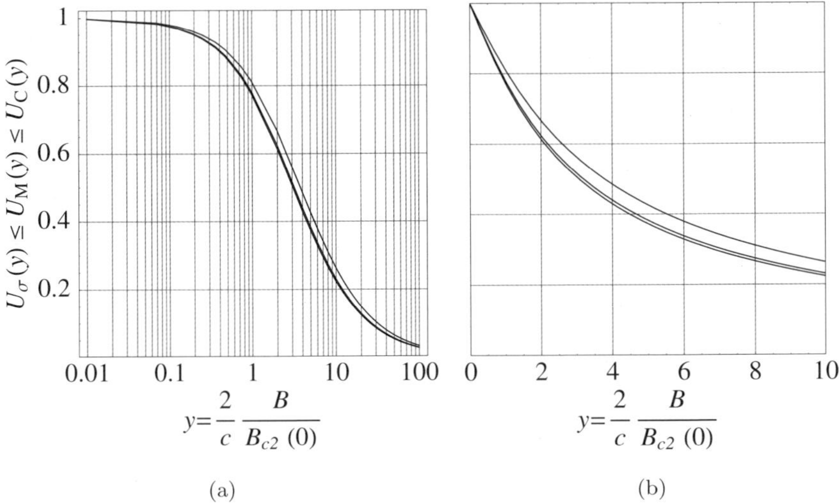

Fig. 5.2 Universal scaling curves ((a) semi-logarithmic plot; (b) linear plot) of a quasi-2D superconductor for the fluctuation conductivity $B\sigma(T_c, B) \propto U_\sigma$, magnetic moment $M(T_c, B) \propto U_M$, and heat capacity $BC(T_c, B) \propto U_C$ versus dimensionless magnetic field $y = 2h/c$. The fit of the scale in horizontal direction gives the GL cutoff parameter c. The scales in vertical directions are related correspondingly to diffusion constant of Cooper pairs $\xi_{ab}^2(0)/\tau_0$, effective inter-layer distance s_{eff}, and 2D Ginzburg number ϵ_{Gi}, cf. Eqs. (5.199), (5.137), and (5.201).

At least for conductivity the experimental confirmation for the quasi-2D superconductors ($h \gg r, \epsilon_{\text{Gi}}$) can be easily achieved. How different are the animals...? Biochemists believed that what is true for *Escherichia coli* holds true for the elephant. Analogously, we consider that $B\sigma(T_c, B)$ versus B will be within 20% accuracy the same for conventional Pb layers and for strongly anisotropic underdoped $Bi_2Sr_2CaCu_2O_8$ in spite of the bunch of sophisticated theories of high-T_c superconductivity. The GL theory gives the scaling law, the notions and notations, and in this sense the language for analysis of the fluctuation phenomena. The precisely measured deviations from the GL scaling low could give the basis for further microscopic consideration using the methods of the statistical mechanics. This is the last example how the development of the Gaussian fluctuation spectroscopy could be of importance not only for the materials science but for the fundamental physics of superconductivity as well.

Chapter 6

Kinetics of fluctuation Cooper pairs

6.1 Introduction

For all high temperature superconductors, the fluctuation phenomena can be observed and their investigation takes a significant part of the complete understanding of these materials; for a contemporary review on the fluctuation phenomena in superconductors, see the review by Larkin and Varlamov [460] and more recent discussion in Ref. [461]. The Ginzburg–Landau (GL) approach of the order parameter is an adequate tool to investigate the low-frequency behavior of fluctuations near to T_c; for a review of the Gaussian GL fluctuations see Ref. [429]. A lot of important papers on the fluctuation phenomena in superconductors and related topics have not been cited in these reviews, see for example Refs. [462–470]. We wish to point out that the GL approach is the standard tool for the investigation of magnetic field penetration in superconductors [471] and even non-Gaussian approach to critical fluctuations [472].

Amidst all kinetic phenomena, the fluctuation conductivity created by the metastable in the normal state Cooper pairs is probably best investigated. The Boltzmann equation is a standard tool for investigation of kinetic phenomena and the purpose of the present chapter is to derive the Boltzmann equation for fluctuation Cooper pairs and to illustrate its work on the example of the fluctuation conductivity; a shortened version of the present research has been presented in preliminary communications [473, 474]. We rederive the frequency dependence of the Aslamazov–Larkin conductivity, fluctuation Hall effect at weak magnetic fields, and magnetoconductivity. We analyze the experimental data for indium oxide films and find significant deviation from the BCS weak coupling prediction. We are coming to the conclusion that a systematic investigation of lifetime of fluctuation Cooper pairs will give important information for our understanding of the physics of superconductivity.

6.2 From TDGL equation via Boltzmann equation to Newton equation

Our starting point is the time-dependent Ginzburg–Landau (TDGL) equation for the superconducting order parameter derived in the classical paper by Abrahams and Tsuneto [411] (see also Refs. [392, 410, 412, 417, 475, 476] and references cited in the review by Larkin and Varlamov [460])

$$\frac{(-i\hbar D_r)^2}{2m^*}\Psi + a\Psi + b|\Psi|^2\Psi = -\hbar\gamma\left(D_t\Psi - \zeta\right), \tag{6.1}$$

where m^* and $|e^*| = 2|e|$ are the mass and charge of the Cooper pairs, parameter γ describes the dissipation, and $\zeta(\mathbf{r}, t)$ is the external noise in TDGL equation. Here

$$\begin{aligned} -i\hbar D_r &= -i\hbar\nabla - e^*\mathbf{A}/c, \\ i\hbar D_t &= i\hbar\partial_t - e^*\varphi, \end{aligned}$$

are the operators of kinetic momentum and energy, $\mathbf{A}$ is a vector-potential, and φ is the potential.

Close to the critical temperature $a(T) \approx (T - T_c)a_0/T_c$, and $b \approx \text{const}$, where $a_0 = \hbar^2/2m^*\xi^2(0)$, and $\xi(0)$ is coherence length.

The correlations of the white noise $\langle\zeta\rangle = 0$,

$$\langle\,\zeta^*(\mathbf{r}_1, t_1)\,\zeta(\mathbf{r}_2, t_2)\,\rangle = \Gamma\,\delta(t_1 - t_2)\,\delta(\mathbf{r}_1 - \mathbf{r}_2), \tag{6.2}$$

are parameterized by fluctuation parameter Γ. The BCS theory gives

$$\gamma_{BCS} = \frac{\pi}{8}\frac{a_0}{T_c},$$

and that is why we parameterize $\gamma = \gamma_{BCS}\tau_{rel}$, by the dimensionless parameter $\tau_{rel} \simeq 1$, which describes the relative life-time of fluctuation Cooper pairs.

The most simple is the case of free particle, which means $\mathbf{A} = 0$, $\varphi = 0$, $b|\Psi|^2 \approx 0$. Introducing the Fourier transform

$$\Psi(\mathbf{r}, t) = \sum_p \frac{e^{i\mathbf{p}\cdot\mathbf{r}/\hbar}}{\sqrt{\mathcal{V}}}\,\psi_p(t), \tag{6.3}$$

$$\zeta(\mathbf{r}, t) = \sum_p \frac{e^{i\mathbf{p}\cdot\mathbf{r}/\hbar}}{\sqrt{\mathcal{V}}}\,\zeta_p(t), \tag{6.4}$$

where

$$\sum_p \approx \mathcal{V} \int \frac{d^D p}{(2\pi\hbar)^D},$$

and

$$\langle\, \zeta_p^*(t_1)\, \zeta_q(t_2)\,\rangle = \Gamma \delta_{p,q}\, \delta(t_1 - t_2),$$

we obtain TDGL equation in momentum representation

$$(\varepsilon_p + a)\psi_p = -\hbar\gamma(d_t\psi_p - \zeta_p). \tag{6.5}$$

Its solution reads

$$\psi_p(t) = e^{-t/2\tau_p} \left(\int_0^t e^{t'/2\tau_p}\, \zeta_p(t')dt' + \psi_p(0) \right), \tag{6.6}$$

where

$$\tau_p = \frac{\hbar\gamma}{2(\varepsilon_p + a(T))}, \qquad \varepsilon_p = \frac{p^2}{2m^*} \tag{6.7}$$

are momentum-dependent lifetime and kinetic energy of fluctuation Cooper pairs. The number of particles for every momentum can be found by noise averaging

$$n_p = \langle\, \psi_p^*(t)\psi_p(t)\rangle = n_p(0)e^{-t/\tau_p} + (1 - e^{-t/\tau_p})\bar{n}_p, \tag{6.8}$$

where $n_p(0) = |\psi_p(0)|^2$ is the initial number. The time differentiation of this solution gives the well-known Boltzmann equation

$$\frac{d}{dt} n_p(t) = -\frac{1}{\tau_p}(n_p(t) - \bar{n}_p), \tag{6.9}$$

which can be considered in this physical situation as a consequence of the TDGL equation. The quantity

$$\bar{n}_p = n_p(t = \infty) = \Gamma\tau_p$$

gives the equilibrium number of particles. The fluctuation parameter Γ is related to dissipation parameter γ by the fluctuation-dissipation theorem, which here takes the form

$$\Gamma = \frac{2T}{\hbar\gamma} = \frac{\bar{n}_p}{\tau_p} = \frac{T}{a_0\tau_0}, \tag{6.10}$$

where

$$\bar{n}_p = \frac{T}{\varepsilon_p + a(T)}$$

is the Rayleigh-Jeans distribution.

Let us now analyze the influence of a weak electric field in the Boltzmann [477] equation

$$\partial_t n_p + e^* \mathbf{E} \cdot \partial_{\mathbf{p}} n_p = -\frac{1}{\tau_p}(n_p - \bar{n}_p). \tag{6.11}$$

Quantum mechanics was born during Halle conference in 1891, when exposed to the ignorant criticism of both statistical methods and atomic physics, Boltzmann suddenly made a remark: "I see no reason why energy shouldn't also be regarded as divided atomically" Ref. [478]. Later on applying the Boltzmann method to the problem of black body radiation Planck found that the constant appearing in the photon spectrum is just the volume of the Boltzmann cells in the phase space. Due to this reason, Planck called the quantity $2\pi\hbar$ after Boltzmann - Boltzmann constant.

For the solution we search in the form

$$n_p(t) = n(\mathbf{p}, t) \approx \bar{n}\left(\mathbf{p} - m^* \mathbf{V}(\mathbf{p}, t)\right), \tag{6.12}$$

and we obtain the Newton equation

$$m^* d_t \mathbf{V}_p(t) = e^* \mathbf{E} - \frac{m^*}{\tau_p} \mathbf{V}_p(t) \tag{6.13}$$

for the field of drift velocity in momentum space. The general formula for the current gives

$$\mathbf{j}_{\mathrm{fl}} = \sum_p e^* \frac{n_p}{\mathcal{V}} \mathbf{v}_p = \boldsymbol{\sigma}_{\mathrm{fl}} \cdot \mathbf{E}, \qquad n_{\mathrm{D}} = \sum_p \frac{n_p}{\mathcal{V}}, \tag{6.14}$$

where n_D is the D-dimensional volume density of the fluctuation Cooper pairs. Substitution here of the shifted equilibrium distribution gives the well-known formula for the conductivity tensor [479]

$$\boldsymbol{\sigma}_{\mathrm{fl}} = e^{*2} \int \frac{d^D p}{(2\pi\hbar)^D} \frac{\mathbf{v}_p \otimes \mathbf{v}_p}{1/\tau_p - i\omega} \left(-\frac{\partial\, n_p}{\partial\, \varepsilon_p} \right), \tag{6.15}$$

where $\mathbf{v}_p = \partial_p \varepsilon_p = \mathbf{p}/m^*$ is the Cooper pairs' velocity. This is only a small fraction of the total conductivity

$$\sigma(T) = \sigma_{\mathrm{N}}(T) + \sigma_{\mathrm{fl}}(\epsilon), \quad \epsilon \equiv \ln \frac{T}{T_{\mathrm{c}}} \approx \frac{T - T_{\mathrm{c}}}{T_{\mathrm{c}}}, \quad \sigma_{\mathrm{fl}} \ll \sigma_{\mathrm{N}}. \tag{6.16}$$

For thin superconducting films $D = 2$ substituting

$$\frac{dp_x dp_y}{(2\pi\hbar)^2} = \frac{d(\pi p^2)}{(2\pi\hbar)^2} = \frac{m^*}{2\pi\hbar^2} d\varepsilon_p, \quad -\frac{\partial \bar{n}}{\partial \varepsilon} = \frac{T}{(\varepsilon + a)^2}, \tag{6.17}$$

we obtain the classical result by Aslamazov and Larkin [405]

$$\sigma_{\mathrm{AL}}(\epsilon) = \frac{e^2}{16\hbar} \tau_{\mathrm{rel}} \frac{T_{\mathrm{c}}}{T - T_{\mathrm{c}}} = \frac{e^2 T}{\pi\hbar^2} \tau(\epsilon), \tag{6.18}$$

where

$$\tau(\epsilon) \equiv \tau(\mathbf{p}=0,\epsilon) = \frac{\pi\hbar}{16T_c}\frac{\tau_{\mathrm{rel}}}{\epsilon} = \frac{\tau_0}{\epsilon} \tag{6.19}$$

is the lifetime for Cooper pairs with zero momentum.

For the two-dimensional (2D) case conductivity is just the inverse resistance $\sigma^{(2D)}(T) = R_\square^{-1}(T)$. For conventional disordered superconductors normal conductivity can be approximated by residual conductivity far above T_c, for example, $T = 3T_c$. In this approximation Aslamazov–Larkin conductivity can be rewritten in a convenient for experimental data processing form

$$\left(\frac{1}{R_\square(T)} - \frac{1}{R_\square(3T_c)}\right)^{-1} \approx \frac{16\hbar}{e^2\tau_{\mathrm{rel}}}\left(\frac{T}{T_c} - 1\right). \tag{6.20}$$

Performing the linear regression fit of the data presented in Ref. [480] we have obtained that for indium oxide films $\tau_{\mathrm{rel}} = 1.15$. This significant 15 % deviation from the weak coupling BCS value is created by strong coupling effects. We conclude that analogous systematic investigations for thin films would be very helpful for our understanding of the dynamics of the order parameter in superconductors. Decreasing the lifetime and τ_{rel} by depairing impurities or disorder for anisotropic gap superconductors definitely deserves great attention.

6.3 Fluctuation conductivity in different physical condition

6.3.1 *High frequency conductivity*

For diagonal components of conductivity taking into account that $\mathrm{Tr}\mathbf{1} = D$, from the general formula Eq. (6.15), we obtain

$$\sigma_{\mathrm{fl}} = \frac{e^{*2}}{D}\int \frac{d^D p}{(2\pi\hbar)^D}\frac{v_p^2}{1/\tau_p - i\omega}\left(-\frac{\partial\, n_p}{\partial\, \varepsilon_p}\right). \tag{6.21}$$

It is convenient to introduce a dimensionless frequency $z = \omega\tau(\epsilon)$. In order to derive the dimensionless complex conductivity $\varsigma(\omega)$ we need to solve the elementary integral

$$\varsigma(z) = \varsigma_1(z) + i\varsigma_2(z) = 2\int_1^\infty \frac{x-1}{x^2(x+y)}dx \tag{6.22}$$

$$= \frac{2}{y}\left[\left(1+\frac{1}{y}\right)\ln(1+y) - 1\right], \tag{6.23}$$

$$= \frac{2}{-iz}\left[\left(1+\frac{1}{-iz}\right)\ln(1-iz) - 1\right],$$

where $x = p^2/2m^*a(\epsilon) + 1$ is the kinetic energy of Cooper pairs taken into account from the "chemical potential" in $a(\epsilon)$ units, and $y = -iz = -i\omega\tau(\varepsilon)$ is the dimensionless Matsubara frequency $\zeta_{\mathrm{M}} = -i\omega$. The integral Eq. (6.22) is solved considering the Matsubara frequency y to be a real variable. Then we can make the analytical continuation to real frequencies substituting $y = -iz$ in the result Eq. (6.23). This method is very popular in the quantum field theory, but works effectively for classical problems as well. In such a way we obtain

$$\varsigma_1(z) = \frac{2}{\pi}\mathcal{P}\int_0^\infty \frac{x\varsigma_2(x)}{x^2 - z^2}dx = \frac{2}{z^2}\left[z\arctan(z) - \frac{1}{2}\ln(1+z^2)\right] \tag{6.24}$$

$$\varsigma_2(z) = -\frac{2z}{\pi}\mathcal{P}\int_0^\infty \frac{\varsigma_1(x)}{x^2 - z^2}dx = \frac{2}{z^2}\left[\arctan(z) - z + \frac{z}{2}\ln(1+z^2)\right]. \tag{6.25}$$

Then the frequency-dependent conductivity reads

$$\sigma_{\mathrm{2D}}(\epsilon,\omega) = \sigma_{\mathrm{AL}}(\epsilon)\,\varsigma(\omega\tau(\epsilon)). \tag{6.26}$$

The integral Eq. (6.21) can be solved for arbitrary dimension

$$\sigma_{\mathrm{D}}(\epsilon,\omega) = \sigma_{\mathrm{D}}(\epsilon)\,\varsigma_{\mathrm{D}}(z), \tag{6.27}$$

cf. the paper by Dorsey [423]

$$\sigma_{\mathrm{D}}(\epsilon) = 4\frac{\Gamma(2 - D/2)}{(4\pi)^{D/2}}\frac{e^2}{\hbar}\left[\xi(\epsilon)\right]^{2-D}\frac{T\tau(\epsilon)}{\hbar}, \tag{6.28}$$

where $\xi(\epsilon) \equiv \xi(0)/\sqrt{\epsilon}$ is the temperature-dependent coherence length. The conductivity in this case has the form

$$\varsigma_{1,\mathrm{D}}(z) = \frac{8}{D(D-2)z^2}\left[1 - (1+z^2)^{D/4}\cos\left(\frac{D}{2}\arctan z\right)\right],$$

$$\varsigma_{2,\mathrm{D}}(z) = \frac{8}{D(D-2)z^2}\left[-\frac{D}{2}z + (1+z^2)^{D/4}\sin\left(\frac{D}{2}\arctan z\right)\right]. \tag{6.29}$$

6.3.2 *Hall effect*

The fluctuation Hall conductivity also can be derived in the framework of the Boltzmann kinetic equation. We have to take into account a small imaginary part α of γ parameter in the TDGL equation, i.e., $\gamma \to \gamma + i\alpha$, and $\alpha \ll \gamma$. The solution [473, 474] of the kinetic equation gives

$$\sigma_{xy}(\epsilon) = \frac{Z}{3}\omega_c\tau(\epsilon)\sigma_{\mathrm{AL}}(\epsilon) \propto \tau^2(\epsilon), \tag{6.30}$$

where $\omega_c = e^* B/m^* c$ is the "cyclotron" frequency and

$$Z = -\mathrm{Im}\frac{1}{\gamma + i\alpha} \approx \frac{\alpha}{\gamma^2} \ll 1. \tag{6.31}$$

This result agrees with microscopic calculations [460]. Due to the small value of the parameter α, fluctuation Hall effect is difficult to observe. With fitting of α and m^* from the experimental data finishes the complete determination of parameters of TDGL theory.

6.3.3 *Magnetoconductivity*

It is interesting to mention that the classical formula for the conductivity Eq. (6.15) correctly works even for strong magnetic fields. We only have to substitute the momentum integration with summation on discrete Landau levels, taking into account the density of Landau magnetic subbands

$$\epsilon_p \to \epsilon_n = \hbar\omega_c \left(n + \frac{1}{2}\right) = a_0(2n+1)h, \tag{6.32}$$

where

$$h = \frac{\hbar\omega_c}{2a_0} = \frac{B_z}{B_{\mathrm{c2}}(0)} \tag{6.33}$$

is the dimensionless magnetic field and

$$B_{\mathrm{c2}}(0) = -T_{\mathrm{c}}\frac{d}{dT}B_{\mathrm{c2}}(T)|_{T_{\mathrm{c}}} \tag{6.34}$$

is the linear extrapolation. In the numerator of Eq. (6.15) we have to substitute the classical velocity with the oscillator matrix elements of the momentum. Analogously for the energy-dependent lifetime we have to average on neighboring levels. Due to the triviality of the oscillator problem these substitutions can be performed in only one way, and Aslamazov–Larkin conductivity Eq. (6.18) is substituted by the magnetoconductivity of Abrahams, Prange and Stephen [420] (APS)

$$\sigma_{\mathrm{AL}}(\epsilon) = \frac{e^2 T\tau_0}{\pi\hbar}\frac{1}{\epsilon} \to \sigma_{\mathrm{APS}}(\epsilon, h) = \frac{e^2 T\tau_0}{\pi\hbar} f(\epsilon, h), \tag{6.35}$$

i.e., $1/\epsilon$ has to be substituted by APS function

$$f(\epsilon, h) = \frac{2}{h^2}\left[\epsilon F\left(\frac{1}{2} + \frac{\epsilon}{2h}\right) - \epsilon F\left(1 + \frac{\epsilon}{2h}\right) + h\right]. \tag{6.36}$$

This two-dimensional result can be easily generalized for layered and bulk superconductors using the layering operator introduced in Sec. 5.2.3.

6.3.4 *Strong electric fields*

Using the optical gauge

$$\varphi = 0, \qquad \mathbf{A} = -t\mathbf{E},$$

the TDGL equation Eq. (6.1) reads

$$d_u\psi_q(u) = -\frac{1}{2}\left[(q+fu)^2+\epsilon\right]\psi_q(u)+\bar{\zeta}_q(u), \tag{6.37}$$

where we are introducing dimensionless variables for the $u = t/\tau_0$ time, $q = p\xi(0)/\hbar$ momentum, $f = e^*E\tau_0\xi(0)/\hbar$ electric field, and $\bar{\zeta}_q(u) = \tau_o\zeta_p(t)$ noise. We have a linear ordinary differential equation which can be solved for arbitrary $f(u)$, i.e., for arbitrary time dependence of the electric field. For constant electric field the TDGL equation has the solution

$$\psi_q(u) = \left\{\int_0^u \exp\left[\frac{1}{2}\int_0^{u_1}\left[(q+fu_2)^2+\epsilon\right]du_2\right]\bar{\zeta}_q(u_1)du_1+\psi_q(0)\right\}$$
$$\times\exp\left[-\frac{1}{2}\int_0^u\left[(q+fu_3)^2+\epsilon\right]du_3\right]. \tag{6.38}$$

In order to obtain the static $[t \gg \tau(\epsilon)]$ momentum distribution we have to perform the noise averaging

$$n_k = \lim_{u\to\infty}\langle|\psi_{q+fu}(u)|^2\rangle \tag{6.39}$$
$$= \frac{T}{a_0}\int_0^\infty \exp\left[-\left(k^2+\epsilon\right)v+fkv^2-\frac{1}{3}f^2v^3\right]dv,$$

where $u_1 = u - v$ and

$$k = q + fu = (p - e^*A)\,\frac{\xi(0)}{\hbar}$$

isthe dimensionless kinetic momentum. This distribution can be directly derived from Boltzmann equation Eq. (6.11) for fluctuation Cooper pairs [481]. In Ref. [481] it was demonstrated that substitution of the momentum distribution Eq. (6.39) in the formula for the current density Eq. (6.14) gives the result which agrees with the formula by Dorsey [423]; cf. also the paper by Gor'kov [482]

$$j(E_x) = \frac{e^2\tau_{\rm rel}E_x}{16\hbar\left[2\pi^{1/2}\xi(0)\right]^{D-2}}\int_0^\infty\frac{\exp\left(-\epsilon u-gu^3\right)}{u^{(D-2)/2}}\,du, \tag{6.40}$$

where

$$g \equiv \frac{f^2}{12}, \quad f = \frac{e^*E_x\xi(0)\tau_0}{\hbar} = \frac{\pi}{8}\,\frac{eE_x\xi(0)}{T_{\rm c}}\,\tau_{\rm rel}.$$

Differentiating the upper expression we obtain differential conductivity

$$\sigma_{\mathrm{diff}} = \frac{dj(E_x)}{dE_x} = \frac{e^2\tau_{\mathrm{rel}}}{16\hbar(2\sqrt{\pi}\xi(0))^{D-2}} \times \int_0^\infty \frac{1-2gu^3}{u^{(D-2)/2}} \exp(-\epsilon u - gu^3)\, du. \tag{6.41}$$

Applying a voltage $U(t) = U_{\mathrm{DC}} + U_{\mathrm{AC}} \cos\omega t$ to the nanowire, the differential conductivity can be easily determined measuring the AC component for the current if $U_{\mathrm{AC}} \ll U_{\mathrm{DC}}$. Cooling the sample the differential conductivity will decrease, then at some temperature it will be annulated and what will happen at further cooling is an interesting experimental question.

6.4 Current functional: self-consistent approximation and energy cut-off

The self-consistent approximation for the reduced critical temperature [423, 429] in the one-dimensional (1D) case reads

$$\epsilon_{\mathrm{ren}} = \ln\frac{T}{T_0} + \frac{b}{a_0} n_{1D} = \ln\frac{T}{T_0} + \epsilon_{1G}\mathcal{N}_1(\epsilon_{\mathrm{ren}}, f), \tag{6.42}$$

where n_{1D} is the bulk density of the fluctuation Cooper pairs when we have 1D fluctuations in a wire with cross section $S \ll \xi^2(\epsilon)$ and

$$\epsilon_{1G} \equiv \frac{\mu_0\lambda^2(0)\xi(0)e^2T_{\mathrm{c}}}{\sqrt{\pi}S\hbar^2} = \frac{k_{\mathrm{B}}}{8\sqrt{\pi}\Delta C\xi(0)S}, \tag{6.43}$$

where $\lambda(\epsilon) = \lambda(0)/\sqrt{-\epsilon}$ is the temperature-dependent penetration depth and ΔC is the jump of the specific heat at T_{c} per unit volume. For numerical calculations the function

$$\mathcal{N}_1(\epsilon, f) \equiv \int_0^\infty \exp(-\epsilon v - gv^3)\frac{dv}{\sqrt{v}} \tag{6.44}$$

has to be programmed as

$$\mathcal{N}_1(\epsilon, f) = 2\int_0^\infty \exp(-gz^6 - \epsilon z^2) dz, \quad z^2 = v. \tag{6.45}$$

Analogously for the thin superconducting film with thickness $d_f \ll \xi(\epsilon)$ the equation for reduced temperature at zero electric field takes the form [429]

$$\epsilon_{\mathrm{ren}} = \ln\frac{T}{T_0} + \frac{b}{a_0} n_{2D} = \ln\frac{T}{T_0} + \epsilon_{2G}\mathcal{N}_2(\epsilon_{\mathrm{ren}}), \tag{6.46}$$

where n_{2D} is the volume density of the fluctuation Cooper pairs having 2D fluctuations,

$$\epsilon_{2G} \equiv \frac{k_B}{4\pi\Delta C\xi^2(0)d_f} = 2\pi\mu_0\frac{T_c}{d_f}\left(\frac{\lambda(0)}{\Phi_0}\right)^2 \tag{6.47}$$

is the 2D Ginzburg number and

$$\mathcal{N}_2(\epsilon) \equiv \ln\left(\frac{c+\epsilon}{\epsilon}\right). \tag{6.48}$$

As simplest possible application of these results we have to mention nanostructured superconductors, e.g., nanowires [483–490], similar to those used for long time for investigation of phase slip centers [491–498] of the superconducting phase. We are pointing out that paraconductivity is a property of the normal phase.

For general (nonparabolic) dispersion we can derive from the TDGL equation the formula for the distribution of fluctuation Cooper pairs

$$\begin{aligned} n_k(u) = {} & \frac{T}{a_0}\int_0^u \exp\left\{-\int_{u_1}^u\left[\frac{\varepsilon(k(u_2))}{a_0}+\epsilon\right]du_2\right\}du_1 \\ & + \bar{n}_k(0)\exp\left\{-\int_{u_1}^u\left[\frac{\varepsilon(k(u_2))}{a_0}+\epsilon\right]du_2\right\}, \end{aligned} \tag{6.49}$$

where the dimensionless kinetic momentum and the vector potential are

$$k(u) = q + \overline{A}(u), \quad \overline{A}(u) = -\frac{e^*\xi(0)}{\hbar}A(t).$$

In the case of parabolic dispersion and arbitrary time dependence of the electric field we can write Eq. (6.14) for the current functional in the form

$$j[A] = \frac{\sqrt{\pi}\hbar e}{m^*\xi(0)}\left[\frac{T}{a_0}\int_0^u F_A[u_1]du_1 + \bar{n}_k(0)F_A[0]\right], \tag{6.50}$$

where for brevity we introduce the functionals

$$F_A[u_1] \equiv \left(\frac{\overline{A}(u)}{\sqrt{u-u_1}} - \frac{B_A[u_1]}{(u-u_1)^{3/2}}\right)\exp\left[\frac{(B_A[u_1])^2}{u-u_1} - G(u_1)\right] \tag{6.51}$$

and

$$B_A[u_1] \equiv \int_{u_1}^u \overline{A}(u_2)du_2, \tag{6.52}$$

$$G_A[u_1] \equiv -\epsilon(u-u_1) + \int_{u_1}^u (\overline{A}(u_2))^2 du_2. \tag{6.53}$$

Local GL theory with an energy cut-off for the kinetic energy $\varepsilon_p < a_0 c$ [or for the kinetic energy taken into account from the chemical potential

$\varepsilon_p + a(T) < a_0 c]$ is very often used for describing the fluctuation phenomena [429]

$$\varepsilon(k) = \begin{cases} a_0 k^2, & |k| < \Lambda \\ a_0 c, & |k| > \Lambda \end{cases} \tag{6.54}$$

where the dimensionless constant $\Lambda \equiv \sqrt{c} \simeq 1$. For $YBa_2Cu_3O_{7-\delta}$ the recent investigations [499, 500] of high-frequency fluctuation conductivity determined $\Lambda \approx 0.5$. See also the recent work by Puica and Lang, Ref. [501], where the effects of energy cut-off and self-consistent interaction are treated in great details. Then the functional which participates in the formula for the current Eq. (6.50) takes the form

$$F_A[u_1] = -\frac{\sinh(2\Lambda B_A[u_1])}{\sqrt{u-u_1}} \exp\left[-(u-u_1)\left(c + \frac{(B_A[u_1])^2}{(u-u_1)^2}\right)\right] \tag{6.55}$$

$$+ \left(\overline{A}(u) - \frac{B_A[u_1]}{(u-u_1)}\right) \exp\left(\frac{(B_A[u_1])^2}{u-u_1} - G_A[u_1]\right)$$

$$\times \int_{-\Lambda - A(u)}^{\Lambda - A(u)} \exp\left[-(u-u_1)\left(q + \frac{B_A[u_1]}{u-u_1}\right)^2 dq\right].$$

These rather complicated formulas are necessary for investigation of paraconductivity in the THz range. In the next section we will give an illustration for the important one dimensional case.

6.5 Fluctuation conductivity in nanowires

The recent development of the technology of the performance of nanowires made it possible and even indispensable for the investigation of fluctuation conductivity. In this section we will analyze in detail the general results in 1D case.

The integrants in the momentum distribution Eq. (6.39) is actually age distribution for fluctuation Cooper pairs

$$\mathcal{F}(v;\, k, \epsilon, f) = \exp\left[-\left(k^2 + \epsilon\right)v + fkv^2 - \frac{1}{3}f^2 v^3\right], \tag{6.56}$$

the variable v is the age in units $\tau(\epsilon)$. Time integration returns us to the momentum distribution, which using the dimensional variables

$$k_\epsilon \equiv \frac{k}{|\epsilon|^{1/2}}, \qquad f_\epsilon \equiv \frac{f}{|\epsilon|^{3/2}} \tag{6.57}$$

reads

$$\overline{n}(k;\,\epsilon,f)=\frac{n_T}{|\epsilon|}\mathcal{F}_{\pm}(k_\epsilon,f_\epsilon), \tag{6.58}$$

where

$$\mathcal{F}_{\pm}(k_\epsilon,f_\epsilon)\equiv\int_0^\infty \exp\left[-(k_\epsilon^2\pm 1)x+f_\epsilon k_\epsilon x^2-\frac{1}{3}f_\epsilon^2x^3\right]dx \tag{6.59}$$

and $x=|\epsilon|u$.

For the case $\epsilon=0$, $T=T_c$ or when $f\to\infty$, i.e.,

$$f_\epsilon=e^*E_x\xi(\epsilon)\tau(\epsilon)/\hbar\gg 1,\qquad \xi(\epsilon)=\xi(0)/|\epsilon|^{1/2} \tag{6.60}$$

we obtain

$$n(k,f)=\frac{n_T}{f^{2/3}}\mathcal{F}_0(k_f),\qquad k_f=\frac{k}{f^{1/3}}, \tag{6.61}$$

where in

$$\mathcal{F}_0(k_f)\equiv\int_0^\infty \exp\left[-k_f^2y+k_fy^2-\frac{1}{3}y^3\right]dy \tag{6.62}$$

we use the transformation $y=f^{2/3}v$.

For 1D case using Eq. (6.40), we express the current

$$j(E_x,\epsilon)=\frac{\sqrt{\pi}e^2}{8\hbar}\tau_{\rm rel}\xi(0)E_x\mathcal{J}(\epsilon,f) \tag{6.63}$$

where

$$\mathcal{J}(\epsilon,f)=\int_0^\infty \exp(-\epsilon v-gv^3)\sqrt{v}dv,\quad g\equiv\frac{f^2}{12}. \tag{6.64}$$

The fluctuational current for Cooper pairs above and under the critical temperature is

$$j=\frac{\pi e^2\tau_{\rm rel}\xi(0)E_x}{16\hbar|\epsilon|^{3/2}}\varsigma_{\pm}(f_\epsilon), \tag{6.65}$$

where

$$\varsigma_{\pm}(f_\epsilon)=\frac{2}{\sqrt{\pi}}\int_0^\infty \exp(\mp v_\epsilon-g_\epsilon v_\epsilon^3)\sqrt{v_\epsilon}dv_\epsilon \tag{6.66}$$

is the dimensionless function, which depends on the strength of the electric field. For convenience we use $g_\epsilon=f_\epsilon^2/12$ and $v_\epsilon=v|\epsilon|$.

We wish to point out the normalization and the asymptotics

$$\varsigma_+(0)=1,\qquad \varsigma_{\pm}(f_\epsilon\to\infty)\sim\frac{4}{\sqrt{3}f_\epsilon}. \tag{6.67}$$

Using the strong field asymptotic $f_\epsilon \gg 1$, we obtain the fluctuational current at T_c

$$j_\infty \equiv j(f_\epsilon \to \infty) = \frac{2}{\sqrt{3}} \frac{eT_c}{\hbar}. \tag{6.68}$$

We wish to point out that $j(f_\epsilon \to \infty)/T_c$ is universal and contains only fundamental physical constants. All material constants like $\xi(0)$, $\lambda(0)$ or the cross section of the nanowire S are cancelled. For experimental data processing it is necessary to perform linear regression of the IV curve

$$j_{\mathrm{tot}} = \frac{U}{R_N(T_c)} + \mathrm{sign}(U) \times 24.22\,\mathrm{nA}\; T_c[\mathrm{K}], \tag{6.69}$$

where the second term is the universal fluctuational current at strong electric fields around T_c. In order to avoid the thermoelectric effect coming from different materials forming the contacts to the nanowire one can analyze the current harmonics predicted by Eq. (6.69) for $U(t) = U_0 \sin \omega t$ at $T = T_c$

$$j_{\mathrm{tot}}(t) = \frac{U_0}{R_N(T_c)} \sin \omega t + \frac{4}{\pi} j_\infty \sum_{l=1}^{\infty} \frac{\sin[(2l+1)\omega t]}{2l+1}. \tag{6.70}$$

This current response is a good approximation even slightly above T_c for voltage amplitudes

$$U_0 \gg \frac{8}{\pi} \frac{T_c}{e} \frac{L}{\xi(0)} |\epsilon|^{3/2}, \tag{6.71}$$

where L is the length of the nanowire. In such a way the investigation of fluctuation current in nanowires can be used as a high accuracy test for applicability of the TDGL equation for nanostructured superconductors. For further references related to harmonic generations close to T_c see Ref. [63, 65].

Using Eq. (6.14) and Eq. (6.39) we derive 1D dimensional density

$$n_{1D} = \frac{n_T}{2\sqrt{\pi}\xi(0)S} \mathcal{N}_1(\epsilon, f), \tag{6.72}$$

where $\mathcal{N}_1(\epsilon, f)$ is previously defined in Eq. (6.44), when we considered without a derivation the self-consistent equation for the reduced temperature. The analysis of temperature and electric field dependence is reduced to two functions of one variable f_ϵ above and below the T_c

$$n_{1D} = \frac{n_T}{2\xi(0)S\sqrt{|\epsilon|}} N_\pm(f_\epsilon), \tag{6.73}$$

where

$$\mathcal{N}_{\pm}(f_\epsilon) = \frac{1}{\sqrt{\pi}} \int_0^\infty e^{\mp v_\epsilon - g_\epsilon v_\epsilon^3} \frac{dv_\epsilon}{\sqrt{v_\epsilon}}, \quad N_+(f_\epsilon = 0) = 1. \tag{6.74}$$

Analogously at T_c, where $\epsilon = 0$ and $f_\epsilon \gg 1$ we obtain

$$n_{1D}(f) = \frac{\Gamma(1/6) n_T}{2^{2/3} 3^{5/6} \sqrt{\pi} \xi(0) S} \frac{1}{f^{1/3}}. \tag{6.75}$$

Choosing a typical set of parameters for the Sn nanowire: $\xi(0) = 1000$ Å, $\lambda(0) = 1000$ Å, $D = 500$ Å, $S = \pi D^2/4$ according Eq. (6.43) we obtain for 1D Ginzburg number $\epsilon_{1G} = 4.3 \times 10^{-5}$. This parameter is essential for the numerical solution of the equation for renormalized reduced temperature Eq. (6.42).

6.6 Discussion

Solving in parallel the TDGL equation and the Boltzmann equation we obtained coinciding results: not only for the linear case of Aslamazov–Larkin conductivity, but for the cases of strong electric fields, arbitrary time dependence of the electric field, nonparabolic momentum dependence of energy of Cooper pairs, energy cut-off, self-consistent equation for the renormalized reduced temperature, frequency dependence of the fluctuation conductivity etc. The number of fluctuation Cooper pairs which participates in the Boltzmann equation and the formulas for the current is actually the diagonal element of the order parameter correlator [499, 500] $n_k(t) = C\left[p_{\mathrm{kin}} = p - e^* A(t); t, t\right]$. One can also easily check that the entropy of fluctuation Cooper pairs η is increasing with the time $d\eta/dt \geqslant 0$; the capital η in the η-theorem by Boltzmann is often spelled as Latin H. Our self-consistent formula for the fluctuation conductivity of a superconducting nanowire can be directly used for the experimental data processing. In such a way we conclude that Boltzmann equation for fluctuation Cooper pairs reproduces the known results of the fluctuation theory in the normal phase and it is a adequate tool to predict new phenomena related to metastable Cooper pairs, like negative differential conductivity in the fluctuation regime predicted in Ref. [481] and strong electric field effect in nanostructured superconductors where the heating effects are reduced. Our universal result Eq. (6.69) for a fluctuational current in a nanowire under strong electric field shows that Boltzmann equation for the fluctuation Cooper pairs will become an indispensable tool for understanding the electronic processes in nanostructured superconductors.

Chapter 7

Fluctuation conductivity in strong electric fields

7.1 Introduction

Fluctuation phenomena in superconductors have been intensively studied for more than three decades since the discovery of the fluctuation smearing of the superconducting transition. The interest in fluctuations was resurrected when the high-temperature superconductors (HTSC) were found. In conventional BCS-like superconductors the transition temperature T_c marks a sharp dividing point between distinct regions of "superconducting" and "normal" behavior, and critical fluctuations are almost unobservable. Largely due to their high temperature, small coherence length and quasi-two-dimensional nature, HTSC show a significant smearing of the critical transition and thus the effect of fluctuations in the critical region of copper-oxides is much more pronounced and experimentally accessible. The conductivity, the specific heat, the diamagnetic susceptibility, the thermopower etc. have been observed to increase considerably in the normal state in the vicinity of the transition temperature in cuprate compounds. Importantly, for high-quality HTSC samples the interactions between fluctuations are negligibly small, fluctuations can be described as Gaussian ones and their theory is rather simple. As a result, nowadays the study of fluctuation phenomena in superconductors occupies an essential and significant part of the whole physics of superconductivity. For a contemporary review we refer the reader to the work of Larkin and Varlamov [460]; cf. also the classical review by Skocpol and Tinkham [377] (Sec. 5.3, Fig. 15, and references therein) and especially Chapter 5, devoted to Gaussian fluctuations in layered superconductors.

In this chapter, we are concerned with the effect of fluctuations of the superconducting order parameter on both the linear and the nonlinear conductivity near and above the bulk superconducting critical temperature. In order to account for the fluctuation conductivity (or paraconductivity) one needs to apply the time-dependent generalization of the Ginzburg–Landau theory (TDGL). The theoretical study of fluctuation conductivity dates back to the 1968 paper of Aslamazov and Larkin [405], based on the microscopic approach. Although a long time has elapsed, nonlinear effects have not been sufficiently investigated and the theoretical results obtained so far require specification.

The aim of the present chapter is to study the kinetics of the superconducting order parameter in a *strong electric field* within a simple phenomenological approach and to provide new experimental methods for determining such important material constants of superconductors as the lifetime of fluctuation Cooper pairs τ_0 and the coherence length $\xi(0)$.

We adopt the ideology of the TDGL theory and show that the velocity distribution law of fluctuation Cooper pairs can be derived from a Boltzmann equation. The cases of superconducting wires, thin films, bulk superconductors and striped and layered superconducting systems are analyzed. We are particularly interested in the effects of nonlinearity of conductivity, which can be observed experimentally. We show that the strong critical behavior in the vicinity of T_c of the third harmonic of the electric field generated by a harmonic current, can be used as an experimental technique for the determination of the metastable Cooper pair lifetime constant.

This chapter is organized as follows. In Sec. 7.3 we discuss the kinetic Boltzmann equation for fluctuation Cooper pairs in strong electric fields and present the momentum distribution law for Cooper pairs in the normal phase and the expression for the fluctuation current. Section 7.4 is mainly technical and devoted to dimensionless notations we introduce to make our calculation less complicated. The derivation of the time evolution of the distribution function using canonical variables is given in the next section. Section 7.5 explains the paraconductivity in layered materials, such as, for example, HTSC. We provide the analytical derivation of the conductivity correction in the normal phase in a Lawrence–Doniach superconductor. In Sec. 7.6 the Aslamazov–Larkin conductivity dependent on the electric field is described in detail for one-, two- and three-dimensional superconductors. We study strong electric field effects on conductivity as well as weak electric field influences below the critical temperature. The interesting case of a striped superconductor, which is becoming very popular nowadays, and that of a thick film can be found in Sec. 7.7. The conductivity of such a superconductor depending on the thickness of the film is obtained. The expressions for the nonlinear conductivity correction can be used for experimental data processing for extracting the lifetime constant of Cooper pairs, as is shown in Sec. 7.8. A new experimental technique for probing τ_0, based on suggested measurements of the third harmonic of the electric field, can be found in Sec. 7.9. Finally, a discussion and conclusions are given in the last section.

7.2 Solution to the Boltzmann equation

In the Boltzmann equation Eq. (7.20) the decay rate of fluctuation Cooper pairs depends on the kinetic energy $\varepsilon(\mathbf{p})$. As an illustration here we will analyze the isotropic GL model with

$$\frac{1}{\tau(\mathbf{p})} = \frac{1}{\tau_0}\frac{\varepsilon(\mathbf{p}) + a_0\epsilon}{a_0}, \qquad \varepsilon(\mathbf{p}) = \frac{\mathbf{p}^2}{2m^*}, \qquad \mathbf{p} = m^*\mathbf{v}, \qquad \mathbf{v} = \frac{\partial\varepsilon(\mathbf{p})}{\partial\mathbf{p}}. \tag{7.1}$$

The Boltzmann equation is a dynamic equation and we consider it natural to trace the time evolution of the velocity distribution starting from some arbitrary initial distribution

$$n(\mathbf{p}, t = 0) = n_0(\mathbf{p}). \tag{7.2}$$

As a rule dynamic problems can be more elegantly analyzed using canonical variables. That is why we will analyze the momentum distribution $N(\mathbf{P}, t)$ defined with respect to the canonical momentum

$$\mathbf{P} = \mathbf{p} + e^*\mathbf{A}. \tag{7.3}$$

For a constant and space-homogeneous electric field $\mathbf{E} = \text{const}$ in the $\varphi = 0$ gauge we have

$$\begin{gathered}\mathbf{A}(t) = -t\mathbf{E}, \quad \mathbf{A}(t=0) = 0, \quad \mathbf{P}(\mathbf{p}, t) \equiv \mathbf{p} - e^*\mathbf{E}t, \\ \left(\frac{\partial\mathbf{P}}{\partial\mathbf{p}}\right)_t = \mathbf{1}_{D\times D}, \quad \left(\frac{\partial\mathbf{P}}{\partial t}\right)_p = -e^*\mathbf{E}.\end{gathered} \tag{7.4}$$

Now we will solve the Boltzmann equation using the distribution $N(\mathbf{P}, t)$. The physical quantity is the same – the number of Cooper pairs living at some momentum point – but mathematically the functions are different and their correspondence is given by

$$N(\mathbf{P}, t) \equiv n(\mathbf{p}, t). \tag{7.5}$$

After this change of the variables the sum of partial derivatives on the left-hand-side of the Boltzmann equation reduces to a usual derivative

$$\left(\frac{\partial}{\partial t}+e^*\mathbf{E}\cdot\frac{\partial}{\partial \mathbf{p}}\right)N(\mathbf{P}(\mathbf{p},t),t)=\frac{\partial N}{\partial t}+\frac{\partial N}{\partial \mathbf{P}}\cdot\frac{\partial \mathbf{P}}{\partial t}+\frac{\partial N}{\partial \mathbf{P}}\cdot\frac{\partial \mathbf{P}}{\partial \mathbf{p}}\cdot e^*\mathbf{E}$$
$$=\frac{\partial N(\mathbf{P},t)}{\partial t}=\frac{dN_P(t)}{dt}. \tag{7.6}$$

The physical interpretation is very simple, in an external electromagnetic field the canonical momentum is conserved and reduces simply to a label with which the distribution can be parametrized. In this derivation it is essential to note that when a partial derivative is taken with respect to one argument, the other argument of the function is kept constant. As a consequence, $(\partial n/\partial t)_p \neq (\partial N/\partial t)_P$. Taking into account that the decay rate and the equilibrium distribution are functions of the kinetic momentum $\mathbf{p}=\mathbf{P}-e^*\mathbf{A}$, the Boltzmann equation takes the form

$$\frac{dN_P(t)}{dt}=-\frac{N_P(t)-\overline{n}(\mathbf{P}-e^*\mathbf{A})}{\tau(\mathbf{P}-e^*\mathbf{A})}=-\left[\frac{(\mathbf{P}+e^*\mathbf{E}t)^2}{2m^*}+a_0\epsilon\right]\frac{N_P(t)}{a_0\tau_0}+\frac{n_T}{\tau_0}. \tag{7.7}$$

Introducing now dimensionless time and dimensionless canonical momentum

$$\tilde{u}=\frac{t}{\tau_0},\quad \mathbf{q}=\xi(0)\mathbf{P}/\hbar,\quad q=k-f\tilde{u}, \tag{7.8}$$

we find that in the 1D case the Boltzmann equation takes the convenient form

$$\frac{dN_P(\tilde{u})}{d\tilde{u}}=-\left[(q+f\tilde{u})^2+\epsilon\right]N_P(\tilde{u})+n_T. \tag{7.9}$$

One can easily check that the function

$$N_q(\tilde{u})=n_T\exp\left\{-\frac{1}{3f}(q+f\tilde{u})^3-\epsilon\tilde{u}\right\}\int_0^{\tilde{u}}\exp\left\{\frac{1}{3f}\left(q+fu'\right)^3+\epsilon u'\right\}du'$$
$$+\,n_0(q)\exp\left\{-\frac{1}{3f}\left[(q+f\tilde{u})^3-q^3\right]-\epsilon\tilde{u}\right\} \tag{7.10}$$

satisfies this ordinary linear differential equation with the initial condition

$$N_q(\tilde{u}=0)=n_0(q). \tag{7.11}$$

Now introducing a new dimensionless variable

$$u\equiv\tilde{u}-u' \tag{7.12}$$

we find for the distribution $n_k(\tilde{u}) \equiv N_q(\tilde{u})$ with respect to the kinetic momentum $k = q + f\tilde{u}$,

$$n_k(\tilde{u}) = n_T \int_0^{\tilde{u}} \exp\left\{-\left(k^2+\epsilon\right)u + kfu^2 - \frac{1}{3}f^2u^3\right\} du$$
$$+ n_0(k - f\tilde{u}) \exp\left\{-\left(k^2+\epsilon\right)\tilde{u} + kf\tilde{u}^2 - \frac{1}{3}f^2\tilde{u}^3\right\}. \tag{7.13}$$

Note that the argument of n_0 is a conserved quantity $k - f\tilde{u}$, cf. also Eq. (7.10).

The transition to the limit $t = \tau_0\tilde{u} \to \infty$ is very simple. After several relaxation times we have the stationary distribution

$$n(k) \equiv n_k(\tilde{u} \to \infty) = n_T \int_0^{\infty} \exp\left\{-\left(k^2+\epsilon\right)u + kfu^2 - \frac{1}{3}f^2u^3\right\} du. \tag{7.14}$$

Restoring $k \to k_x$, $\epsilon \to \epsilon + w$, and $n(k) \to n(k_x; w, f, \epsilon)$ we return to the solution of the static Boltzmann equation Eq. (7.34). Consequently the dummy parameter $t = \tau_0 u$ in Eq. (7.28) has the meaning of the time interval between the birth of fluctuation Cooper pairs and the moment when we measure the current. Hence the integrand in the formula for the current Eq. (7.30) has the meaning of the part of the current given by Cooper pairs born a time span t *before* the moment of the measurement. We consider it interesting that according to Eqs. (7.80) and (7.81) this "age distribution" has a sharp δ-like maximum at $t_0 = \tau_0 u_0$ for the supercooled normal phase in small electric fields.

Finally, it is amusing to note that although the fluctuation Cooper pairs are neither particles nor quasi-particles the conservation of canonical momentum implies Newton's equation of motion

$$\frac{d\mathbf{p}}{dt} = e^*\mathbf{E}. \tag{7.15}$$

7.3 Boltzmann equation and formula for the current

As was already mentioned in Sec. 7.1, the Ginzburg–Landau theory serves as an adequate tool for describing the fluctuation conductivity phenomena in superconductors. The important fundamental constant of the TDGL theory is τ_0, which is proportional to the lifetime of metastable Cooper pairs in the normal state:

$$\tau(\epsilon) = \frac{\tau_0}{\epsilon}, \tag{7.16}$$

where $\epsilon \equiv (T - T_c)/T_c$.

Within the weak-coupling BSC theory in the case of negligible depairing mechanisms the temperature-independent constant τ_0 satisfies the relation:

$$\tau_0^{(\mathrm{BCS})} = \frac{\pi}{16} \frac{\hbar}{k_{\mathrm{B}} T_{\mathrm{c}}}. \tag{7.17}$$

At present the experimental value for τ_0, obtained for the layered cuprate superconductors, is in good agreement with the BCS theory [429]. From an experimental point of view it is convenient to introduce the dimensionless ratio:

$$\tau_{\mathrm{rel}} = \frac{\tau_0}{\tau_0^{(\mathrm{BCS})}}, \tag{7.18}$$

which just characterizes the deviations of the experimental value from that which is theoretically derived from the microscopic theory for the case of small coupling and negligible depairing. Before establishing the relation between experimental results on fluctuation conductivity and the lifetime constant τ_0 the important question to be clarified is what the momentum or velocity distribution of the metastable Cooper pair actually is.

The charge of the Cooper pair is $|e^*| = 2\,|e|$ and its effective mass m^* is unambiguously experimentally accessible through many principally different methods: electrostatic charge modulation of the kinetic inductance [6,370], surface Hall effect [359], magnetoplasma waves and cyclotron resonance [502, 503], Doppler effect of Cooper pair plasmons [504] and, which is most reliable, determination of the thermodynamic equilibrium electric potential related to the Bernoulli potential [358]. The velocity of a pair along the x-direction reads

$$v_x = \frac{p_x}{m_a^*} \tag{7.19}$$

and along the y and z directions the velocities are correspondingly $v_y = p_y/m_b^*$, and $v_z = p_z/m_c^*$. Note that we allow for the presence of mass anisotropy.

It was shown [473, 474] within the general TDGL theory that, surprisingly, the momentum distribution law for the metastable Cooper pair is described by the classical Boltzmann equation introduced in physics in 1872 shortly before the electron was theorized (1874, and experimentally discovered in 1897). Nowadays this equation is discussed in many well-known textbooks on kinetics and solid state physics [271, 272, 479, 505, 506]. For a detailed description of the relaxation time approximation to the Boltzmann equation see also the well-known textbook [507]. The Boltzmann equation in our context is given by

$$\left(\frac{\partial}{\partial t} + e^* \mathbf{E} \cdot \frac{\partial}{\partial \mathbf{p}}\right) n(\mathbf{p}, t) = -\frac{n(\mathbf{p}, t) - \overline{n}_p}{\tau_p} = -\frac{n_p(t)}{\tau_p} + \frac{n_T}{\tau_0}. \tag{7.20}$$

This kinetic equation for fluctuation Cooper pairs has been applied to the fluctuation Hall effect in thin superconducting films [473] and to the fluctuation paraconductivity within the framework of time-dependent Ginzburg–Landau theory [474], cf. also Ref. [460]. In Eq. (7.20) $n_p(t) = n(\mathbf{p}, t)$, and we introduce the dimensionless number $n_T = k_{\mathrm{B}}T/a_0$, where

$$a_0 = \frac{\hbar^2}{2m_a^*\xi_a^2(0)} = \frac{\hbar^2}{2m_b^*\xi_b^2(0)} = \frac{\hbar^2}{2m_c^*\xi_c^2(0)} \tag{7.21}$$

is in fact proportional to the first coefficient in the Ginzburg–Landau expansion

$$a(\epsilon) = a_0\epsilon \tag{7.22}$$

and

$$\overline{n}_p = \frac{k_{\mathrm{B}}T}{\varepsilon_p - \mu} \tag{7.23}$$

is a standard equilibrium distribution, where the energy spectrum of the layered superconductor with layer spacing s is given by

$$\varepsilon_p = \frac{p_x^2}{2m_a^*} + \frac{p_y^2}{2m_b^*} + \frac{\hbar^2}{m_c^* s^2}\left[1 - \cos\left(\frac{sp_z}{\hbar}\right)\right], \tag{7.24}$$

and the chemical potential is defined through $-\mu(T) = a(\epsilon)$. The momentum-dependent relaxation time τ_p obtained [473,474] by TDGL theory reads as

$$\tau_p = \frac{\tau_0 a_0}{\varepsilon_p + a(\epsilon)}. \tag{7.25}$$

Let us assume that the electric field is constant and applied along the x-direction in the ab-plane: $\mathbf{E} = (E_x, 0, 0) = \text{const}$. In this case from the Boltzmann equation (7.20) follows the stationary momentum distribution law

$$\begin{aligned} n(p_x;\varepsilon_\perp, E_x, T) &= \\ \frac{k_{\mathrm{B}}T}{\hbar^2/2m_a^*\xi_a^2(0)} &\int_0^\infty \exp\left\{-\left[\left(\frac{\xi_a(0)p_x}{\hbar}\right)^2 + \frac{\varepsilon_\perp}{\hbar^2/2m_a^*\xi_a^2(0)} + \epsilon\right]\frac{t}{\tau_0}\right\} \\ \times \exp&\left\{+\frac{e^*E_x\xi_a(0)}{\hbar/\tau_0}\,\frac{\xi_a(0)p_x}{\hbar}\left(\frac{t}{\tau_0}\right)^2 - \frac{1}{3}\left(\frac{e^*E_x\tau_0}{\hbar/\xi_a(0)}\right)^2\left(\frac{t}{\tau_0}\right)^3\right\}\frac{dt}{\tau_0}, \end{aligned} \tag{7.26}$$

where

$$\varepsilon_\perp(p_y, p_z) = \frac{p_y^2}{2m_b^*} + \frac{\hbar^2}{m_c^* s^2}\left[1 - \cos\left(\frac{sp_z}{\hbar}\right)\right], \tag{7.27}$$

and the physical meaning of the dummy parameter t as time is revealed in Sec. 7.2. For zero field $E_x = 0$ from (7.26) we naturally obtain

$$\overline{n}(p_x, \varepsilon_\perp) = \frac{k_{\mathrm{B}} T/a_0}{\left(\xi_a(0) p_x/\hbar\right)^2 + \varepsilon_\perp/a_0 + \epsilon}. \tag{7.28}$$

Once we have the momentum distribution law for the fluctuation Cooper pair, we can immediately proceed to the expression for the stationary fluctuation current for arbitrary dimension D:

$$\mathbf{j} = \sum_p e^* \mathbf{v}_p \frac{n_p}{\mathcal{V}} = \int e^* \mathbf{v}_p \, n(\mathbf{p}) \frac{d^D p}{(2\pi\hbar)^D}, \qquad \mathbf{v}_p = \frac{\partial \varepsilon_p}{\partial \mathbf{p}}, \tag{7.29}$$

where $\mathcal{V} = L_x L_y L_z$ is the volume of a system. For the current along the x-direction in a layered superconductor we have

$$\begin{aligned} j_x &= e^* \frac{k_{\mathrm{B}} T}{\hbar^2/2m_a^* \xi_a^2(0)} \int_0^\infty \frac{dt}{\tau_0} \exp\left\{ -\frac{1}{3} \left(\frac{e^* E_x \tau_0}{\hbar/\xi_a(0)} \right)^2 \left(\frac{t}{\tau_0} \right)^3 \right\} \\ &\times \int_{-\pi\hbar/s}^{\pi\hbar/s} \frac{dp_z}{2\pi\hbar} \int_{-\infty}^{\infty} \frac{dp_y}{2\pi\hbar} \exp\left\{ -\left[\epsilon + \left(\frac{p_y \xi_b(0)}{\hbar} \right)^2 + \left(\frac{2\xi_c(0)}{s} \sin \frac{s p_z}{2\hbar} \right)^2 \right] \frac{t}{\tau_0} \right\} \\ &\times \int_{-\infty}^{\infty} \frac{p_x}{m_a^*} \left\{ -\frac{t}{\tau_0} \left(\frac{\xi_a(0) p_x}{\hbar} \right)^2 + \frac{e^* E_x \xi_a(0)}{\hbar/\tau_0} \left(\frac{t}{\tau_0} \right)^2 \frac{\xi_a(0) p_x}{\hbar} \right\} \frac{dp_x}{2\pi\hbar} \\ &= \sigma_{xx} E_x. \end{aligned} \tag{7.30}$$

A similar formula for the current was presented by Gor'kov [482] and modified for layered superconductors by Varlamov and Reggiani [508].

7.4 Dimensionless variables

To make our further calculations a bit less cumbersome we introduce a number of dimensionless variables: $k_x = \xi_a(0) p_x/\hbar$ and $k_y = \xi_b(0) p_y/\hbar \in (-\infty, \infty)$ are dimensionless momenta, $\theta = s p_z/\hbar \in (-\pi, \pi)$ is the Josephson phase and $u = t/\tau_0 \in (0, \infty)$ is in fact the renormalized time. We also introduce

$$w \equiv \frac{\varepsilon_\perp}{a_0} = \left(\frac{p_y \xi_b(0)}{\hbar} \right)^2 + \left(\frac{2\xi_c(0)}{s} \sin \frac{s p_z}{2\hbar} \right)^2 = k_y^2 + \omega(\theta), \tag{7.31}$$

where for the well-known Lawrence–Doniach model $\omega(\theta) = \varepsilon_z(p_z)/a_0 = \frac{1}{2} r (1 - \cos\theta) = r \sin^2 \frac{\theta}{2}$. The factor $r = (2\xi_c(0)/s)^2$ originates from the parametrization of the effective mass in c-direction, m_c^*, for an anisotropic GL model.

It is convenient to deal with some dimensionless variable f, proportional to the electric field

$$f = \frac{e^* E_x \tau_0}{\hbar/\xi_a(0)}, \tag{7.32}$$

and with the parameter g, proportional to the nonlinear electric field correction

$$g = \frac{f^2}{12} = \frac{1}{12}\left(\frac{e^* E_x \xi_a(0)}{\hbar/\tau_0}\right)^2. \tag{7.33}$$

With the above notations we can now rewrite our expression (7.26) for the momentum distribution of the metastable Cooper pair in the layered superconductor:

$$n(k_x; w, f, \epsilon) = n_T \int_0^\infty \exp\left[-\left(k_x^2 + \epsilon + w\right) u + f k_x u^2 - \frac{1}{3} f^2 u^3\right] du. \tag{7.34}$$

This function is a solution of the static Boltzmann equation which can be written in the form ($k = k_x$)

$$f \frac{dn(k)}{dk} = -\left(k^2 + \epsilon + w\right) n(k) + n_T. \tag{7.35}$$

Let us mention some details concerning the derivation of expression (7.34). Direct solution of the Boltzmann equation gives

$$\begin{aligned} n(k) = &\exp\left(-\frac{k^3/3 + (\epsilon + w)k}{f}\right)\left[n_T \int_{k_0}^{k} \exp\left(\frac{\tilde{k}^3/3 + (\epsilon + w)\tilde{k}}{f}\right)\frac{d\tilde{k}}{f}\right. \\ &\left. + n_0 \exp\left(\frac{k_0^3/3 + (\epsilon + w)k_0}{f}\right)\right]. \end{aligned} \tag{7.36}$$

From here it is easy to see that the boundary condition is $n(k_0) = n_0$. Then writing $\tilde{k} = k - uf$, introducing $u = (k - \tilde{k})/f$ and using

$$\frac{\tilde{k}^3 - k^3}{3f} = -k^2 u + f k u^2 - \frac{1}{3} f^2 u^3, \tag{7.37}$$

we obtain from Eq. (7.36)

$$\begin{aligned} n(k) = n_T &\int_0^{(k-k_0)/f} \exp\left[-\left(k^2 + \epsilon + w\right) u + f k u^2 - \frac{1}{3} f^2 u^3\right] du \\ &+ n_0 \exp\left(-\frac{k^3/3 + (\epsilon + w)k}{f}\right) \exp\left(\frac{k_0^3/3 + (\epsilon + w)k_0}{f}\right). \end{aligned} \tag{7.38}$$

In the limit $k_0 \to -\infty$ for $f > 0$ or $k_0 \to \infty$ for $f < 0$ we have $n(-\infty) = n(\infty) = 0 = n_0$ and we arrive at Eq. (7.34). Finally the expression for the current Eq. (7.30) in our notations for the layered superconductor takes the form

$$\begin{aligned} j_x = e^* \frac{2k_{\rm B}T}{\hbar}\xi_a(0) \int_0^\infty du \, \exp\left\{-\frac{1}{3}f^2u^3\right\} \int_{-\pi}^{\pi} \frac{d\theta}{2\pi s} \int_{-\infty}^{\infty} \frac{dk_y}{2\pi\xi_b(0)} \\ \times \exp\left\{-\left[\epsilon + k_y^2 + \frac{r}{2}(1-\cos\theta)\right] u\right\} \\ \times \int_{-\infty}^{\infty} k_x \exp\left\{-uk_x^2 + fu^2k_x\right\} \frac{dk_x}{2\pi\xi_a(0)} = \sigma_{xx}E_x. \end{aligned} \tag{7.39}$$

It is informative to compare this result with previously proposed expressions [482,508]. Our further analysis of the fluctuation conductivity is based on this general formula.

7.5 Paraconductivity in a layered metal

As we already pointed out in Sec. 7.1 the fluctuations should be more important in high-temperature materials due in part to their high transition temperatures and extremely short coherence lengths (on the order of 10 Å). Thus the study of Gaussian fluctuation effects on the conductivity in layered compounds is indispensable and of primary importance. In this section we examine in detail the electric field influence on paraconductivity in layered superconductors and derive analytically the correction to the conductivity in the normal phase.

The integration over k_x in the current expression (7.39) can be easily performed:

$$\int_{-\infty}^{\infty} k_x \exp\left\{-uk_x^2 + fu^2k_x\right\} dk_x = \frac{1}{2}\sqrt{\pi u}\, f \exp\left(\frac{1}{4}f^2u^3\right). \tag{7.40}$$

Hence we have for the current in the layered superconductor the expression

$$\begin{aligned} j_x = \frac{e^* k_{\rm B}T}{2\pi^{1/2}\hbar} f \int_{-\pi}^{\pi} \frac{d\theta}{2\pi s} \int_{-\infty}^{\infty} \frac{dk_y}{2\pi\xi_b(0)} \int_0^\infty du\sqrt{u} \\ \times \exp\left\{-\left[\epsilon + k_y^2 + \frac{r}{2}(1-\cos\theta)\right] u - \frac{1}{12}f^2u^3\right\} = \sigma_{xx}E_x \end{aligned} \tag{7.41}$$

and, consequently, the fluctuation conductivity reads

$$\begin{aligned} \sigma_{xx} = \frac{2}{\pi^{1/2}} \frac{e^2 k_{\rm B} T\tau_0\xi_a(0)}{\hbar^2} \int_{-\pi}^{\pi} \frac{d\theta}{2\pi s} \int_{-\infty}^{\infty} \frac{dk_y}{2\pi\xi_b(0)} \int_0^\infty du\sqrt{u} \\ \times \exp\left\{-\left[\epsilon + k_y^2 + \omega(\theta)\right] u - gu^3\right\} = j_x/E_x. \end{aligned} \tag{7.42}$$

After integrating over k_y we obtain from (7.42):

$$\sigma_{xx} = \frac{e^2 k_{\mathrm{B}} T \tau_0 \xi_a(0)}{\pi \hbar^2 s \xi_b(0)} \int_{-\pi}^{\pi} \frac{d\theta}{2\pi} \int_0^{\infty} du \exp\left\{-\left[\epsilon + \omega(\theta)\right] u - g u^3\right\}$$
$$= j_x / E_x. \tag{7.43}$$

Thus the conductivity in the Lawrence–Doniach (LD) model can be expressed in the way

$$\sigma_{\mathrm{LD}} = \frac{e^2 k_{\mathrm{B}} T \tau_0 \xi_a(0)}{\pi \hbar^2 s\, \xi_b(0)} \int\limits_0^{\infty} du \exp\left\{-\left(\epsilon + \frac{r}{2}\right) u - g u^3\right\}$$
$$\times \int\limits_{-\pi}^{\pi} \exp\left\{\frac{r}{2} \cos(\theta) u\right\} \frac{d\theta}{2\pi} \tag{7.44}$$

or, more conveniently,

$$\sigma_{\mathrm{LD}}(\epsilon, g; r) = \frac{e^2 k_{\mathrm{B}} T \tau_0 \xi_a(0)}{\pi \hbar^2 s\, \xi_b(0)} \int_0^{\infty} du \, \exp\left\{-\left(\epsilon + r/2\right) u - g u^3\right\} I_0\left(r u / 2\right), \tag{7.45}$$

where

$$I_0(x) = \frac{1}{\pi} \int_0^{\pi} \exp\left(x \cos\theta\right) d\theta = \sum_{n=0}^{\infty} \left[\frac{(x/2)^n}{n!}\right]^2 = J_0(ix) \tag{7.46}$$

is the Bessel function of imaginary argument.

In the case of zero electric field $E_x = 0$ the integral over u in Eq.(7.45) is easy to compute since

$$\int_{-\pi}^{\pi} \frac{d\theta}{2\pi} \int_0^{\infty} du \exp\left\{-\left[\epsilon + \frac{r}{2}\left(1 - \cos\theta\right)\right] u\right\}$$
$$= \int_{-\pi}^{\pi} \frac{d\theta}{2\pi} \frac{1}{\left(\epsilon + \frac{r}{2}\right) - \frac{r}{2}\cos\theta} = \frac{1}{\sqrt{\epsilon(\epsilon + r)}}, \tag{7.47}$$

and we obtain the Lawrence–Doniach formula for the fluctuation conductivity of a layered superconductor

$$\sigma_{\mathrm{LD}}(\epsilon; r) = \frac{e^2 k_{\mathrm{B}} T \tau_0 \xi_a(0)}{\pi \hbar^2 s\, \xi_b(0)} \frac{1}{\sqrt{\epsilon(\epsilon + r)}}. \tag{7.48}$$

Hereafter we define, for simplicity, $\sigma(\epsilon; r) \equiv \sigma(\epsilon, 0; r)$.

It is simpler to deal with the dimensionless function (cf. Varlamov and Reggiani [508] Eq. (7))

$$\varsigma_{\mathrm{LD}}(\epsilon, g; r) = \frac{\sigma_{\mathrm{LD}}(\epsilon, g; r)}{\sigma_{\mathrm{LD}}(\epsilon; r)} \tag{7.49}$$
$$= \sqrt{\epsilon(\epsilon + r)} \int_0^{\infty} \exp\left\{-\left(\epsilon + r/2\right) u - g u^3\right\} I_0\left(r u / 2\right) du.$$

Then Eq. (7.45) reads

$$\sigma_{\mathrm{LD}}(\epsilon, g; r) = \sigma_{\mathrm{LD}}(\epsilon; r)\, \varsigma_{\mathrm{LD}}(\epsilon, g; r). \tag{7.50}$$

It is important to note that the product of two exponents in (7.49) can be expressed in terms of the third-derivative operator acting on the function of ϵ

$$\exp\left(-gu^3\right)\exp\left(-\epsilon u\right) = \exp\left(g\frac{\partial^3}{\partial\epsilon^3}\right)\exp\left(-\epsilon u\right). \tag{7.51}$$

This key observation immediately leads us to the simple general relation between the nonlinear fluctuation conductivity and the linear one in evanescent field, $\sigma(\epsilon)$,

$$\sigma(\epsilon > 0, g) = \exp\left(g\frac{\partial^3}{\partial\epsilon^3}\right)\sigma(\epsilon), \tag{7.52}$$

which is similar to the relation for the magnetoconductivity, derived in Ref. [429]. For our dimensionless function (7.49) we get

$$\varsigma(\epsilon, g) = \frac{1}{\sigma(\epsilon)}\exp\left(g\frac{\partial^3}{\partial\epsilon^3}\right)\sigma(\epsilon). \tag{7.53}$$

In the limit $g \to 0$ the electric-field-dependent conductivity can be written in a simple form

$$\sigma(\epsilon, f) = \sigma(\epsilon) + \Delta\sigma_f(\epsilon, f) \approx \left(1 + g\frac{\partial^3}{\partial\epsilon^3}\right)\sigma(\epsilon), \tag{7.54}$$

and thus the nonlinear electric-field correction to the conductivity reads

$$\Delta\sigma_f(\epsilon, f) \approx \frac{f^2}{12}\frac{\partial^3}{\partial\epsilon^3}\sigma(\epsilon). \tag{7.55}$$

Now we can derive the conductivity, dependent on the electric field in the Lawrence–Doniach model of a layered superconductor. In principle we may start either from the right-hand-side of Eq. (7.47) (i.e., we can take the third derivative after averaging over the Josephson phase), or we can perform the differentiation before integrating over θ. The latter is a useful method for the evaluation of complicated integrals necessary for experimental data processing and leads to

$$\oint \frac{d\theta}{2\pi}\frac{1}{\left[\epsilon + \frac{r}{2}\left(1 - \cos\theta\right)\right]^4} = -\frac{1}{6}\frac{\partial^3}{\partial\epsilon^3}\frac{1}{\sqrt{\epsilon(\epsilon + r)}} = \frac{\epsilon^3 + \frac{3}{2}r\epsilon^2 + \frac{9}{8}r^2\epsilon + \frac{5}{16}r^3}{\left[\epsilon(\epsilon + r)\right]^{7/2}}. \tag{7.56}$$

Finally for the Lawrence–Doniach conductivity we have

$$\sigma_{\rm LD}(\epsilon, g; r) \approx \left[1 - \frac{1}{2}\frac{\epsilon^3 + \frac{3}{2}r\epsilon^2 + \frac{9}{8}r^2\epsilon + \frac{5}{16}r^3}{(\epsilon + r)^3} f_\epsilon^2\right] \sigma_{\rm LD}(\epsilon; r), \qquad f_\epsilon \ll 1, \tag{7.57}$$

where

$$f_\epsilon = \frac{|e^* E_x| \xi_a(\epsilon)\tau(\epsilon)}{\hbar} = \frac{|f|}{\epsilon^{3/2}} = \frac{|E_x|}{E_c(\epsilon)}, \qquad g_\epsilon = \frac{1}{12} f_\epsilon^2 = \frac{g}{\epsilon^3}, \tag{7.58}$$

and other notations are

$$E_c(\epsilon) = \frac{\hbar}{|e^*|\xi(\epsilon)\tau(\epsilon)} = E_c(0)\epsilon^{3/2}, \qquad E_c(0) = \frac{\hbar}{|e^*|\xi(0)\tau_0}, \tag{7.59}$$

$$\xi_{a,b,c}(\epsilon) = \frac{\xi_{a,b,c}(0)}{\epsilon^{1/2}}, \qquad r_\epsilon = \frac{r}{\epsilon} = \left(\frac{2\xi_c(\epsilon)}{s}\right)^2. \tag{7.60}$$

Our dimensionless ς function can now be rewritten in the scaled variables, as can be seen from (7.49),

$$\varsigma_{\rm LD}(\epsilon, g; r) = \varsigma_{\rm LD}(1, g_\epsilon; r_\epsilon). \tag{7.61}$$

To conclude, we obtain the formula for the electric-field-dependent correction to the conductivity in the Lawrence–Doniach model for a layered superconductor:

$$\begin{aligned}\Delta\sigma_f(\epsilon, E_x) &= -\frac{4k_{\rm B}Te^4\left[\xi_a(0)\tau_0\right]^3}{\pi\hbar^4 s\xi_b(0)} \frac{\left[\epsilon^3 + \frac{3}{2}r\epsilon^2 + \frac{9}{8}r^2\epsilon + \frac{5}{16}r^3\right]}{\left[\epsilon(\epsilon + r)\right]^{7/2}} E_x^2 \\ &\equiv \Delta j_x / E_x.\end{aligned} \tag{7.62}$$

7.6 Aslamazov–Larkin conductivity for D-dimensional superconductors

In this section we are concerned with the derivation of the Aslamazov–Larkin electric-field-dependent conductivity for bulk superconductors ($D = 3$), thin films ($D = 2$) and wires ($D = 1$). The results can be generalized for the case of arbitrary dimension.

We start with a one-dimensional superconductor. In order to derive the expression for the one-dimensional fluctuation conductivity we should set $p_y = p_z = 0$ in (7.24) and exclude from Eq. (7.42) the integration over the perpendicular component of the momenta

$$\mathbf{p}_\perp = (p_y, p_z) = \left(\frac{\hbar k_y}{\xi_b(0)}, \frac{\hbar\theta}{s}\right), \tag{7.63}$$

(the electric field is as usual parallel to the x-direction). We immediately get the final expression

$$\sigma_{1\mathrm{D}} = \frac{2}{\pi^{1/2}} \frac{e^2 k_{\mathrm{B}} T \tau_0 \xi_a(0)}{\hbar^2} \int_0^\infty \sqrt{u} \exp\left\{-\epsilon u - g u^3\right\} du = \frac{I_x}{E_x}. \tag{7.64}$$

For the derivation of a two-dimensional conductivity, induced by the fluctuations, we have to cancel averaging with respect of the Josephson phase θ in Eq. (7.43) and we find

$$\sigma_{2\mathrm{D}} = \frac{e^2 k_{\mathrm{B}} T \tau_0 \xi_a(0)}{\pi \hbar^2 \xi_b(0)} \int_0^\infty \exp\left\{-\epsilon u - g u^3\right\} du = \frac{j_x^{(2\mathrm{D})}}{E_x}. \tag{7.65}$$

The case of a layered superconductor turns into the bulk (3D) one in the limit $r \to \infty$, when the distance between the layers tends to zero. Taking into account the asymptotics for the Bessel function in Eq. (7.45), cf. Ref. [508]:

$$I_0(x \gg 1) \approx \frac{\mathrm{e}^x}{\sqrt{2\pi x}}, \qquad \exp\left(-\frac{r}{2}u\right) I_0\left(\frac{r}{2}u\right) \approx \frac{1}{\sqrt{\pi r u}}, \tag{7.66}$$

we obtain the bulk fluctuation conductivity expression

$$\sigma_{3\mathrm{D}} = \frac{e^2 k_{\mathrm{B}} T \tau_0 \xi_a(0)}{2\pi^{3/2} \hbar^2 \xi_b(0) \xi_c(0)} \int_0^\infty \exp\left\{-\epsilon u - g u^3\right\} \frac{du}{\sqrt{u}} = \frac{j_x^{(3\mathrm{D})}}{E_x}. \tag{7.67}$$

These results bring us naturally to the general expression for the fluctuation conductivity for arbitrary dimension D and T above T_{c}:

$$\sigma_{\mathrm{D}}(\epsilon, g) = \frac{e^2 k_{\mathrm{B}} T \tau_0 \xi_a^2(0)}{2^{D-2} \pi^{D/2} \hbar^2 \xi^D(0)} \int_0^\infty \exp\left\{-\epsilon u - g u^3\right\} \frac{du}{u^{(D-2)/2}}, \tag{7.68}$$

(cf. Dorsey [423], Eq. (47), for $\epsilon > 0$).

One can express Eq.(7.68) in a form similar to Eq. (7.50):

$$\sigma_{\mathrm{D}}(\epsilon, f) = \sigma_{\mathrm{D}}(\epsilon) \varsigma_{\mathrm{D}}(g_\epsilon), \tag{7.69}$$

where

$$\sigma_{\mathrm{D}}(\epsilon) = \frac{4\Gamma\left(\frac{4-D}{2}\right)}{(4\pi)^{D/2}} \frac{e^2}{\hbar^2} k_{\mathrm{B}} T \frac{\tau(\epsilon) \xi_a^2(\epsilon)}{\xi^D(\epsilon)}, \tag{7.70}$$

is the fluctuation conductivity in zero electric field, and

$$\xi^D(\epsilon) \equiv \begin{cases} \xi_a(\epsilon) & \text{for } D = 1, \\ \xi_a(\epsilon) \xi_b(\epsilon) & \text{for } D = 2, \\ \xi_a(\epsilon) \xi_b(\epsilon) \xi_c(\epsilon) & \text{for } D = 3, \\ \ldots \end{cases} \tag{7.71}$$

The function

$$\varsigma_{\rm D}(g_\epsilon) = \frac{1}{\Gamma\left(\frac{4-D}{2}\right)} \int_0^\infty v^{1-D/2} {\rm e}^{-v-g_\epsilon v^3}\, dv, \qquad v = \epsilon u \tag{7.72}$$

is a dimensionless function containing the electric-field dependence of the fluctuation conductivity.

For small electric field $f_\epsilon \ll 1$, in accordance with Eq (7.54), we have:

$$\varsigma_{\rm D}(f_\epsilon) \approx 1 - \left(\frac{4-D}{2}\right)\left(\frac{6-D}{2}\right)\left(\frac{8-D}{2}\right)\frac{f_\epsilon^2}{12}. \tag{7.73}$$

7.6.1 *Strong electric field expansion*

In this subsection we study the conductivity analytically in the limit of electric fields that are large compared with the reduced temperature distance to the critical point. To this end we introduce the scaled variable $\epsilon_f = 12^{1/3}\,\epsilon/f^{2/3}$ and focus on the regime of small ϵ_f. The sign of ϵ_f is arbitrary. Our analysis applies to $T \leqslant T_{\rm c}$ as well as $T > T_{\rm c}$. Changing the integration variable u in Eq. (7.68) to $f^{2/3}u$ allows us to extract the dominant field dependence at the critical point. Subsequently we use the series expansion of $\exp(-\epsilon u)$ and integrate it term by term for obtaining a series in ϵ_f. We arrive at the result

$$\begin{aligned} J_{\rm D}(\epsilon, g) &\equiv \int_0^\infty \exp\left\{-\epsilon u - g u^3\right\} \frac{du}{u^{(D-2)/2}} \\ &= 4\frac{f^{-(4-D)/3}}{12^{(D+2)/6}} \sum_{n=0}^{\infty} \frac{(-\epsilon_f)^n}{n!}\Gamma\left(\frac{2n+4-D}{6}\right). \end{aligned} \tag{7.74}$$

The convergence is very good for $|\epsilon_f| \leqslant 1$. For instance, for $\epsilon_f = 0.2$ the errors in zeroth, first, second and third order are, respectively, 12%, 1%, 0.07% and 40 ppm (parts per million). This series leads to the following expression for the conductivity

$$\sigma_{\rm D}(\epsilon, f) = \sigma_{\rm D}(f)\Sigma_{\rm D}\left(\epsilon_f\right) \tag{7.75}$$

with

$$\Sigma_{\rm D}(\epsilon_f) \equiv 1 + \sum_{n=1}^{\infty} \frac{(-\epsilon_f)^n}{n!} \frac{\Gamma\left(\frac{2n+4-D}{6}\right)}{\Gamma\left(\frac{4-D}{6}\right)} \tag{7.76}$$

and

$$\begin{aligned} \sigma_{\rm D}(f) &= \frac{\Gamma\left(\frac{4-D}{6}\right) e^2 k_{\rm B} T \tau_0 \xi_a^2(0) f^{-(4-D)/3}}{3^{(D+2)/6} 2^{(4D-10)/3} \pi^{D/2} \hbar^2 \xi^D(0)} \\ &= \frac{4\Gamma\left(\frac{4-D}{6}\right) k_{\rm B} T \tau_0^{(D-1)/3}}{\left[2\sqrt{\pi}\xi(0)\right]^D E_x^{(4-D)/3}} \left(\frac{e\xi_a(0)}{\sqrt{3}\hbar}\right)^{(D+2)/3} \end{aligned} \tag{7.77}$$

is the fluctuation conductivity at $T_{\rm c}$; cf. [423, 508, 509].

7.6.2 Weak electric fields below T_c

In order to study the fluctuation conductivity for temperatures slightly below T_c and in small electric fields we must take into account that the limit of zero field is singular due to the occurrence of bulk superconductivity. The contribution to the conductivity that we calculate in this subsection must be well separated physically from that due to the onset of bulk order. The approximation scheme we develop here is a good one for relatively weak electric fields, which satisfy at $T < T_c$ the condition

$$f_\epsilon \equiv \frac{|f|}{|\epsilon|^{3/2}} = \frac{2\,|eE_x|\,\xi_a(0)\tau_0}{\hbar|\epsilon|^{3/2}} < 1. \tag{7.78}$$

Especially below T_c we have to take into account that the temperature distance to the critical point ϵ is renormalized as can be seen in self-consistent mean-field-like approximations which we will briefly consider later. The origin of this effect lies in the non-linear character of the TDGL equation, which cannot be neglected for high densities of fluctuation Cooper pairs. The self-consistent approximation decouples the non-linearity, resulting in a linear problem with a modified ϵ, to be denoted by ϵ_r. To alleviate the notation we will postpone this substitution until the end of this subsection.

For $T < T_c$ we write $\epsilon = -|\epsilon|$ and to the integral in Eq. (7.74) we apply the Gaussian saddle point approximation for weak electric fields. This amounts to looking for the maximum of the argument of the exponential function, since the remaining factor is an algebraic function of u and therefore slowly varying. Defining

$$F(u) = -|\epsilon|u + \frac{f^2u^3}{12} \tag{7.79}$$

the saddle-point approximation can be written as

$$\mathrm{e}^{-F(u)} \approx \mathrm{e}^{-F(u_0)}\sqrt{\frac{2\pi}{F''(u_0)}}\,\delta(u-u_0), \tag{7.80}$$

where u_0 is the minimum of $F(u)$, given by

$$u_0 = 2\frac{\sqrt{|\epsilon|}}{|f|}; \tag{7.81}$$

the time interval $t_0 = \tau_0 u_0$ has a transparent physical meaning, $|e^*E_x|\,t_0 = 2\hbar/\xi_a(\epsilon)$. In order for this approximation to be accurate, the condition

$$F''(u_0)u_0^2 \gg 1 \tag{7.82}$$

must be satisfied, which is equivalent to

$$f_\epsilon \ll 1. \tag{7.83}$$

This condition can be seen to arise from two requirements. Firstly, already in $D = 2$, in the absence of the power of u in the integral, the validity of the saddle-point approximation requires

$$\exp\left(-\frac{1}{2}F''(u_0)u_0^2\right) \ll 1 \tag{7.84}$$

in order for the integration interval to be extendable to $(-\infty, \infty)$. Secondly, in the presence of slowly varying additional factors in the integrand, we must check the consistency of the approximation by performing a Taylor expansion of the algebraic function of u about u_0. If we denote this function by $G(u)$, we may approximate this by the constant $G(u_0)$ in the integral, provided $G(u)$ deviates only weakly from linearity in a neighbourhood of width w_F around u_0, where w_F is the standard deviation of the Gaussian function. This is fulfilled when

$$G''(u_0)w_F^2/G(u_0) \ll 1. \tag{7.85}$$

Since $w_F^2 = 1/F''(u_0)$ and $G(u)$ is simply a power of u, this condition coincides with Eq. (7.82).

Within the range of validity of the saddle-point approximation we thus arrive at the following result for the integral

$$J_{\mathrm{D}}(-|\epsilon|, f) \approx \sqrt{\pi}\,|\epsilon|^{-\frac{4-D}{2}}\left(\frac{2}{f_\epsilon}\right)^{\frac{3-D}{2}}\exp\left(\frac{4}{3f_\epsilon}\right), \quad \text{for } \mathrm{e}^{2/f_\epsilon} \gg 1. \tag{7.86}$$

According to Eq. (7.68) and taking into account the renormalization of ϵ to ϵ_r we obtain

$$\begin{aligned}\sigma_{\mathrm{D}}(-|\epsilon|, f) \approx & \frac{e^2 k_{\mathrm{B}} T}{2^{(3D-7)/2}\pi^{(D-1)/2}\hbar^2} \\ & \times \frac{\tau(\epsilon_r)\xi_a^2(\epsilon_r)}{\xi^D(\epsilon_r)} f_\epsilon^{(D-3)/2}\exp\left(\frac{4}{3f_\epsilon}\right).\end{aligned} \tag{7.87}$$

The replacement $\epsilon \to \epsilon_r$ derives from the use of a Maxwell-type self-consistent approach (cf. Ref. [429]) for solving the Boltzmann equation. The implicit equation relating the renormalized and bare parameters is given by

$$\epsilon_r - \epsilon = \frac{\mu_0}{m_{ab}^*}\left[e^*\lambda_{ab}(0)\right]^2 \int \frac{d^D p}{(2\pi\hbar)^D} n(\mathbf{p}, \mathbf{E}, \epsilon_r), \tag{7.88}$$

with

$$\frac{1}{\lambda_{ab}^2(0)} \equiv -T_{\mathrm{c}} \left.\frac{d}{dT}\frac{1}{\lambda_{ab}^2(T)}\right|_{T_{\mathrm{c}}^-}, \tag{7.89}$$

where details of the procedure of ultraviolet regularization will be presented elsewhere.

The most significant aspect of our result is the dramatic exponential increase in the fluctuation conductivity for small electric fields. Therefore, for $f_\epsilon \ll 1$, the fluctuation part becomes of the same order of magnitude as the normal-state background, $\sigma_D(-|\epsilon|, f) \simeq \sigma_N$, and the fluctuation conductivity will no longer be just a perturbation but a significant part of the total conductivity

$$\sigma_{\text{tot}} = \sigma_D(\epsilon, f) + \sigma_N(T) = j_{\text{tot}}/E_x. \tag{7.90}$$

This extraordinary increase of the conductivity naturally leads to a *minimum* in the current as a function of the applied field, in agreement with the Gor'kov analysis [482] that the current-voltage characteristic must have a perfectly noticeable section corresponding to negative differential conductivity. At a suitable potential difference between the ends of the film, the generation of radiation should be observed. Indeed, for every $\epsilon < 0$ the generation of radiation will start for electric fields lower than the critical one, $E_{\text{gen}}(\epsilon = -|\epsilon|)$, determined by the criterion

$$\left.\frac{dj_{\text{tot}}(E_x)}{dE_x}\right|_{E_{\text{gen}}} = 0, \tag{7.91}$$

which is equivalent to

$$f_{\text{gen}} \left.\frac{d}{df}\sigma(-|\epsilon|, f)\right|_{f_{\text{gen}}} + \sigma(-|\epsilon|, f_{\text{gen}}) = -\sigma_N(T). \tag{7.92}$$

Within the self-consistent approximation Eq. (7.89) it is easy to obtain theoretical formulae for the case when the fluctuations are nonlinear

$$E_{\text{gen}}(-|\epsilon_r|) = E_c(0) f_{\text{gen}}(-|\epsilon_r|), \tag{7.93}$$

where the unit of electric field $E_c(0)$ is defined in Eq. (7.59), and this will be a nontrivial test of the validity of the self-consistent approximation applied to the TDGL equation and, following from it, the Boltzmann equation for fluctuation Cooper pairs. Now we are addressing cases important for the applications, in which the fluctuation superconductivity can be easily investigated.

7.7 Striped superconductors and thick films

The latest achievements in nanotechnology provide us in principle with a tool for a practical realization of so called striped superconductors with controlled parameters. For this it is necessary to cut stripes from a superconducting film using some appropriate lithographic technology. The amazing observation about the striped materials is that they are of "intermediate" dimensionality, i.e., they are neither one-dimensional systems nor two-dimensional ones. The closer the striped superconductor is to its critical temperature T_c, the more "perfectly" two-dimensional material it becomes, because the stripes become increasingly coherent. Analogously to the Lawrence–Doniach model for the layered superconductor we can describe this situation in terms of "dimensional crossover". In the present section we show that the fluctuation conductivity longitudinal to the stripes can be derived following the standard procedure for layered superconductors. We also have to mention that probably some underdoped cuprates are naturally striped and this phenomenon has been at the center of the attention attracted by HTSC during the last few years.

It is obvious from Sec. 7.6, for example, that the LD conductivity for a layered superconductor can be naturally derived from a two-dimensional AL conductivity just by integrating that over the momentum in the direction perpendicular to the plane

$$\sigma_{\mathrm{LD}}(\epsilon, f) = \int \frac{dp_z}{2\pi\hbar}\sigma_{2\mathrm{D}}\left(\epsilon + \frac{\varepsilon_z(p_z)}{a_0}, f\right)$$
$$= \int_{-\pi}^{\pi} \sigma_{2\mathrm{D}}\left(\epsilon + \frac{r}{2}(1-\cos\theta), f\right)\frac{d\theta}{2\pi s}. \tag{7.94}$$

Thus the fluctuation conductivity for a striped superconductor reads

$$\sigma_{\mathrm{striped}}(\epsilon, f) = \frac{1}{s}\oint \sigma_{1\mathrm{D}}\left(\epsilon + \frac{r}{2}(1-\cos\theta), f\right)\frac{d\theta}{2\pi}, \tag{7.95}$$

where s now stands for the period of the stripes.

For a thick film with thickness d_{film} we have to sum over the discrete spectrum of the energy associated with the motion in z-direction:

$$\varepsilon_z(p_z) = \frac{p_z^2}{2m_c^*}, \qquad \frac{\varepsilon_z(p_z)}{a_0} = \left(\frac{\pi\xi_c(0)}{d_{\mathrm{film}}}\right)^2 n_z^2, \tag{7.96}$$

since

$$p_z = \frac{\pi\hbar}{d_{\mathrm{film}}}n_z, \qquad n_z = 0, 1, 2, 3, \ldots. \tag{7.97}$$

In the expression for the fluctuation conductivity of a thick film we have now summation instead of integration

$$\sigma_{\mathrm{film}}(\epsilon, f) = \frac{1}{d_{\mathrm{film}}}\sum_{p_z}\sigma_{2\mathrm{D}}\left(\epsilon + \varepsilon_z(p_z)/a_0, f\right)$$
$$= \frac{1}{d_{\mathrm{film}}}\sum_{n_z}^{\infty}\sigma_{2\mathrm{D}}\left(\epsilon + \left(\pi\xi_c(0)/d_{\mathrm{film}}\right)^2 n_z^2, f\right), \tag{7.98}$$

which for zero electric field can be readily performed using

$$\sum_{n=0}^{\infty} \frac{1}{a^2 + b^2 n^2} = \frac{\pi}{2ab \tanh(\pi a/b)} + \frac{1}{2a^2}. \tag{7.99}$$

Finally we obtain the expression for the 3D conductivity of the film, which interpolates between 2D and 3D behaviour analogously to the conductivity for a layered superconductor:

$$\begin{aligned}\sigma_{\mathrm{film}}(\epsilon) &= \sigma_{2D}(\epsilon) \left\{ \frac{1}{2d_{\mathrm{film}}} + \frac{1}{2\xi_c(\epsilon) \tanh\left(d_{\mathrm{film}}/\xi_c(\epsilon)\right)} \right\} \\ &= \frac{e^2 \tau_{rel}}{16\hbar\epsilon} \left\{ \frac{1}{2d_{\mathrm{film}}} + \frac{\sqrt{\epsilon}\coth\left(d_{\mathrm{film}}\sqrt{\epsilon}/\xi_c(0)\right)}{2\xi_c(0)} \right\},\end{aligned} \tag{7.100}$$

$$\sigma_{LD}(\epsilon) = \frac{\sigma_{\mathrm{2D}}(\epsilon)}{\sqrt{s^2 + (2\xi_c(\epsilon))^2}} = \frac{e^2 \tau_{\mathrm{rel}}}{16\hbar\sqrt{\epsilon^2 s^2 + \epsilon\,(2\xi_c(0))^2}}. \tag{7.101}$$

Note that the thick film becomes "rightly" *two*-dimensional in the vicinity of T_{c}, thus the dimensionality of it decreases, whereas for the layered superconductor the dimensionality goes up from $D = 2$ to $D = 3$ as we approach the critical temperature. In such a way a thick film of a strongly anisotropic layered superconductor, $\xi_c(0) \ll s \ll d_{\mathrm{film}}$, can have two [510] dimensional crossovers.

7.8 Determination of the lifetime constant τ_0

In the current section we show that conductivity measurements in a strong electric field can serve as a method for probing fundamental properties of superconductors such as the lifetime constant of metastable Cooper pairs τ_0 and the coherence length $\xi(0)$. We demonstrate that our theoretical results can be effectively used for experimental data processing and determination of both τ_0 and $\xi(0)$.

As a rule in an experiment the temperature dependence of a resistivity $\rho_{\mathrm{exp}}(T)$ is examined. The experimentally measured conductivity is consequently $\sigma_{\mathrm{exp}} = 1/\rho_{\mathrm{exp}}(T)$. The in-plane current, if we take into account the first nonlinear correction, can be written in the general form

$$j_x = \sigma_{\mathrm{exp}} E_x - \mathcal{A}(\epsilon) E_x^3, \tag{7.102}$$

since the nonlinear correction to the fluctuation conductivity is $\Delta\sigma_f = -\mathcal{A}E_x^2$.

Of primary interest for us are the superconducting films. Let us consider the two-dimensional BCS-like superconductor. In this case the parameter r

that determines the effective dimensionality of the superconductor is zero, and the periodicity of the Lawrence–Doniach model $s \to d_{\text{film}}$ (d_{film} is the thickness of the superconducting film). Thus the two-dimensional current is $j_x^{(2D)} = d_{\text{film}} j_x$.

From the expression for the LD fluctuation conductivity Eq (7.62) it follows:

$$j_x^{(2D)} = \left(\sigma_N(T) + \frac{e^2}{16\hbar} \frac{\tau_{\text{rel}}}{\epsilon} \right) E_x - \frac{4\pi^2}{\hbar} \left[\xi_{ab}(0)/k_{\text{B}}T \right]^2 \frac{e^4 \tau_{\text{rel}}^3}{(8\epsilon)^4} E_x^3. \tag{7.103}$$

In order to study the fluctuation effect on conductivity one should plot first of all the paraconductivity contribution to the resistance $1/(1/\rho_{\text{exp}}(T) - 1/\rho_N(T))$ as a function of T (see for example [480], Fig. 3, where indium oxide films were examined). For InO_x films the value of $\tau_{\text{rel}} = 1.16$ can be used as a tool for the determination of the in-plane coherence length $\xi_{ab}(0)$.

In the general case of a layered LD superconductor the coefficient in the nonlinear correction to fluctuation conductivity according to Eq. (7.62) reads

$$\mathcal{A}(\epsilon) = \frac{4k_{\text{B}}Te^4 \left[\xi_a(0)\tau_0\right]^3}{\pi\hbar^4 s\xi_b(0)} \frac{\left[\epsilon^3 + \frac{3}{2}r\epsilon^2 + \frac{9}{8}r^2\epsilon + \frac{5}{16}r^3\right]}{\left[\epsilon(\epsilon+r)\right]^{7/2}}. \tag{7.104}$$

In this case we have three coherence lengths $\xi_a(0)$, $\xi_b(0)$, $\xi_c(0)$, and the current anisotropy $J_{\text{max}}/J_{\text{min}}$.

In this Chapter we develop a model-free method for the determination of the lifetime of Cooper pairs τ_0 and the coherence length from the experimental results for the fluctuation conductivity. The final result is derived on the basis of the Eq. (7.54) after three-fold integration over some time interval (ϵ_1, ϵ_2). We obtain the following expression for the lifetime

$$\tau_0 = \sqrt{\frac{3}{2}} \frac{\hbar}{|eE_x|\xi_a(0)} \tag{7.105}$$

$$\times \left\{ \frac{\int_{\epsilon_1}^{\epsilon_2} (\tilde{\epsilon} - \epsilon_1)^2 \left[-\Delta\sigma_f(\tilde{\epsilon}, f)\right] d\tilde{\epsilon}}{\sigma(\epsilon_1) - \left[\sigma(\epsilon_2) + (\epsilon_1 - \epsilon_2)\,\sigma'(\epsilon_2) + \frac{1}{2}(\epsilon_1 - \epsilon_2)^2 \sigma''(\epsilon_2)\right]} \right\}^{1/2}.$$

In (7.105) we have the notations

$$\sigma'(\epsilon) = \frac{\partial}{\partial\epsilon}\sigma(\epsilon), \qquad \sigma''(\epsilon) = \frac{\partial^2}{\partial\epsilon^2}\sigma(\epsilon), \tag{7.106}$$

and $\epsilon_2 \simeq 0.2$ determines the upper bound on the temperature below which the fluctuation conductivity can be reliably measured. In order to simplify the formula (7.105) we can make the approximation $\epsilon_2 \to \infty$. Since the fluctuation phenomena are negligible already for $T - T_\mathrm{c} \approx 0.15 T_\mathrm{c}$ the following asymptotic conditions can be imposed in the limit $\epsilon_2 \to \infty$,

$$\sigma(\epsilon_2) \to 0, \qquad \epsilon_2 \sigma'(\epsilon_2) \to 0, \qquad \epsilon_2^2 \sigma''(\epsilon_2) \to 0. \tag{7.107}$$

Finally, we obtain the formulae for the lifetime of metastable Cooper pairs and the in-plane coherence length, which can be applied for experimental data processing:

$$\tau_0 = \sqrt{\frac{3}{2}} \frac{\hbar}{|eE_x|\xi_a(0)} \left\{ \frac{1}{\sigma(\epsilon)} \int_\epsilon^\infty (\tilde{\epsilon} - \epsilon)^2 \left[-\Delta\sigma_f(\tilde{\epsilon}, f)\right] d\tilde{\epsilon} \right\}^{1/2}, \tag{7.108}$$

$$\tau_\mathrm{rel} = \sqrt{\frac{3}{2}} \frac{16 k_\mathrm{B} T}{\pi |eE_x|\xi_a(0)} \left\{ \frac{1}{\sigma(\epsilon)} \int_\epsilon^\infty (\tilde{\epsilon} - \epsilon)^2 \left[-\Delta\sigma_f(\tilde{\epsilon}, f)\right] d\tilde{\epsilon} \right\}^{1/2}, \tag{7.109}$$

$$\xi_{ab}(0) = \sqrt{\frac{\Phi_0}{\pi |B_z|}} \left\{ \frac{1}{\sigma(\epsilon)} \int_\epsilon^\infty (\tilde{\epsilon} - \epsilon) \left[-\Delta\sigma_h(\tilde{\epsilon}, h)\right] d\tilde{\epsilon} \right\}^{1/4} \tag{7.110}$$

cf. Ref. [429], Eqs. (201)–(-203). Here

$$-\Delta\sigma_h(\epsilon, h) = \frac{h^2}{4} \frac{\partial^2}{\partial \epsilon^2} \sigma(\epsilon), \qquad h \ll \epsilon \tag{7.111}$$

is a nonlinear correction to the magnetoconductivity and the magnetic field

$$h = \frac{B_z}{B_{c2}(0)} \tag{7.112}$$

is oriented perpendicular to the *ab*-plane, where

$$B_{c2}(0) = -T_\mathrm{c} \left. \frac{dB_{c2}(T)}{dT} \right|_{T_\mathrm{c}} = \frac{\Phi_0}{2\pi \xi_{ab}^2(0)} \tag{7.113}$$

is the slope of the upper critical field and

$$\Phi_0 = \frac{2\pi\hbar}{|e^*|} \tag{7.114}$$

is the magnetic flux quantum.

7.9 Conductivity correction by detection of 3rd harmonics

As an alternative method for probing the fundamental constants of the BCS theory we suggest the systematic investigation of the third harmonic of the electric field generated by a harmonic current. Third-harmonic measurements are easier to perform than those of resistivity, and, moreover, the effect arising from fluctuations is exceptionally pronounced.

In general AC current response and investigation of higher harmonics is a standard method for investigation of nonlinear effects on superconductivity. A homogeneous electric field

$$E_x(t) = E_0 \cos(\omega t), \tag{7.115}$$

for example, creates a small nonlinear response for the first harmonic and a cubic field dependence of the 3rd harmonic of the current

$$\begin{aligned} j_x(t) &= \sigma E_x(t) - \mathcal{A}E_x^3(t) \\ &= j_{1f}\cos(\omega t) + j_{3f}\cos(3\omega t), \end{aligned} \tag{7.116}$$

where for the amplitudes of the harmonics we have

$$j_{1f}/E_0 = \sigma - \frac{3}{4}\mathcal{A}E_0^2, \qquad j_{3f} = -\frac{1}{4}\mathcal{A}E_0^3. \tag{7.117}$$

If necessary, a smooth analytical normal part of the nonlinear coefficient $\mathcal{A}$ can be subtracted from the experimental data $\mathcal{A}_{\mathrm{exp}}$

$$\mathcal{A}_{\mathrm{exp}} = \mathcal{A}_N(T) + \mathcal{A}(\epsilon), \qquad \mathcal{A}_N(T) = A + BT + CT^2, \tag{7.118}$$

in order to extract pure fluctuation behaviour from the nonlinear coefficient of the conductivity correction

$$\Delta\sigma_f = -\mathcal{A}(\epsilon)E_x^2. \tag{7.119}$$

If we use the so defined conductivity correction $\Delta\sigma_f$ the electric field E_x is actually cancelled in Eq. (7.108) and we have to use the coefficient $\mathcal{A}(\epsilon)$ in the expression for τ_0, i.e.,

$$\tau_0 = \sqrt{\frac{3}{2}}\frac{\hbar}{|e|\xi_a(0)}\left\{\frac{1}{\sigma(\epsilon)}\int_\epsilon^\infty (\tilde{\epsilon}-\epsilon)^2\,\mathcal{A}(\tilde{\epsilon})d\tilde{\epsilon}\right\}^{1/2}. \tag{7.120}$$

There is no doubt that the electric field is a useful tool for a theoretical analysis but for the experimental realization of the suggested method we have to apply a harmonic current and to measure the harmonics of the voltage

$$I(t) = I_0\cos\omega t, \qquad U(t) = U_{1f}\cos\omega t + U_{3f}\cos 3\omega t + U_{5f}\cos 5\omega t + \dots. \tag{7.121}$$

For small current amplitudes used to avoid heating of the sample the voltage response is in first approximation linear and we have Ohm's law for the resistance of the superconductor strip with length L, width w and thickness d_{film},

$$\begin{aligned} U_{1f} &= R(T)I_0, \qquad R(T) = \rho(T)\frac{L}{wd_x\mathrm{film}}, \\ \rho(T) &= \frac{1}{\sigma_{\mathrm{exp}}(T)}, \qquad E_0 = \frac{U_{1f}}{L} = \frac{\rho(T)I_0}{wd_{\mathrm{film}}}. \end{aligned} \tag{7.122}$$

Then the absence of the 3rd harmonic of the current, $j_{3f} = 0$, according to Eq. (7.116) with $E_x(t) = U(t)/L$ gives

$$\frac{U_{3f}}{\rho(T)L} \approx \frac{1}{4}\mathcal{A}(\epsilon)\left(\frac{U_{1f}}{L}\right)^3 \tag{7.123}$$

to lowest order in I_0, and finally we obtain

$$\mathcal{A}(\epsilon) \approx 4\frac{L^2}{\rho(T)}\frac{U_{3f}}{(U_{1f})^3} = 4\frac{L^3}{wd_{\mathrm{film}}}\frac{I_0 U_{3f}}{(U_{1f})^4}. \tag{7.124}$$

In this way the nonlinear coefficient necessary for the determination of the lifetime constant in Eq. (7.120) can be expressed through the electronically measured current I_0, voltage amplitudes U_{3f} and U_{1f}, and the geometrical parameters of the strip L, w and d_{film}. So the suggested experiment can be performed in every laboratory involved in investigations of superconductivity.

Let us describe qualitatively the temperature dependence of the intensity of the 3rd harmonic when the temperature is increased. In the superconducting state the voltage response is negligible and the $3f$ signal will appear abruptly when we reach the critical temperature. After a sharp maximum at T_c the $3f$ signal will decrease with smaller slope and in the normal region the $3f$ signal will be small again and created only by the $2f$ oscillations of the temperature and the temperature dependence of the resistivity. In short we predict that U_{3f} will have a λ-shaped asymmetric critical singularity. The location of this λ-point provides a new method for the determination of the critical temperature of superconductors based on the properties of fluctuation phenomena. Our self-consistent theoretical calculation is applicable above T_c where U_{3f} is much smaller than the value at the λ-point, but still clearly detectable experimentally.

7.10 Discussion

Let us start analyzing the results derived with the momentum distribution Eqs. (7.26) and (7.28). We have a characteristic velocity related to the equilibrium distribution

$$v_c(\epsilon) = \frac{\hbar}{m_a^* \xi_a(\epsilon)} = v_c(0)\sqrt{\epsilon}, \qquad m_a^* v_c(0) = \frac{\hbar}{\xi_a(0)} \tag{7.125}$$

and Eq. (7.28) can now be rewritten as

$$n(v_x, \varepsilon_\perp) = \frac{n_T}{\left[v_x/v_c(\epsilon)\right]^2 + \epsilon + \varepsilon_\perp/a_0}. \tag{7.126}$$

According to the Drude consideration, a small electric field creates a drift velocity

$$v_{\text{drift}}(\epsilon) = \frac{e^* E_x \tau(\epsilon)}{m_a^*} = \frac{e^* E_x \tau_0}{m_a^* \epsilon} \tag{7.127}$$

and the dimensionless electric field is just the ratio of those two velocities,

$$f_\epsilon = \frac{|v_{\text{drift}}(\epsilon)|}{v_c(\epsilon)}, \qquad v_c(\epsilon) \simeq \sqrt{\epsilon}\, v_{\text{Fermi}} \exp\left(-\frac{1}{\rho_{\text{Fermi}} V_{\text{pairing}}}\right). \tag{7.128}$$

From a microscopic point of view the characteristic thermal velocity $v_c(\epsilon)$ is proportional to the Fermi velocity times the small parameter of the BCS theory, the famous exponent which contains the density of states at the Fermi level and the matrix element of the pairing interaction. This order of magnitude estimation is applicable to anisotropic gaps as well. In addition we have a critical slowing down multiplier $\sqrt{\epsilon}$. Those two factors significantly decrease the characteristic velocity and make possible the experimental observation of the electric-field correction to the fluctuation conductivity proportional to f_ϵ^2. For normal metals the ratio $v_{\text{drift}}^2/v_{\text{Fermi}}^2 \ll 1$ is extremely small and only AC oscillations of the temperature $T(t)$, mainly $2f$, can create harmonics in the voltage response. This effect should also be carefully taken into account for cuprate films for which the thermal resistance between the substrate and the film can be very high; this will be the subject of another work.

Let us also consider in short the $\omega\tau$-quasiparticle criterion to check whether or not Cooper pairs are quasiparticles in the usual sense of condensed matter physics, i.e., $\tau_p\left(\epsilon_p - \mu\right)/\hbar \gg 1$. For $\mathbf{p} = 0$ taking the microscopic value for the lifetime we have

$$\tau(\epsilon)\frac{a(\epsilon)}{\hbar} = \tau_0^{(\text{BCS})}\frac{a_0}{\hbar} = \frac{\pi}{16}\frac{a_0}{k_{\text{B}} T_{\text{c}}} = \frac{\pi}{16}\frac{1}{n_T} \ll 1. \tag{7.129}$$

This strong inequality means that fluctuation Cooper pairs are not quasiparticles. The notion of Cooper pairs is only a language to describe the properties of slowly decaying diffusion modes of the superconducting order parameter above T_{c}. Analogously the "mean free path" $l(\epsilon) = v(\epsilon)\tau(\epsilon)$ is also very short relative to the correlation radius:

$$\frac{l(\epsilon)}{\xi(\epsilon)} = \frac{v(\epsilon)\tau(\epsilon)}{\xi(\epsilon)} = \frac{\hbar}{m^*\xi(\epsilon)}\frac{\tau(\epsilon)}{\xi(\epsilon)} = 2\frac{a_0\tau_0}{\hbar} = \frac{\pi}{8}\frac{1}{n_T} \ll 1. \tag{7.130}$$

As an illustration let us take a set of parameters corresponding to a high-T_{c} cuprate: $T_{\text{c}} = 90$ K, $k_{\text{B}} T_{\text{c}} = 7.76$ meV, $m_{ab}^* = 11 m_e$, where m_e is the mass of a free electron, $\xi_{ab}(0) = 11$ Å. Then $a_0 = \hbar^2/2m_{ab}^*\xi_{ab}^2(0) = 2.86$ meV and

$n_T = k_B T_c / a_0 = 2.71 > 1$. The last inequality ensures that $n(\epsilon) = n_T/\epsilon \gg 1$ and this condition of applicability of Rayleigh-Jeans statistics justifies the treatment of the Ginzburg–Landau Ψ-function as a classical complex field.

In view of the new effects which can be predicted using the derived velocity distribution we consider the possibility of supercooling of the normal phase, cf. Ref. [482], to be very interesting. In this case the fluctuation conductivity could create a negative differential conductivity, which opens perspectives for many technical applications. In order to prevent the nucleation of superconductivity from regions where the current densities and electric fields are very small, depairing impurities should be introduced in the contact area of the microbridge. This could be realized, for example, by evaporation of Ni on the wide area of the Al microbridge or by Mn ions in cuprate films. In both cases the central narrow region of the microbridge should be protected. In a sample prepared under these conditions the criterion of negative differential conductance can be easily satisfied and so the predicted generation of oscillations would probably be the best example of significant fluctuation effects in superconductors.

Whether the threshold for the generation regime can be described within the nonlinear theory or whether we need to calculate the fluctuation density and the renormalized temperature $T_r = T_c(\epsilon_r + 1)$ in a self-consistent way, depends on the numerical value of the Ginzburg number. In any case the analyzed solution of the Boltzmann equation suffices to predict a cross-over from positive to negative differential conductivity, as the field decreases. As a precursor of oscillations, when the electric field is decreased, due to strong nonohmic behaviour and low dissipation, the sample will be an excellent frequency mixer. The incipient bulk conductivity should always be taken into account, because the supercooled normal state is metastable and applying a voltage to the superconducting state leads as a rule to a space- and time-inhomogeneous phase.

Some words should be added concerning the history of the kinetic equation introduced by Boltzmann. This was the first use of probability concepts in a dynamical theory but the real recognition of the Boltzmann equation was stimulated by the electronic industry in the second half of the 20th century. Indeed, the Boltzmann equation is an essential tool for understanding how electronic devices work.

In summary, in this chapter we have used the Boltzmann equation not only as a didactical instrument but also as a means of deriving new results. The Boltzmann equation for fluctuation Cooper pairs is a consequence of the time-dependent Ginzburg–Landau equation. This equation has been

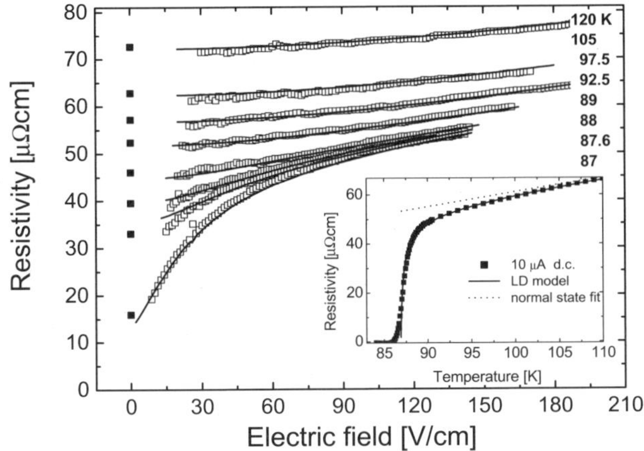

Fig. 7.1 Electric field dependence of the resistivity ρ of a $YBa_2Cu_3O_{7-x}$ film (□: experimental data) for different temperatures (reprinted with permission from Ref. [511]; Copyright © 2004 by IOP Publishing). The more pronounced electric field dependence for lower temperatures is a fluctuations effect. At higher T and electric fields, the normal-phase behavior is nearly Ohmic. The inset shows the temperature dependence of ρ at vanishing current; the dotted line gives the extrapolated ρ of the normal phase. Solid lines represent the self-consistent TDGL fit for the LD model.

derived from microscopic theory and in this sense the Boltzmann equation is a tool for the application of the microscopic theory of superconductivity. That is why the Boltzmann equation can be used to predict the results of new experiments, to help in their interpretation, and even to correct some previously obtained results derived in the frame-work of microscopic theory.

In a detail study by Puica and Lang [511], the self-consistent treatment of the fluctuation conductivity was applied to fit the experimental data for $YBa_2Cu_3O_{7-x}$, Fig. 7.1. The excellent agreement between theory and experiment, observed in this figure, is an evidence for the quantitative applicability of the self-consistent treatment of the TDGL theory and the Boltzmann equation to high-T_c superconductors.

Chapter 8

Linear-T electrical resistivity and normal phase properties

8.1 Introduction

The linear temperature dependence of the electrical resistivity $\rho_{ab} \propto T$ [512–514] is one of the most important properties of the normal-phase kinetics of high-T_c layered cuprates [422, 515–517]. However, despite the intensive investigations over a period of more than ten years and the numerous theoretical models proposed [518–523] this simple law does not yet have a unique explanation. In the physics of conventional metals it is well established that Pt, for instance, has a linear resistivity but the underlying physics is completely different. A parallel between similar behaviour for Cu and the data for $La_{1.825}Sr_{0.175}CuO_4$ and $YBa_2Cu_3O_7$ is available as well [512]. The linear temperature dependence of the resistivity is one of the most discussed problems in the physics of high-T_c superconductors [396].

The aim of this chapter is to present a simple model estimation explaining the $\rho_{ab} \propto T$ behaviour. Our model is based on the strong anisotropy of the electrical resistivity in layered cuprates. In the c-direction, perpendicular to the conducting CuO_2 planes (ab-planes), the electrical resistivity ρ_c is significantly lower than the in-plane one ρ_{ab}. For the c-polarised electric fields also the plasma frequency ω_c can be lower than the critical temperature

$$\hbar\omega_c < k_{\mathrm{B}} T_{\mathrm{c}}. \tag{8.1}$$

Plasma oscillations exist only in the superconducting phase. Plasmons in superconductors are observable only in the superconducting phase while in the normal phase they are overdamped. Therefore the criterion for applicability of the model is the lack of coherent transport in c-direction and almost frequency independent electromagnetic response for $\hbar\omega < k_{\mathrm{B}}T$ (in Ref. [524], for instance, the author finds that coherence is not very important for the normal resistivity; for the applicability of our model it would suffice that only the mean free path in c-direction be not significantly larger than the lattice constant c_0 in the same direction). In this case the electric field between the CuO_2 planes can be considered as a ilassical field and Boltzmann's distribution for the energy of the plane capacitors formed by these planes is applicable. Such a picture is probably most appropriate for $YBa_2Cu_3O_{7-\delta}$ and $Bi_2Ca_2SrCu_2O_8$ which contain double CuO_2 layers spaced by distance d_0 almost corresponding to the diameter of the oxygen ions. Every plaquete within a double CuO_2 plane is considered here as an independent plane capacitor with capacitance $C = \varepsilon_0 a_0^2/d_0$, where $1/4\pi\varepsilon_0 \approx 9 \times 10^9\ \mathrm{J\,m/C^2}$, ε_0 being the dielectric permeability of vacuum and a_0—the lattice parameter of the CuO_2 plane; the distance between the copper and oxygen ions is thus given by $\frac{1}{2}a_0$.

8.2 Qualitative picture

The capacitors defined in Sec. 8.1 are the main ingredient of the proposed mechanism for creation of resistivity. According to the equipartition theorem their average energy can be written as $\frac{1}{2C}\langle Q^2\rangle = \frac{1}{2}k_{\mathrm{B}}T$, which gives for the averaged square of the electric charge

$$\langle Q^2\rangle = \varepsilon_0 \frac{a_0^2}{d_0} k_{\mathrm{B}} T. \tag{8.2}$$

Consider now the scattering of one nearly free charge carrier (e.g., a hole moving in a CuO_2 plane) by a localized charge Q. For definiteness the hole is assumed to move in the x-direction and pass by the charge Q at minimum distance r to it. The trajectory of the hole is approximated by a straight line and its velocity is nearly constant and corresponds in our model estimation to the Fermi velocity v_{F}. The time needed for the charge carrier to pass by the scatterer (the flypast-time) we evaluate as $\tau_Q \simeq 2r/v_{\mathrm{F}}$ [525]. The maximal Coulomb force acting perpendicular to the trajectory is $F_\perp = eQ/4\pi\varepsilon_0 r^2$. Hence for the perpendicular momentum gained by the scattered hole one has $\Delta p_\perp \simeq \tau_Q F_\perp$. The latter quantity is much smaller than the Fermi momentum $p_{\mathrm{F}} = mv_{\mathrm{F}}$, and for small scattering angles $\theta_r \ll 1$ one has

$$\theta_r \simeq \frac{\Delta p_\perp}{p_{\mathrm{F}}} = \frac{A}{r}, \tag{8.3}$$

where $A \equiv eQ/4\pi\varepsilon_0 E_{\mathrm{F}}$, $E_{\mathrm{F}} = \frac{1}{2}mv_{\mathrm{F}}^2$ is the Fermi energy, $m = m_{\mathrm{eff}}m_0$ is the effective mass in the CuO_2 plane, m_{eff} is the dimensionless one, and

m_0 is the free electron mass. We note that the Rutherford scattering is the same both in classical and quantum mechanics and the Coulomb logarithms in the theory of plasma exceed the accuracy of such model estimations. In principle the classical and quantum results might slightly differ upon taking into account prefactors containing Coulomb logarithms, however, the difference would be much smaller than the uncertainty introduced by the lack of knowledge of material and/or model parameters. Let us stress that any mechanism for electrical resistivity must incorporate in an essential way some mechanism for transmission of the electron quasimomentum to the lattice. The capacitor model does this implicitly. The thermally excited charges in the capacitors play the role of the defects in the metal. Strictly speaking, the model considered is not purely electronic. It contains implicitly some weak inelastic electron-phonon interaction. In spite of its large relaxation time the latter ensures the equipartition theorem for the thermal energy of the independent capacitors.

The picture outlined above can be easily generalized to account for the influence of all scattering plaquetes along the $x = 0$ line. Accordingly, the charge carrier travels at distance $r = \pm a_0, \pm 2a_0, \pm 3a_0, \ldots$. Since the charges of the capacitors are independent random variables the average square of the scattering angle is an additive quantity [312]

$$\langle\theta^2\rangle_{\mathrm{line}} = \sum_r \langle\theta_r^2\rangle = 2\left(\frac{A}{a_0}\right)^2\left(1+\frac{1}{2^2}+\frac{1}{3^2}+\ldots\right). \tag{8.4}$$

The field outside the plane capacitors is essentially a dipole field, however, we will not discuss such details because the corresponding correction gives a factor of order of one: $\zeta(2) = 1 + 1/2^2 + 1/3^2 + \cdots = \pi^2/6 \simeq 1$.

Now let us apply the discrete lattice model in order to address the diffusion of the charge carrier momentum on the Fermi surface. The mean free path l is the distance after which the charge carrier "forgets" the direction of its earlier motion and scatters by $90^\circ = \pi/2$ having travelled a distance equal to l/a_0 lattice constants, i.e.,

$$\langle\theta^2\rangle_{l/a_0} = \frac{l}{a_0}\langle\theta^2\rangle_{\mathrm{line}} = \left(\frac{\pi}{2}\right)^2. \tag{8.5}$$

Consequently, for the mean free path[1] we get finally

$$l = 3\pi\frac{4\pi\varepsilon_0}{e^2}\frac{E_{\mathrm{F}}^2 d_0}{k_{\mathrm{B}}T}a_0. \tag{8.6}$$

[1] This notion was introduced in physics in 1858 by R. Clausius, 14 years before the Boltzmann equation.

Inserting the transport lifetime of the carriers defined for metals as $\tau_{\rm tr} = l/v_{\rm F}$ in Drude's formula for the conductivity

$$\sigma = \frac{ne^2\tau_{\rm tr}}{m} = \frac{1}{\rho}, \tag{8.7}$$

where n is the number of charge carriers per unit volume and e is the electron charge, one recovers the linear temperature dependence of the resistivity (distinguishing feature for the classical statistics [526])

$$\rho(T) = \frac{p_{\rm F}}{ne^2 l} = \frac{1}{4\pi\varepsilon_0}\frac{p_{\rm F}}{3\pi n d_0 a_0 E_{\rm F}^2}k_{_{\rm B}}T = \frac{k_{_{\rm B}}T}{3\pi^2\varepsilon_0 n d_0 a_0 m v_{\rm F}^3}. \tag{8.8}$$

It is remarkable that the squared electron charge e^2 is cancelled and $\hbar$ does not appear explicitly as well. The electrical resistivity is by definition a property of the normal state whereas the cuprates have attracted attention because of their high $T_{\rm c}$. That is why we consider it useful to perform a comparison with the experiment employing parameters of the superconducting phase. For clean superconductors when the mean free path $l(T_{\rm c})$ is much larger than the Ginzburg-Landau coherent length $\xi_{ab}(0)$ we can evaluate the effective mass $m_{\rm eff}$ as half of the effective mass of the Cooper pairs. For thin cuprates films, $d_{\rm film} \ll \lambda_{ab}(0)$, the effective mass of the Cooper pairs is determined by the electrostatic modulation of the kinetic inductance [6, 369], but in principle it is also accessible from the Bernoulli effect [358], the Doppler effect for plasmons [504], magnetoplasma resonances [502], or the surface Hall effect [359]. Further, the electron density can be extracted from the extrapolated to zero temperature in-plane penetration depth

$$\frac{1}{\lambda_{ab}^2(0)} = \frac{\mu_0 n e^2}{m}, \qquad \mu_0 = 4\pi\times 10^{-7}, \qquad \epsilon_0 = 1/\mu_0 c^2, \qquad c = 299792458\ \mathrm{m/s}. \tag{8.9}$$

The Fermi momentum $p_{\rm F}$ can be determined on the basis of the model of a two-dimensional (2D) electron gas which for bilayered cuprates gives

$$n = \frac{2}{c_0}n^{(2D)}, \qquad n^{(2D)} = \frac{p_{\rm F}^2}{2\pi\hbar^2}. \tag{8.10}$$

So after some elementary algebra the Eq. (8.8) takes the form

$$\frac{\sqrt{m_{\rm eff}}}{\lambda_{ab}^5(0)}\frac{d\rho}{dT} \approx C_{\rho\lambda} \equiv \frac{8}{3\pi^{1/2}}\frac{k_{_{\rm B}}e^5}{m_0^{1/2}(2\pi\hbar)^3}\frac{\mu_0^{5/2}}{\epsilon_0}\frac{1}{a_0 d_0 c_0^{3/2}} = \mathrm{const.} \tag{8.11}$$

We expect a weak doping dependence of the left-hand side of the above equation, while

$$T_{\rm c} \propto E_{\rm F} \propto n^{(2D)} \propto 1/\lambda_{ab}^2(0), \quad \text{for } n \ll n_{\rm opt} \tag{8.12}$$

can vary significantly upon going from underdoped to optimally doped regime $n_{\rm opt}$. In the overdoped regime $n > n_{\rm opt}$ the resistivity often displays non-linear temperature dependence. Simultaneously, if the doping dependence of the effective mass is negligible for $n \ll n_{\rm opt}$, i.e., $m_{\rm eff} \approx$ const, the model predicts

$$\frac{d\rho}{dT} \propto \lambda_{ab}^5(0) \propto \frac{1}{n^{5/2}} \propto \frac{1}{T_{\rm c}^{2.5}}, \quad \text{for } n \ll n_{\rm opt}. \tag{8.13}$$

Let us note also that for a single-plane material ($d_0 \equiv c_0$) only the 2D density $n^{(2D)} = nc_0$, i.e., the number of electrons per unit area, is relevant for the bulk 3D resistivity. The cancellation of the lattice constant c_0 can by easily understood inspecting the expression for the bulk conductivity of a system with equidistant conducting planes $\sigma = c_0^{-1} n^{(2D)} \tau_{\rm tr}(c_0)/m$. According to the plane capacitor model, cf. Eq. (8.2), the scattering rate is proportional to c_0^{-1} and, according to Eq. (8.6), the transport lifetime $\tau_{\rm tr} \propto c_0$. As a result, for single-plane materials σ does not depend on the interplane distance c_0, assuming $n^{(2D)} =$ const.

8.3 Quantitative estimate

Let us provide now an estimate for $v_{\rm F}$ and l based on the proposed model for the set of parameters $d\rho/dT = 0.5\ \mu\Omega$cm/K [422], $n^{(2D)} = \frac{1}{2}a_0^{-2} = 3.37 \times 10^{14}$ cm^{-2}, $n = 1/(a_0^2 c_0) = 5.72 \times 10^{21}$ cm^{-3}, $2\pi\hbar/p_{\rm F} = 3.54a_0$, $a_0 = 3.85$ Å, $c_0 = 11.8$ Å, $d_0 = 3.18$ Å, $m = m_{\rm eff} m_0$, $m_{\rm eff} = 3$, cf. Ref. [6], and $m_0 = 9.11 \times 10^{-31}$ kg. Substituting these parameters into Eq. (8.8) and Eq. (8.6) we get an acceptable value for the Fermi velocity (cf. Table 3 of Ref. [396], where 31, 140, 200 and 220 km/s estimates are cited)

$$v_{\rm F} = \left(\frac{k_{\rm B}}{3\pi^2 m} \frac{a_0 c_0}{d_0 \varepsilon_0} \frac{dT}{d\rho} \right)^{1/3} = 1.76 \times 10^5 \text{ m s}^{-1} = 176 \text{ km/s}, \tag{8.14}$$

and for the mean free path, respectively

$$l(T = 300 \text{ K}) = 5.67a_0 = 22 \text{ Å}, \tag{8.15}$$

$$\pi \left(\frac{l}{a_0} \right)^2 \approx 101.$$

For $T_{\rm c} = 90$ K and $\xi_{ab}(0) = 12$ Å we have $\xi_{ab}(0)/l(T_{\rm c}) = 16\%$, $\lambda_{ab}(0) = 122$ nm, $\lambda_{ab}(0)/\xi_{ab}(0) = 101$, $8/(3\sqrt{\pi}) \approx 1.50$, this numerical prefactor slightly changes should we apply more sophisticated approach to treat electric field fluctuations and the electron scattering by the random electric potential. Finally, for the $C_{\rho\lambda}$ constant, introduced in Eq. (8.11), describing the ρ-λ correlations, we get $C_{\rho\lambda} = 320\,\Omega\mathrm{K}^{-1}(\mu\mathrm{m})^{-4}$. These numerical estimates lead us to conclude that the suggested model does not contradict the experimental data, cf. Refs. [396,512]. Furthermore, we consider that a detailed state-of-the-art derivation of the charge density fluctuation of the plasma in layered cuprates could be an adequate quantitative model for the theory of their electrical resistivity. Along the same line, a systematic study of the ρ-λ correlations would provide an efficient tool to analyse the scattering mechanisms in layered perovskites.

The mechanism of the electrical resistivity is qualitatively fairly simple and we present it schematically in Fig. 8.1. The conducting CuO_2 planes constitute plates of plane capacitors and one has to take into account the Boltzmann (or Rayleigh-Jeans) statistics of the electrostatic energy of the capacitors. The last criterion for applicability of the model is the significant low-frequency reflection coefficient for an electromagnetic wave from a single CuO_2 plane. It exists only for high two-dimensional conductivity $\sigma c_0 > \varepsilon_0 c_{\mathrm{light}}$ [513], where $\rho_{ab}(4/c_0) = 300\ \Omega$ sheet resistance is evaluated just above T_c for $Bi_2Ca_2SrCu_2O_8$. While moving in the conducting CuO_2 planes the charge carriers are scattered by charge density fluctuations in the same planes. In fact, it is a self-consistent problem for collisionless plasma. The electrical resistivity appears upon taking into account the charge density fluctuations. This scattering mechanism is analogous to the Rayleigh's blue-sky law [527, 528] where the light is scattered by fluctuations of the air density. Only the "electron" sky is rather red. In the case of Coulomb scattering, the faster charges of smaller wavelength are scattered less intensively while for light the effect is opposite.

8.4 Discussion

The numerical example presented in Sec. 8.3 demonstrates that scattering by charge density fluctuations can explain the total resistivity of layered cuprates or at least constitutes a significant part of it. The plane capacitor is an important structural detail of the scenario. Now we want to address two interesting issues. (i) Whether the existence of high-T_c structures having linear resistivity is possible without the plane capacitor detail and vice versa? (ii) Why other layered structures do not display linear resistivity? The answers have qualitative character:

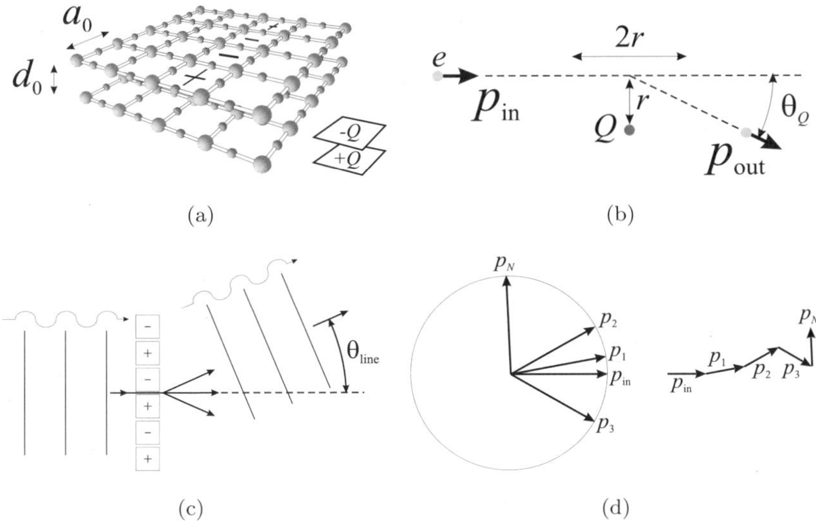

Fig. 8.1 The plane capacitor scenario for the resistivity of layered cuprates in a nutshell: (a) Two conducting CuO_2 planes (spaced by distance d_0) of a layered perovskite are considered as an array of independent plain capacitors; $\frac{1}{2}a_0$ is the Cu-O ion distance. In every capacitor we have thermally fluctuated charge Q for the avaraged square of which the equipartition theorem gives $\langle Q^2 \rangle/2C = \frac{1}{2}k_{\rm B}T$, where C is the capacitance; (b) In the model evaluation the charges are considered as point ones. A charge carrier (a hole with charge e) with momentum $p_{\rm in}$ passes near the scattering centre Q at distance r and deviates at a small angle $\theta_Q(r)$, with $p_{\rm out}$ being its momentum after the scattering event; (c) The electron waves are further scattered by a line of plane capacitors. The capacitor charges are independent random variables, thus the averaged square of the scattering angle $\theta_{\rm line}$ is an additive quantity. The Rutherford cross section is the same in classical and quantum mechanics; (d) Having crossed $N = l/a_0$ plaquete lines, the charge carrier "forgets" the direction of its initial momentum $p_{\rm in}$. This is employed to evaluate the mean free path l. The diffusion of momentum on the Fermi surface is presented also as a trajectory containing segments in the real space $\{p_{\rm in}, p_1, \ldots, p_N\}$.

(i) High-$T_{\rm c}$ superconductivity can exist even in artificial structures having a CuO_2 monolayer and the resistivity is yet linear. In this case the plane capacitors are missing; however, one has to take into account the electric field fluctuations close to the two-dimensional (2D) CuO_2 layer. The 2D plasmons are gapless and overdamped above $T_{\rm c}$. Therefore we have to calculate 2D charge density fluctuations and the corresponding thermodynamic fluctuations of the electric field. Due to the equipartition theorem the linear resistivity is recovered again, with only the prefactor being different.

(ii) There are various artificial metal-insulator structures where the temperature dependence of the in-plane conductivity is specific and non universal. When a metallic layer contains even a few monolayers we should take into account that the electric field does not penetrate through the metal. Thus due to trivial electrostatic reasons the thermodynamic fluctuations of the electric field in the insulator layers would be an inefficient mechanism for charge scattering in thick metallic layers. Metallic monolayers have residual resistivity related to defects, significant electron-phonon coupling etc. According to Mattissen's rule we could search for the charge density fluctuation part of the scattering, but it is unlikely that this mechanism dominates. The theory of fluctuations of the electromagnetic field between metallic layers (this is the geometry of the Casimir effect) is a typical problem in statistical physics, however, the difficult task will be the experimental separation of the linear term provided many other mechanisms contribute to the resistivity.

To summarise, we came to the conclusion that an important hint for the applicability of the suggested model will be the existence of linear resistivity in other layered structures containing no CuO_2 planes but having almost 2D charge carriers and of course good quality. If our simple electrostatic explanation is correct any layered material having good 2D metallic layers should display the same behaviour regardless of its electronic structure. In this case it is ensured that linear resistivity is not related to some specific subtle properties of electron band structure of the CuO_2 plane or some sophisticated non-Fermi-liquid-like strongly correlated electron processes due to Cu3d electrons, but rather to such a universal cause as the omnipresent fluctuations of the electromagnetic field and related to them charge density fluctuations—*who could be blind to the blue sky.*

The linear resistivity of the layered ruthenates [529–531] could be considered as such an example and crucial experiment. In Sr_2RuO_4 the temperature dependence of the Hall coefficient is similar to the one measured in cuprates and the striking linear dependence of the conductivity persists over the whole temperature range 1–1000 K [529–531], Fig. 8.2. It is impressive to observe any physical quantity exhibiting linear behaviour over three orders of magnitude change of the temperature although it is a simple consequence of the conventional transport theory of metallic solids. The authors of Ref. [529–531] note that they were unaware of any model that specifically predicts or can convincingly account for essentially linear behaviour of resistivity over three decades of temperatures although most theories of high-T_c superconductivity discuss linear resistivity. Nevertheless they suggest that linear resistivity is not an exclusive feature of the normal state of high-T_c cuprates, but rather of all layered oxides especially

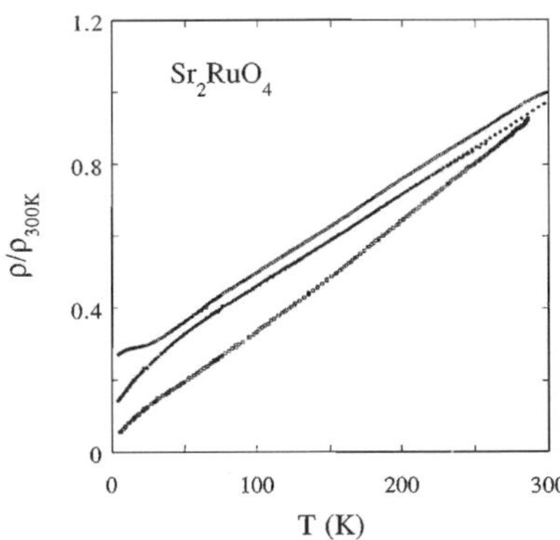

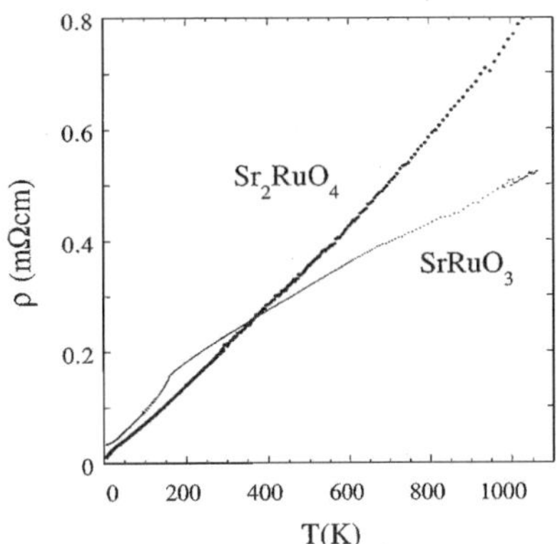

Fig. 8.2 (left) Temperature dependence of resistivity for three different single crystals of $Sr_2RuO_{4-\delta}$ that show linear slope of resistivity up to room temperature. (right) Resistivity of $Sr_2RuO_{4-\delta}$ and $SrRuO_3$ up to $\sim$ 1050 K (reprinted with permission from Ref. [529]; Copyright © 1998 by EDP Sciences).

perovskites, possibly even independently of the magnitude of T_c [529–531]. Indeed, Sr_2RuO_4 ($T_c < 1$ K) presents an unequivocal demonstration that its linear resistivity is not related to processes involved in the pairing mechanism. The linear resistivity is created by thermally activated electric fields while the pairing originates in the *s*-*d* exchange considered in Chapter 2.

Concluding, we believe that there is nothing mysterious [396] in the linear dependence of the in-plane resistivity of high-T_c layered cuprates—it is just a consequence of well-known physical laws dating back to the end of XIX and the beginning of XX century.

8.5 Outlook: relation between the normal state transport properties and the pairing mechanism

Let us address again the question of what is common between the CuO_2 and RuO_2 planes. Apparently, it is not the paring mechanism, but rather the electric field fluctuations between the conducting 2D layers. Linear T-dependence of the resistivity demonstrates the linear T-dependence of the in-plane electron density fluctuations.

We shall attempt now to qualitatively see how these density fluctuations drive the transport lifetime τ_{tr} anisotropy within the *s*-*d* pairing interaction, Sec. 2.2.2. In a self-consistent picture, the electron density fluctuations can be considered as quenched impurity scattering centers. In calculating the resistivity due to such centers, backscattering is especially important. This is accounted for by the $(1 - \cos\theta)$ factor in the expression for the transport cross-section

$$\sigma_{\rm tr} = \int (1 - \cos\theta)\sigma(\theta)\, 2\pi \sin\theta\, d\theta \propto \frac{1}{\tau_{\rm tr}}, \tag{8.16}$$

where $\sigma(\theta)$ is the impurity differential cross-section—a standard result in the theory of bulk normal metals [271, 272]. Let us now analyze the backscattering ($\theta = 180°$) anisotropy along the Fermi contour. A Coulomb scattering would be essentially isotropic due to small $v_{\rm F}$ variation. However, a backscattering resulting from the *s*-*d* exchange interaction, Eq. (2.9), would have an amplitude proportional to the hybridization factor $\chi_{\mathbf{p}} = S_{\mathbf{p}} D_{\mathbf{p}}$, Eq. (2.32), which is a consequence of the scattering matrix element between the initial and final states. But according to Eq. (2.33) the backscattering amplitude is proportional to the superconducting gap anisotropy function $\chi_{\mathbf{p}}$. Thus, the hypothesis that the same interaction is responsible for both Cooper pairing and Ohmic resistivity is qualitatively validated,

$$\tau_{\rm tr}^{-1}(T) \propto T(|\chi_{\mathbf{p}}|^2 + \text{const}), \qquad \Delta_{\mathbf{p}}(T) = \Xi(T)\chi_{\mathbf{p}}. \tag{8.17}$$

Charge carriers in vicinity of the gap nodes are characterized by longer transport lifetimes and, like stealth aircrafts, are invisible to the other carriers. In our view, the above qualitative picture explains the anisotropic scattering in $Tl_2Ba_2CuO_{6+\delta}$ [532]. Carriers with momenta corresponding to maximal gap in the superconducting state experience the strongest scattering. The scattering-gap correlation shows up for conventional superconductors as well but it is not related to the pairing mechanism: Pb is a bad conductor with high $T_{\rm c}$, whereas Al is a perfect conductor with very low $T_{\rm c}$. To make an analogy with the CuO_2 Fermi contour we can assign Al to the nodal directions, and the Pb to those with the maximal gap.

For the further development of the theory of superconductivity in general, similar investigations need to be carried out for Sr_2RuO_4. The history of cuprate superconductivity has set, at least, the right direction to be followed. We believe that the $\tau_{\rm tr}^{-1}(\mathbf{p}) \propto |\Delta_{\mathbf{p}}|^2$ correlation is common to all superconductors. The lessons learned from the cuprates tells us how the properties of conventional superconductors can be generalized to new materials, thus allowing many anomalies of the normal phase to get the traditional explanation. As already alluded to in Chapter 2, space inhomogeneities and stripes may require technically more elaborate theories. We believe, however, that the pseudogap problem can be reduced to some type of statistics of density waves.

We shall conclude this chapter with a remark on the mid-infrared maximum of the CuO_2-plane conductivity. The importance of this common for the cuprates feature was aptly pointed out in the monograph [81]. Within our traditional approach, it is just the transition between the conduction and the empty Cu4s bands. It is an additional argument in favor of the assumption that the Cu4s lies only slightly above the Fermi level.

Chapter 9

Terahertz electric oscillations in supercooled superconductors

9.1 Introduction

Due to their low ohmic dissipation, superconductors find a lot of technical applications. Superconductors can be used in resonators, cables, electromagnets, transformers, electrical engines and generators. As another important technical application we wish to mention the SQUIDs based on Josephson effect. However, up to now there is a little progress in the use of superconductors as active elements in electronic circuits such as amplifiers and generators integrated in superconducting electronics. The present work falls in this still uncompleted technical niche. We are suggesting how superconductors, and especially high-T_{c} superconductors, can be used as generators of high frequency electric oscillations including the THz range. When the superconductor is implemented as a thin layer the work of the generator can be influenced by the electrostatic charge modulation as in a gate transistor, by heat, light or simply by the change of the DC bias voltage. Such a way oscillations can be modulated, the generator can be used as an optoelectronic device or as a bolometer. The electric oscillations are generated by negative differential conductivity (NDC) of a superconductor in non equilibrium condition. The superconductor is supercooled in the normal phase below the critical temperature T_{c} under a constant electric field. The electric field prevents transition of the superconductor to the superconducting phase state. NDC is coming from so called fluctuation conductivity which is strongly expressed for cuprate superconductors having smaller coherence length.

9.2 Physical model

9.2.1 *Qualitative consideration and analogies*

First we will describe the idea how the negative differential conductivity (NDC) can be created in a superconductor. Let us consider the physical processes which happen in a supercooled in the normal state superconductor under voltage. The material is in a normal state and in his volume thermally activated Cooper pairs are continuously created. This stochastic process is analogous to the Brownian motion but is related to the wave function of the metastable Cooper pairs, the Ginzburg–Landau order parameter $\Psi(\mathbf{r}, t)$ which for the normal phase is a time and space-vector dependent stochastic function. When the temperature is below the critical one $T < T_c$ both the amplitude of the wave function and the number of fluctuation Cooper pairs increase with the time. Such wave amplification is analogous to the lasing process in lasers or to the dynamics of the Bose condensation. This is the precursor of the transition of the superconductor in the superconducting phase which has infinite conductivity; the electric current can flow without an external voltage in the superconducting phase. However, the applied DC electric field prevents fluctuation Cooper pairs to condense in a coherent superconducting phase; superconducting phase cannot exist under external voltage. Once born, the electric field accelerates the fluctuation Cooper pairs and their kinetic energy increases. However, the decay rate of the Cooper pairs is energy dependent and increases with the energy. Like on a highway big velocity increases the probability of accidents. Roughly speaking, the life of faster Cooper pairs is shorter and the electric field finally destroys the accelerated Cooper pairs. During their life metastable Cooper pairs carry significant electric current (fluctuation current) comparable and even bigger than the current of normal charge carriers electrons or holes (normal current); the total current is the sum of fluctuation and normal current.

We consider as very instructive to give also popular explanations using the analogy with ideas from two atmospheric phenomena which have no direct link in real meteorology: rains and meteors; as we stressed it will be only analogy. (1) Imagine a very humid atmosphere supercooled below the condensation temperature. Small droplets continuously grow in size and it begins raining. We have different droplets with different sizes and velocities that create the current density of the rain – a total amount of water passing through a square meter per one second. (2) Second phenomenon is related to meteorites falling from the cosmic space; moving with high velocities stones burn in the atmosphere due to the heating by the friction force. Imagine now that Earth acceleration is many orders of magnitude higher (for the real meteorology it is unrealistic) that makes big rain droplets moving with high velocity evaporate by heating from the friction force in bigger extent. This sets limits to the maximal size for falling droplets and the rain could be a continuous process with a current density determined by the acceleration force. The subtle question here is what will happen if acceleration will slightly decrease. The smaller acceleration will lead to a bigger size of the largest droplets and as a consequence the debit of the rain will increase. This is an example of how NDC appears – small decrease in the driving force cause increasing of the current. The analog of the droplets is the square of modulus of the Fourier components of the effective Ginzburg–Landau function $|\Psi_{\mathbf{k}}(t)|^2$. When a superconductor is cooled below the critical temperature $T_{\rm c}$ in the bulk of the superconductor starts a process similar to Bose condensation. Electrons start to condensate in Cooper pairs and many Cooper pairs can have a common momentum $\hbar\mathbf{k}$—this is the analog of a rain droplet having a velocity proportional to its momentum.

After this analogy and a qualitative description of the main processes we can analyze in detail the appearance of the NDC—the current increasing when the electric field decreases. Such a behavior is opposite to the ohmic conductivity.

When the electric field is smaller, the acceleration is slower, the decay rate of fluctuation Cooper pairs is slower, and as a result the volume density of the fluctuation Cooper pair is higher and the current is higher too. The acceleration, creating the decay of Cooper pairs, is an analog of the evaporation of droplets by the heating force from our above-mentioned example. So, the electric field accelerates and destroys Cooper pairs and the superconductor cannot move to superconducting state. Thus it is logically that the electric current density $j(E)$ increases when the electric field E decreases, i.e., the differential conductivity of the material is negative

$$\varsigma_{\mathrm{diff}}(E) = \frac{dj(E)}{dE}. \tag{9.1}$$

In such a non equilibrium situation supercooled superconductor will have NDC due to the electric field dependence of the density of metastable Cooper pairs.

Here, by way of illustration we will consider another analogy of the creation of electric oscillations by supercooled superconductors. In some sense, the active part of a superconductor with NDC operates as a laser. However, there is a need to pump the energy in the active medium in order to make lasers operating. Now, let us consider how the superconductor is lasing. Below T_c, the ground state of the material is the superconducting state having the lowest free energy and the normal state (nonsuperconducting state with an Ohmic resistivity) is the "excited state" having higher energy and generating electric oscillations. The energy for these oscillations comes from the constant electric field but in some sense the excited state is reached by cooling. For high-T_c superconductors the cooling can be done by liquid nitrogen that is substantially cheaper. The initial demonstrations of the effect of the NDC, however, can be realized at helium temperatures by technologically more convenient conventional superconductors.

Like the current, the total differential conductivity is a sum of its fluctuation and normal parts. The normal part of the differential conductivity $\varsigma_N(T)$ weakly depends on the electric field. In the next subsection we will describe the state-of-the-art theory of the differential conductivity of the fluctuation Cooper pairs.

9.2.2 *Formulas for the differential conductivity*

For a comprehensive contemporary review on the properties of superconductors see the book edited by Benneman and Ketterson which starts with a review article by Larkin and Varlamov on the fluctuation phenomena in superconductors [460]. There can be found detailed explanations of the Bardeen, Cooper and Schrieffer (BCS) theory of superconductivity and also of the time dependent Ginzburg-Landau (TDGL) theory for the order parameter of superconductors. Ref. [429] is another review especially devoted to the Gaussian fluctuation in superconductors. The Boltzmann equation for the fluctuation Cooper pairs was derived [473, 474] in the framework of TDGL theory. For the case of strong electric fields Boltzmann equation was solved in Ref. [481], and a general formula for the fluctuation current was derived as well; see also the references therein. The same result was rederived [533, 534] directly from the TDGL theory. Our formula for small electric fields below T_c is similar to the formula by Gor'kov which was, however, directly derived [482] within the framework of the BCS theory; see also the work by Tucker and Halperin [535]. Differentiating the formula for the fluctuation current [481, 533, 534]

$$j_{\mathrm{fl}}(E_x) = \frac{e^2 \tau_{\mathrm{rel}} E_x}{16\hbar \left[\pi^{1/2}\xi(0)\right]^{D-2}} \int_0^\infty \frac{\exp\left(-\epsilon u - g u^3\right)}{u^{(D-2)/2}}\, du, \qquad (9.2)$$

we obtain the formula for the total differential conductivity

$$\varsigma_{\mathrm{diff}}(E) = \varsigma_N(T) + \frac{e^2 \tau_{\mathrm{rel}}}{16\hbar \left[\pi^{1/2}\xi(0)\right]^{D-2}} \times \int_0^\infty \frac{\exp\left(-\epsilon u - g u^3\right)}{u^{(D-2)/2}} \left(1 - 2gu^3\right) du, \qquad (9.3)$$

where D is the dimension of the space, e is the electron charge, $\xi(0)$ is the Ginzburg–Landau (GL) coherence length of the superconductor, τ_{rel} is a dimensionless constant which describes how long the fluctuation Cooper pairs live in comparison with the prediction of the BCS theory, U is the voltage difference, L is the length of the sample, and

$$\epsilon = \frac{T - T_{\mathrm{c}}}{T_{\mathrm{c}}}, \quad g = \frac{f^2}{12}, \quad f = \frac{\pi}{8} \frac{eE\xi(0)}{k_{\mathrm{B}} T_{\mathrm{c}}} \tau_{\mathrm{rel}}, \quad E = \frac{U}{L}.$$

The analysis of Eq. (9.3) shows that below T_{c} where $\epsilon < 0$ the differential conductivity is really negative as it is necessary for the work of the suggested current oscillator. The GL theory is formally applicable only close to T_{c} for $|\epsilon| \ll 1$, but qualitatively its results can be used even far from T_{c}. In other words, the differential conductivity will remain negative even if the accuracy of the TDGL formula Eq. (9.3) is not very high. We have to mention that TDGL equation is derived from BCS theory as a result of some approximations and for some cases it could be only a convenient model equation. We also wish to point out that the dimension of the current density depends on the dimension of the space $[j_D] = \mathrm{A/m}^{D-1}$: for a bulk sample $[j_3] = \mathrm{A/m^2}$, for thin films with a thickness $d_{\mathrm{film}} \ll \xi(\epsilon)$ $[j_2] = \mathrm{A/m}$, and for a wire with a cross-section $\ll \xi^2(\epsilon)$ the current density is just the current $[j_1] = \mathrm{A}$. Here

$$\xi(\epsilon) = \frac{\xi(0))}{\sqrt{|\epsilon|}} \qquad (9.4)$$

is the temperature dependent coherence length. It is also convenient to introduce a temperature dependent Cooper pair life-time $\tau(\epsilon)$

$$\tau(\epsilon) = \frac{\tau(0)}{|\epsilon|}, \qquad \tau(0) = \frac{\pi}{16} \frac{\hbar}{k_{\mathrm{B}} T_{\mathrm{c}}} \tau_{\mathrm{rel}}, \qquad (9.5)$$

where the numerical coefficient $\pi/16$ is a result of the microscopic BCS theory. Analogously, it is convenient to introduce a dimensionless temperature dependent electric field

$$f_\epsilon = \frac{f}{|\epsilon|^{3/2}} = |e^* E|\xi(\epsilon)\tau(\epsilon)/\hbar \ll 1, \tag{9.6}$$

where $|e^*| = 2|e|$ is the charge of the Cooper pair. The dummy parameter of the integration in Eq. (9.3) has a physical meaning of a dimensionless time $u = t/\tau(0)$, and analogously, one can introduce another dimensionless time $v = t/\tau(\epsilon)$.

The present theory is applicable for every superconductor which is homogeneous enough in order to avoid nucleation of the superconducting phase. However, we consider as the most promising the cuprate high-T_c superconductors containing as a main structural detail superconducting CuO_2 planes, such as $YBa_2Cu_3O_{7-\delta}$ and $Bi_2Sr_2CaCu_2O_8$ superconductors which have $T_c \approx 90$ K and can be cooled by liquid nitrogen using working temperatures $T = 80$ K and reduced temperature $\epsilon \simeq -0.1$. The coherence lengths in CuO_2 plane are typical for other 90 K cuprates $\xi_{ab}(0) \simeq 2$ nm. All high-T_c cuprates have a significant anisotropy but $Bi_2Sr_2CaCu_2O_8$ is extremely anisotropic. Even for small reduced temperatures $|\epsilon| \simeq 0.1$ the coherence length perpendicular to the CuO_2 plane can be smaller than the distance s between double planes CuO_2. In this case, every double planes operate approximately as an independent two dimensional (2D) layer and the number of layers N_l depends on the film thickness $N_l = d_{\rm film}/s$. If the superconductor is a strip with a width w, patterned from a layered superconductor we have for the total current

$$I = \frac{w d_{\rm film}}{s} j_2(E). \tag{9.7}$$

Such a way for the differential conductance of the sample we obtain

$$\sigma_{\rm diff}(U) = \frac{di(U)}{dU} = \sigma_N + \frac{e^2 \tau_{\rm rel} w d_{\rm film}}{16\hbar s|\epsilon|} S(g_\epsilon), \tag{9.8}$$

where the universal function

$$S_{\rm diff}(g_\epsilon) = \int_0^\infty \left(1 - 2g_\epsilon v^3\right) \exp\left(\mathrm{sgn}(-\epsilon) - g_\epsilon v^3\right) dv \tag{9.9}$$

have to be calculated only once for the $\mathrm{sgn}(-\epsilon) = \pm 1$. The negative differential conductivity arises only for supercooled below T_c superconductors, and in this case $\mathrm{sgn}(-\epsilon) = 1$. In Eq. (9.9) the electric field is parameterized by the dimension parameter

$$g_\epsilon = \frac{g}{|\epsilon|^3} = \frac{1}{12} f_\epsilon^2 = \frac{1}{12|\epsilon|^3} \left(\frac{\pi e U \xi(0) \tau_{\rm rel}}{8 k_{\rm B} T_c L}\right)^2. \tag{9.10}$$

In order to have a significant negative differential conductivity this parameter should be small enough $g_\epsilon \ll 1$. This means that close to the critical region the applied DC voltage U should be small enough.

The fluctuations are stronger in low dimensional systems, that is why another realization of the negative differential conductivity in superconductors could be a nanostructured stripe of conventional superconductor on nanowires. Consequently, we have to use one dimensional (1D) formula for the current. Analogously, for layered superconductors we can use Lawrence-Doniach theory which can be interpolated by some space dimension $2 \leq D \leq 3$. It is necessary to use disordered conventional superconductors or cuprates in order to have smaller normal current.

9.3 Description of the oscillations

In order to illustrate how electric oscillations can be generated by a supercooled superconductor we will use the simplest possible electric scheme used in generators with tunnel diodes [536]. For a pedagogical explanation of this scheme and the Van der Pol equation see also the excellent textbook [537].

The superconductor is connected in parallel with one resistor having resistance R and with one capacitor having capacity C. Those 3 elements are sequentially connected in a circuit with one inductance L and a battery with electromotive force $\mathcal{E}$. For a static current the voltage on the superconductor, the capacitor and the resistor is just the voltage of the battery $U = \mathcal{E}$. In this static case, the voltage of the inductance is zero.

Let us now consider what will happen if the superconductor is supercooled. The whole circuit is cooled below T_c, but we consider that in the beginning superconductor is in the normal state. Imagine that it is heated by a short current or a laser impulse. We will analyze the fluctuations of the voltage of the superconductor taking into account the static solution $U(t) = \mathcal{E} + x(t)$. The deviation from the static solution $x \equiv U(t) - \mathcal{E}$ obeys the differential equation

$$C\frac{d^2}{dt^2}x + \left[\frac{1}{R} + \sigma_N + \sigma_{\mathrm{diff}}(\mathcal{E} + x)\right]\frac{d}{dt}x + \frac{1}{L}x = 0, \tag{9.11}$$

cf. Ref. [537]. After introducing the auxiliary variable

$$y(t) \equiv \frac{d}{dt}x(t) \tag{9.12}$$

the second equation of this system of ordinary differential equations reads as

$$\frac{d}{dt}y(t) = -\nu(v)y - \omega^2 x, \tag{9.13}$$

where

$$\nu(x) \equiv \frac{1}{C}\left[\frac{1}{R} + \sigma_N + \sigma_{\mathrm{diff}}(\mathcal{E} + x)\right], \quad \omega = \frac{1}{\sqrt{LC}}. \tag{9.14}$$

For a moderate accuracy necessary for the modelling of electronic circuits we can use some adaptive Runge–Kutta method [538], or a simple empirical formula for the time step which follows the characteristic frequencies of the circuit

$$\Delta t = 0.1/\sqrt{\nu^2(x) + 4\omega^2}. \tag{9.15}$$

The physical restrictions for the high frequencies are related only to applicability of the TDGL equation and the static formulae for the current response. The static response approximation used for the derivation of Eq. (9.3) is applicable for $\omega\tau(\epsilon) \ll 1$. This means that for a high-T_c superconductor this generator can operate for the whole radio frequency range and even in the far infrared region

$$\omega \ll |\epsilon| k_{\mathrm{B}} T_{\mathrm{c}}/\hbar \simeq 1/\tau(\epsilon). \tag{9.16}$$

Let us now describe the appearance of the oscillations when the superconductor is supercooled. The total differential conductivity of the circuit becomes zero at the temperature T_g, determined by solution of the equation

$$-\sigma_{\mathrm{diff}}(\mathcal{E}, T_g) = \frac{1}{R} + \sigma_N(T_g). \tag{9.17}$$

Further cooling leads to appearance of NDC, the static solution $x(t) = 0$ looses stability, and the voltage in the circuit starts to oscillate. The main difference with the tunnel diode devices is that for superconductors we have no definite region of negative conductivity as a function of the voltage. The current continuously increases when the voltage decreases down to zero voltage. It is quite possible that amplitude of the oscillations will be limited only by ohmic heating of the sample. For thin films the heat current is determined mainly by the boundary resistance $\mathcal{R}_h$ of the interface of the superconductor and the insulator substrate. For high frequencies we can average the dissipated power and calculate the local increasing of the temperature of the superconductor above the ambient temperature

$$\Delta T = \frac{\langle I(t)U(t)\rangle_t}{\mathcal{R}_h}. \tag{9.18}$$

Such a way we obtain a self-consistent correction for the reduced temperature

$$\epsilon \to \epsilon + \Delta T/T_{\mathrm{c}}, \tag{9.19}$$

which have to be substituted in the formula for the differential conductivity Eq. (9.8). The complex problem of the temperature and electric field oscillations can be easily simulated on a computer in order to optimize the parameters of the device and the initial stage of the experimental research.

One can also speculate what will happen if we start with a superconducting sample. Applied voltage will destroy the superconductivity but it is also possible that a space inhomogeneous state will appear. It is difficult to predict the behaviour of the system when having a problem related to the domain structure. That is why an experimental investigation is needed.

9.4 Performance of the generator

The most important prerequisite for the realization of NDC by supercooled superconductors is to keep the superconductor in the metastable normal state by preventing its active part from transition to the superconducting state which is thermodynamically stable below T_{c}. The sample has to be clean from defects, for instance pin holes, that can nucleate locally the superconductivity. Special efforts have to be applied to the contacts of the superconductor sample where the current density is low and these regions could be a source of nucleation of superconducting domains. We consider that only the central working region of the sample should be superconducting. The contact area should have inserted depairing defects. Such defects could be the magnetic impurities for conventional superconductors, Zn in the CuO_2 plane, etc. Oxygenation of $YBa_2Cu_3O_{7-\delta}$ superconductors or changing stoihiometry are also a tool to change T_{c} locally. Overdoped and underdoped cuprate thin films will have opposite behaviour with respect to the stability of appearance of space inhomogeneous domains of the superconducting and normal phases.

The amplitude of the oscillations can also be restricted by a current limiter with a maximal current $I_c(T)$, a narrow superconducting wire, Josephson junction, or narrow superconducting strip sequentially switched to the inductance of the resonance circuit. In a rough approximation the resistance of the limiter $R_{\mathrm{i}} = R_0\theta(I_c(T) - I)$ is switched when the current $I(t)$ becomes bigger than the critical $I_c(T)$. Such additional amplitude dependent dissipation will prevent the sample to pass into normal state when

$U(t) = \mathcal{E} + x(T) = 0$. For such a small modification of the circuit we have to solve the system of equations

$$\mathcal{E} = L\frac{dI}{dt} + R_0\,\theta(I_c(T) - I(t))\,I(t) + U(t), \tag{9.20}$$
$$I(t) = \left(\frac{1}{R} + \frac{1}{R_N}\right)U(t) + C\,\frac{dU(t)}{dt} + I_{\mathrm{fl}}[U(t)],$$

where, if necessary, the Boltzmann equation can be solved in the general case in order to obtain the high frequency functional for the fluctuation current $I_{\mathrm{fl}}[U(t)]$ and eventually the self-interaction between fluctuation Cooper pairs has to be taken into account.

In the paper by Gor'kov [482] were mentioned some early experiments for observation of oscillations in superconductors close to T_c. Those experiments give a hint that creation of oscillations in supercooled superconductor is also possible. But the problem requires detailed experimental investigation. Very often NDC leads to space inhomogeneities instead of time oscillations and we consider that both regimes can be realized. To avoid the appearance of space inhomogeneities one can also use narrow (nanostructured) region of supercooled superconductor which will be analogous to the tunnelling region of a tunnel diode. The length of the superconducting area should be comparable with the coherence length of the superconductor $\xi(\epsilon)$.

Fluctuation effects in conventional superconductors are very weak, with the exception of some very disordered films with a high Ohmic resistance. They cannot be effectively used in THz region due to their relatively smaller critical temperature T_c, the superconducting gap and the energy scale in general. However, systematic investigations of conventional superconducting nanostructures can be very important step for creation of electric oscillators with a superconductor as an active element. In experiments with gaseous plasma and semiconductors the NDC can be reached by various viable methods and it is not strange that the same could happen for superconductors. The investigation of current-voltage characteristics of superconducting nanowires [483] is promising. We consider that for superconducting nanowires could be taken into account the phase slip centers, the domains of different temperature, or even some possible modification of time-dependent Ginzburg-Landau theory might be done. A detailed physical picture can be drawn only after detailed investigations. We suggest conventional superconductors to be used to demonstrate that NDC can be reached and thus to stimulate further investigation of cuprate nanostructures. Successful experiments with high-T_c superconductors can trigger significant applications in the THz range as mentioned above. When the current oscillator is realized by a cuprate the resonance circuit (the capacitor, the induction coil and the current limiter) can be patterned in a single chip from the same submicron cuprate layer. We can generate electric oscillations and NDC to be well inside the THz region only by nanostructured cuprates because the maximal frequency at which NDC can exist is proportional to the maximal value of the superconducting gap and critical temperature T_c.

9.5 Possible applications

The described principle of generation of current oscillations can be realized in superconducting electronics. In general, the negative differential conductivity (NDC) could be a useful tool in the electronics using superconductors as an active element.

We consider that the generation of submillimeter electromagnetic waves by high-T_c superconductors is quite possible. Thin superconducting layers can be electrostatically doped and in this context some preliminary research on superconducting field effect transistors can be extremely helpful for modulation of the oscillations. There are good prospectives for application in the wireless communications. We consider also that in the regime of supercooling some old samples could become working transistors.

Fluctuations are more important for the low dimensional systems. The fact that NDC can be observed in conventional nanostructured superconductors is promising. In this case the 1D theory can be directly used or one can easily perform summation on the perpendicular modes of the Cooper pair wave guide. Shortly said, the combination of nanostrip, capacitor and inductance patterned into one chip or the plasma modes of the superconducting stripe, could be considered as the smallest current oscillator performed for demonstration purposes.

The device constructed upon our instructions will be very sensitive to the temperature and might be stabilized to oscillate near the critical temperature T_c. It should not only prove our theory but could be used as well as a superconducting bolometer. The sensitivity to temperature variations around T_c opens opportunities for optoelectronic applications. The negative differential conductivity can be investigated in the case of zero temperatures $T \ll T_c$ and the critical behaviour in small electric field is appropriate for investigation of the quantum criticality. This is an evidence that development of the applied research can in return stimulate further development of the initializing theory.

The described idea can be realized not only for the in-plane conductivity of MgB_2 and CuO_2 containing high-T_c superconductors but also for the currents in the so called c-direction perpendicular to the CuO_2 layers. For $Bi_2Sr_2CaCu_2O_8$ the NDC can be coupled with plasma resonances having frequencies lower than the superconducting gap. This coupling leads to natural realization of voltage induced FIR oscillations with low dissipation.

The nonlinear conductivity in supercooled regime can be used also for frequency mixers and transistors. The possibility to operate in terahertz range (loosely defined by the frequency range of 0.1 to 10 THz) using hight-T_c superconductors looks very promising. Recently Ferguson and Zhang [539] mentioned that the lack of high-power, low-cost, portable room-temperature THz source is the most significant limitation of the modern THz systems. They consider that the narrow band THz sources are crucial for high-resolution spectroscopy applications. In addition, the authors are stressing that this kind of sources have broad potential applications in telecommunications and are particularly attractive for extremely high bandwidth intersatellite links. Our invention is designed to fill this gap with its simple theoretical explanation and easy for performance practical proposal.

The THz region of the electromagnetic waves (the frequencies between 100 GHz and 10 THz or wavelengths between 3 mm and 30 μm lies between radiofrequencies and optics and up to now is not so well developed for applications and research as neighbouring frequency ranges. The state of the art review on the application and sources of THz radiation and the need of improvement and new sources is given in a recent review article by Mueller [540]. The author points out that practical application of THz radiation is in initial stage and only some reliable sources are available. That is why every new principle of THz wave generation can be not only significant theoretical input but will have important technical applications in the near future. In the review [540] it is mentioned also that recently the researchers are pursuing potential THz-wavelength applications in many fields: quality control; biomedical imaging; THz tomographic imaging in mammography; passengers' screening for explosives at the airports;[1] detecting the

[1]While we were preparing the final draft of this book, on March, 10th 2008 BBC News reported that 'A camera that can "see" explosives, drugs and weapons hidden under clothing from 25 metres has been invented. The ThruVision system could be deployed at airports, railway stations or other public spaces. It is based on so-called "terahertz", or T-ray, technology,... ' [541]. We believe that the next step in this direction could be the realization of much more compact similar high-T_c devices.

presence of cancerous cells; complex dynamics involved in condensed matter physics; molecular recognition and protein folding; environmental monitoring; plasma diagnostics; Antarctic submillimeter telescope which will be used to measure interstellar singly ionized nitrogen and carbon monoxide during the polar winter; significant part of the photons emitted since the Big Bang fall also in THz region - continuous-wave THz sources can be used to help study these photons; THz imaging using time domain spectroscopy developed in Lucent Technologies' Bell Laboratories — it uses the greatly varying absorption characteristics from material to material; NASA's AURA satellite measuring the concentration and distribution of hydroxyl radical (OH^-) in the stratosphere, a crucial component in the ozone cycle, etc.

At the end, we would like to draw reader's attention to another possible application. In a semiconductor nanostructure with two dimensional electron gas and an appropriate grating coupler THz electromagnetic waves can be transmitted in hyper-sound phonons — in such a way we will have an ersatz phonon laser useful for phonon spectroscopy as well [542].

In conclusion we consider that investigation of voltage biased conductivity of nanostructured superconductors is very perspective theme of the fundamental science promising viable variety of technical applications.

9.6 Initial experimental success in the THz range

Plasmons due to Cooper pair motion in thin films, Eq. (4.43), were predicted in the HTS era [5] (see discussion in Chapter 4). Far infrared transparency and plasmon propagation in the bulk superconductor $Bi_2Sr_2CaCu_2O_8$ are one of few new phenomena predicted theoretically for high-T_c superconductors. Since ω_{pl} for this material, Eq. (4.1), falls into the THz range, perhaps the simplest possible idea toward realization of superconductor THz generators would be to excite Cooper pair plasmons by applying a current or an electric field [543]. When an external magnetic field is applied parallel to the CuO_2 planes, an electric field in direction normal to the planes drives a flow of the Josephson lattice. The first technological break-through has already been achieved. Successful continuous and tunable emission (0.6–1 THz) was generated by a sample of slightly overdoped $Bi_2Sr_2CaCu_2O_{8+x}$ [8]. Figure 9.1 shows the experimental setup of Ref. [8] along with the obtained I-V curves of the oscillator and the differential conductivity of the detector demonstrating the appearance of THz plasmon oscillations.

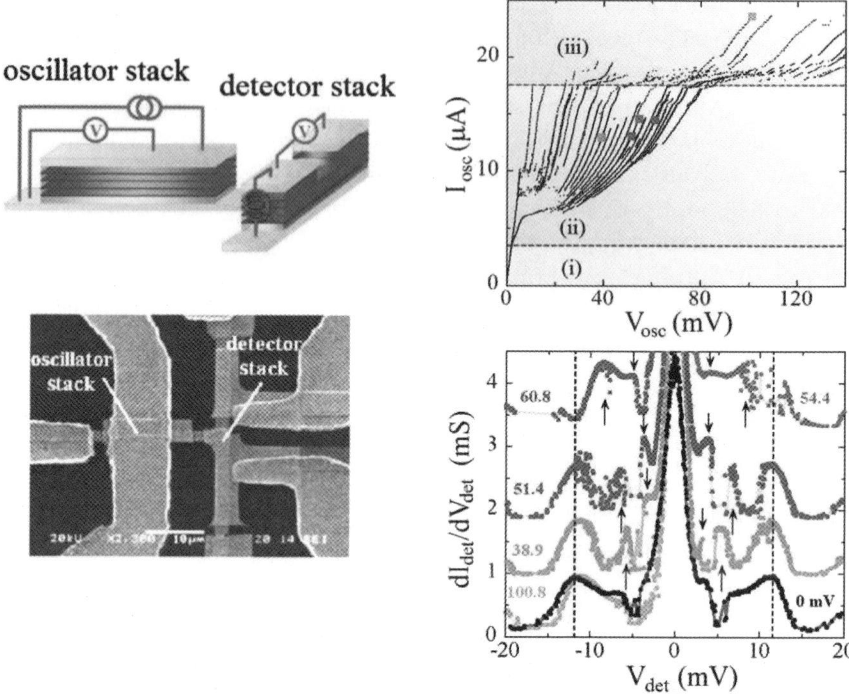

Fig. 9.1 (left) Schematic of the generator and detector and their scanning electron microscopy image. (right) I-V curve of the oscillator and the differential conductivity of the detector (reprinted with permission from Ref. [8]; Copyright © 2007 by the American Physical Society).

Thus, the long research path starting with the microscopic theory has reached new technical applications. We know the electronic structure, pairing mechanism, we have a quantitative theory of the thermodynamic and electrodynamic behavior, and can predict new devices.

Acknowledgments, retrospect

This book is a result of more than thirty years of focused work on the problem of high temperature superconductivity and many people contributed to the scientific ambiance and to the success of this research. The first person to thank is Tsvetan Sariyski with whom in 1979 we challenged the basic concepts such as methods of quantum field theory, the theory of dielectric screening, exchange interaction, the importance of the two-dimensional and layered systems, plasma waves in low-dimensional systems and others that in the following years appeared to be of great importance for further developments in the field. In these early years, superconductivity has been considered in the academic circles as a solved scientific problem and the leader in the field of high-temperature superconductivity (HTSC), V. L. Ginzburg was considered as a supporter of outdated ideas. T. Mishonov had, since 1979, very successful scientific collaboration with Atanas Groshev (1963-1996) which resulted in prediction of plasmons in thin superconducting films, confirmed experimentally later on; his collaboration and friendship are deeply missed. Tihomir Hristov has also contributed to the "biofield" of this early enthusiastic challenging research with his co-authorship and sense of humor. The work of T. Mishonov with some of his students in the 90s - D. Damyanov, I. Genchev, R. Koleva, R. Danev, A. Pachov, N. Zahariev, Y. Maneva, M. Mishonov, M. Stoev has been a constant source of inspiration, and many new ideas were brought into existence. T. Mishonov is thankful to his former students for the critical reading of different parts of the book.

The authors appreciate the comments by J. Brankov, V. Kovachev and N. Tonchev on the early version of the book.

Due to the restricted volume we cannot mention the names of all other people we are grateful to. Special thanks go to our colleagues all over the world, working actively in this field, with whom we had many interesting and fruitful discussions on the subject and on the results published in this book. In particular, we would like to express our gratitude to Joseph Indekeu for his significant support to our research, for his hospitality during our collaboration in Leuven and for his interest in this project. The hospitality of N. Plakida is also highly appreciated by T. Mishonov. T. Mishonov is thankful to M. Angelova for her help. The completion of this book was possible thanks to the indispensable and cordial support by some sponsors. During all these years, the help of V. Mishonova both in preparation of the manuscripts and proofreading as well as her continued support has been invaluable. T. Mishonov is particularly grateful to the late M. Mateev for his continued attention and interest in the development of HTS theory and for his indispensable help in critical moments.

We would like to thank the production staff at World Scientific for their collaboration. We hope that the book will be a stimulating tool for researchers, university professors and students alike.

Bibliography

[1] Bednorz, J. G. & Müller, K. A. Possible high T_c superconductivity in the Ba-La-Cu-O system. *Z. Phys. B* **64**, 189–193 (1986).

[2] Schrieffer, J. R. & Brooks, J. S. (eds.) *Handbook of High-Temperature Superconductivity: Theory and Experiment* (Springer, New York, 2007).

[3] Annett, J. F. *Superconductivity, Superfluids, and Condensates* (Oxford University Press, Oxford, 2004).

[4] Mishonov, T. Predicted plasma oscilations in the $Bi_2Sr_2CaCu_2O_8$ high-temperature superconductor. *Phys. Rev. B* **44**, 12033–12034 (1991).

[5] Mishonov, T. & Groshev, A. Plasmon excitations in Josephson arrays and thin superconducting layers. *Phys. Rev. Lett.* **64**, 2199–2201 (1990).

[6] Mishonov, T. M. Comment on "Metallic and superconducting surfaces of $YBa_2Cu_3O_7$ probed by electrostatic charge modulation of epitaxial films". *Phys. Rev. Lett.* **67**, 3195 (1991).

[7] Bulaevskii, L. N. & Koshelev, A. E. Radiation from a Single Josephson Junction into Free Space due to Josephson Oscillations. *Phys. Rev. Lett.* **97**, 267001 (2006).

[8] Bae, M.-H., Lee, H.-J. & Choi, J.-H. Josephson-Vortex-Flow Terahertz Emission in Layered High-T_c Superconducting Single Crystals. *Phys. Rev. Lett.* **98**, 027002 (2007).

[9] Pickett, W. E. Electronic structure of the high-temperature oxide superconductors. *Rev. Mod. Phys.* **61**, 433–512 (1989).

[10] Shen, Z.-X. & Dessau, D. S. Electronic structure and photoemission studies of late transition-metal oxides – Mott insulators and high-temperature superconductors. *Phys. Rep.* **253**, 1–162 (1995).

[11] Lynch, D. W. & Olson, C. G. *Photoemission studies of High-Temperature Superconductors* (Cambridge University Press, 1999).

[12] King, D. M., Shen, Z.-X., Dessau, D. S., Wells, B. O., Spicer, W. E., Arko, A. J., Marshall, D. S., DiCarlo, J., Loeser, A. G., Park, C. H., Ratner, E. R., Peng, J. L., Li, Z. Y. & Greene, R. L. Fermi surface and electronic structure of $Nd_{2-x}Ce_xCuO_{4-\delta}$. *Phys. Rev. Lett.* **70**, 3159–3162 (1993).

[13] Yu, J. & Freeman, A. J. Electronic structure Fermi liquid theory of high T_c superconductors: Comparison of predictions with experiments. In Bennett,

L. H. (ed.) *Proc. of a Conf. on "Advances in Materials Science and Applications of High Temperature Superconductors" held at Goddard Space Flight Center, Greenbelt, Maryland, April 2–6, 1990*, 365–371 (US Government Printing Office, 1991).

[14] Goringe, C. M., Bowler, D. R. & Hernández, E. Tight-binding modelling of materials. *Rep. Prog. Phys.* **60**, 1447–1512 (1997).

[15] Bullett, D. W. The Renaissance and Quantitative Development of the Tight Binding Method. In *Solid State Physics*, vol. 35, 129–215 (Academic Press, New York, 1980).

[16] Eschrig, H. *Optimized LCAO Method and the Electronic Structure of Extended Systems* (Springer, Berlin, 1989).

[17] Aebi, P., Osterwalder, J., Schwaller, P., Schlapbach, L., Shimoda, M., Mochiku, T. & Kadowaki, K. Complete Fermi surface mapping of $Bi_2Sr_2CaCu_2O_{8+x}$(001): Coexistence of short range antiferromagnetic correlations and metallicity in the same phase. *Phys. Rev. Lett.* **72**, 2757–2760 (1994).

[18] Aebi, P., Osterwalder, J., Schwaller, P., Schlapbach, L., Shimoda, M., Mochiku, T. & Kadowaki, K. Aebi *et al.* Reply. *Phys. Rev. Lett.* **74**, 1886 (1995).

[19] Osterwalder, J., Aebi, P., Schwaller, P., Schlapbach, L., Shimoda, M., Mochiku, T. & Kadowaki, K. Angle-resolved photoemission experiments on $Bi_2Sr_2CaCu_2O_{8+\delta}$(001). *Appl. Phys. A* **60**, 247–254 (1995).

[20] Aebi, P., Osterwalder, J., Schwaller, P., Schlapbach, L., Shimoda, M., Mochiku, T., Kadowaki, K., Berger, H. & Lévy, F. Complete Fermi surface mapping of $Bi_2Sr_2Ca_{n-1}Cu_nO_{2n+4}$(001) with $n = 1, 2$. *Physica C* **235–240**, 949 (1994).

[21] Schwaller, P., Aebi, P., H.Berger, Beeli, C., Osterwalder, J. & Schlapbach, L. Structure and Fermi surface mapping of a modulation-free Pb-Bi-Sr-Ca-Cu-O high-temperature superconductor. *J. Electron Spectrosc. Relat. Phenom.* **76**, 127–135 (1995).

[22] Lu, D. H., Schmidt, M., Cummins, T. R., Schuppler, S., Lichtenberg, F. & Bednorz, J. G. Fermi Surface and Extended van Hove Singularity in the Noncuprate Superconductor Sr_2RuO_4. *Phys. Rev. Lett.* **76**, 4845–4848 (1996).

[23] Maeno, Y., Hashimoto, H., Yoshida, K., Nashizaki, S., Fujita, T., Bednorz, J. G. & Lichtenberg, F. Superconductivity in a layered perovskite without copper. *Nature* **372**, 532–534 (1994).

[24] Bednorz, J. G. & Müller, K. A. Perovskite-type oxides—The new approach to high-T_c superconductivity. *Rev. Mod. Phys.* **60**, 585–600 (1988).

[25] Andersen, O. K., Jepsen, O., Liechtenstein, A. I. & Mazin, I. I. Plane dimpling and saddle-point bifurcation in the band structures of optimally doped high-temperature superconductors: A tight-binding model. *Phys. Rev. B* **49**, 4145–4157 (1994).

[26] Andersen, O. K., Savrasov, S. Y., Jepsen, O. & Liechtenstein, A. I. Out-of-plane instability and electron-phonon contribution to *s*- and *d*-wave pairing in high-temperature superconductors; LDA linear-response calculation for

doped $CaCuO_2$ and a generic tight-binding model. *J. Low Temp. Phys.* **105**, 285–304 (1996).

[27] Andersen, O. K., Liechtenstein, A. I., Jepsen, O. & Paulsen, F. LDA energy bands, low-energy hamiltonians, t', t'', $t_{\perp}(k)$, and $J_{\perp}$. *J. Phys. Chem. Solids* **56**, 1573–1591 (1995).

[28] Poole, J., C. P., Datta, T. & Farach, H. *Copper Oxide Superconductors* (Wiley, Chichester, 1988).

[29] Datta, T. Oxide Superconductors: Physical Properties. In Evetts, J. (ed.) *Concise Encyclopedia of Magnetic and Superconducting Materials*, Advances in Materials Sciences and Engineering, 415 (Pergamon, Oxford, 1992), 1st edn.

[30] Massidda, S., Hamada, N., Yu, J. & Freeman, A. J. Electronic structure of Nd-Ce-Cu-O, a Fermi liquid superconductor. *Physica C* **157**, 571–574 (1989).

[31] Yu, J. & Freeman, A. J. Coulomb correlated band structure and Fermi surfaces of high T_c superconductors. *J. Phys. Chem. Solids* **52**, 1351–1362 (1991).

[32] Radtke, R. J., Levin, K., Schüttler, H.-B. & Norman, M. R. Role of Van Hove singularities and momentum-space structure in high-temperature superconductivity. *Phys. Rev. B* **48**, 15957–15965 (1993).

[33] Radtke, R. J. & Norman, M. R. Relation of extended Van Hove singularities to high-temperature superconductivity within strong-coupling theory. *Phys. Rev. B* **50**, 9554–9560 (1994).

[34] Markiewicz, R. S. A survey of the Van Hove scenario for high-T_c superconductivity with special emphasis on pseudogaps and striped phases. *J. Phys. Chem. Solids* **58**, 1179–1310 (1997). Sec. 5.

[35] Campuzano, J. C. *et al.* Electronic Spectra and Their Relation to the (π, π) Collective Mode in High-T_c Superconductors. *Phys. Rev. Lett.* **83**, 3709–3712 (1999).

[36] Mishonov, T. M., Genchev, I. N., Koleva, R. K. & Penev, E. LCAO analysis of Sr_2RuO_4 band structure. *J. Low Temp. Phys.* **105**, 1611–1616 (1996).

[37] Mishonov, T. M., Genchev, I. N., Koleva, R. K. & Penev, E. Simple tight-binding model for the Fermi surface of Sr_2RuO_4. *Czech. J. Phys.* **46**, (Suppl. S2) 953–954 (1996).

[38] Noce, C. & Cuoco, M. Energy bands and Fermi surface of Sr_2RuO_4. *Phys. Rev. B* **59**, 2659–2666 (1999).

[39] Oguchi, T. Electronic band structure of the superconductor Sr_2RuO_4. *Phys. Rev. B* **51**, 1385–1388 (1995).

[40] Singh, D. J. Relationship of Sr_2RuO_4 to the superconducting layered cuprates. *Phys. Rev. B* **52**, 1358–1361 (1995).

[41] Mackenzie, A. P., Julian, S. R., Diver, A. J., McMullan, G. J., Ray, M. P., Lonzarich, G. G., Maeno, Y., Nishizaki, S. & Fujita, T. Quantum Oscillations in the Layered Perovskite Superconductor Sr_2RuO_4. *Phys. Rev. Lett.* **76**, 3786–3789 (1996).

[42] Puchkov, A. V., Shen, Z.-X., Kimura, T. & Tokura, Y. ARPES results on Sr_2RuO_4: Fermi surface revisited. *Phys. Rev. B* **58**, R13322–R13325 (1998).

[43] Yu, J., Massidda, S., Freeman, A. J. & Koeling, D. D. Bonds, bands, charge transfer excitations and superconductivity of $YBa_2Cu_3O7 - \delta$. *Phys. Lett. A* **122**, 203–208 (1987).

[44] Gofron, K., Campuzano, J. C., Abrikosov, A. A., Lindroos, M., Bansil, A., Ding, H., Koelling, D. & Dabrowski, B. Observation of an "Extended" Van Hove Singularity in $YBa_2Cu_4O_8$ by Ultrahigh Energy Resolution Angle-Resolved Photoemission. *Phys. Rev. Lett.* **73**, 3302–3305 (1994).

[45] Marshall, D. S., Dessau, D. S., King, D. M., Park, C.-H., Matsuura, A. Y., Shen, Z.-X., Spicer, W. E., Eckstein, J. N. & Bozovic, I. Angle-resolved photoemission spectroscopy study of $Bi_2Sr_2CaCu_2O_{8+\delta}$ thin films. *Phys. Rev. B* **52**, 12548–12551 (1995).

[46] Ding, H., Bellman, A. F., Campuzano, J. C., Randeria, M., Norman, M. R., Yokoya, T., Takahashi, T., Katayama-Yoshida, H., Mochiku, T., Kadowaki, K., Jennings, G. & Brivio, G. P. Electronic Excitations in $Bi_2Sr_2CaCu_2O_8$: Fermi Surface, Dispersion, and Absence of Bilayer Splitting. *Phys. Rev. Lett.* **76**, 1533–1536 (1996).

[47] Massidda, S., Yu, J. & Freeman, A. J. Electronic structure and properties of $Bi_2Sr_2CaCu_2O_8$, the third high-Tc superconductor. *Physica C* **152**, 251–258 (1988).

[48] Singh, D. J. & Pickett, W. E. Structural modifications in bismuth cuprates: Effects on the electronic structure and Fermi surface. *Phys. Rev. B* **51**, 3128–3133 (1995).

[49] Novikov, D. L., Gubanov, V. A. & Freeman, A. J. Electronic structure and Fermi surface topology of the infinite-layered superconductor $Sr_{1x}Ca_xCuO_2$. *Physica C* **210**, 301–307 (1993).

[50] Singh, D. J. & Pickett, W. E. Unconventional Oxygen Doping Behavior in $HgBa_2Ca_2Cu_3O_{8+\delta}$. *Phys. Rev. Lett.* **73**, 476–479 (1994).

[51] Novikov, D. L. & Freeman, A. J. Electronic structure and Fermi surface of the $HgBa_2CuO_{4+\delta}$ superconductor: Apparent importance of the role of van Hove singularities on high T_c. *Physica C* **212**, 233–238 (1993).

[52] Novikov, D. L. & Freeman, A. J. Van Hove singularities and the role of doping in the stabilization, synthesis and superconductivity of $HgBa_2Ca_{n-1}Cu_nO_{2n+2+\delta}$. *Physica C* **216**, 273–283 (1993).

[53] Novikov, D. L., Freeman, A. J. & Jorgensen, J. D. Electronic structure of superconductors without apical oxygen: $Sr_2CuO_2F_2$, $Sr_2CuO_2Cl_2$, and $Ca_2CuO_2Cl_2$. *Phys. Rev. B* **51**, 6675–6679 (1995).

[54] Bouvier, J. & Bok, J. Van hove singularity and "pseudo-gap" in HTSC. *J. Superconductivity* **10**, 673–675 (1997).

[55] Bouvier, J., Bok, J., Kim, H., Trotter, G. & Osofsky, M. Coulomb interaction in disordered metals and HTSC. *Physica C* **364–365**, 471–474 (2001).

[56] Labbé, J. & Bok, J. Superconductivity in Alcaline-Earth-Substituted La_2CuO_4: A Theoretical Model. *Europhys. Lett.* **3**, 1225–1230 (1987).

[57] Bouvier, J. & Bok, J. The van Hove scenario of high T_c superconductors: the effect of doping. *Physica C* **288**, 217–225 (1997).

[58] Friedel, J. The high-T_c superconductors: a conservative view. *J. Phys.: Condens. Matt.* **1**, 7757–7794 (1989).

[59] Markiewicz, R. S. Phase separation near the Mott transition in $La_{2-x}Sr_xCuO_4$. *J. Phys.: Condens. Matt.* **2**, 665–676 (1990).

[60] Mishonov, T. M., Indekeu, J. O. & Penev, E. S. Superconductivity of overdoped cuprates: the modern face of the ancestral two-electron exchange. *J. Phys.: Condens. Matter* **15**, 4429–4456 (2003). `cond-mat/0209191`.

[61] Newns, D. M., Tsuei, C. C. & Pattnaik, P. C. Van Hove scenario for d-wave superconductivity in cuprates. *Phys. Rev. B* **52**, 13611–13618 (1995).

[62] Tsuei, C. C., Chi, C. C., Newns, D. M., Pattnaik, P. C. & Däumling, M. Thermodynamic evidence for a density-of-states peak near the Fermi level in $YBa_2Cu_3O_{7-y}$. *Phys. Rev. Lett.* **69**, 2134–2137 (1992).

[63] Mishonov, T. M., Chéenne, N., Robbes, D. & Indekeu, J. O. Generation of 3rd and 5th harmonics in a thin superconducting film by temperature oscillations and isothermal nonlinear current response. *Eur. Phys. J. B* **26**, 291–296 (2002). `cond-mat/0109478`.

[64] Mishonov, T. M. & Stoev, M. V. Determination of the density of states in high-T_c thin films using FET-type microstructures. *Eur. Phys. J. B* **54**, 419–421 (2006).

[65] Chéenne, N., Mishonov, T. & Indekeu, J. O. Observation of a sharp lambda peak in the third harmonic voltage response of high-T_c superconductor thin films. *Eur. Phys. J. B* **32**, 437–444 (2003). `cond-mat/0110632`.

[66] Landau, L. D. & Lifshitz, E. M. *Statistical Physics Part I*, vol. 5 of *Course on Theoretical Physics* (Pergamon Press, Oxford, 1978). Chapter 5.

[67] Lifshitz, I. M., Azbel, M. Y. & Kaganov, M. I. *Electron Theory of Metals* (Consultants Bureau, New York, 1973).

[68] Zhou, X. J. *et al.* Dichotomy between Nodal and Antinodal Quasiparticles in Underdoped $(La_{2-x}Sr_x)CuO_4$ Superconductors. *Phys. Rev. Lett.* **92**, 187001 (2004).

[69] Yoshida, T., Zhou, X. J., Sasagawa, T., Yang, W. L., Bogdanov, P. V., Lanzara, A., Hussain, Z., Mizokawa, T., Fujimori, A., Eisaki, H., Shen, Z.-X., Kakeshita, T. & Uchida, S. Metallic Behavior of Lightly Doped $La_{2-x}Sr_xCuO_4$ with a Fermi Surface Forming an Arc. *Phys. Rev. Lett.* **91**, 027001 (2003).

[70] Zhou, X. J., Yosh ida, T., Kellar, S. A., Bogdanov, P. V., Lu, E. D., Lanzara, A., Nakamura, M., Noda, T., Kakeshita, T., Eisaki, H., Uchida, S., Fujimori, A., Hussain, Z. & Shen, Z.-X. Dual Nature of the Electronic Structure of $(La_{2-x-y}Nd_ySr_x)CuO_4$ and $La_{1.85}Sr_{0.15}CuO_4$. *Phys. Rev. Lett.* **86**, 5578–5581 (2001).

[71] Yoshida, T. *et al.* Electronlike Fermi surface and remnant $(\pi, 0)$ feature in overdoped $La_{1.78}Sr_{0.22}CuO_4$. *Phys. Rev. B* **63**, 220501 (2001).

[72] Ginzburg, V. L. Superconductivity: The Day Before Yesterday, Yesterday, Today, and Tomorrow. *J. Supercond. Novel Magn.* **13**, 665–677 (2000).

[73] Cooper, L. N. Bound Electron Pairs in a Degenerate Fermi Gas. *Phys. Rev.* **104**, 1189–1190 (1956).

[74] Bardeen, J., Cooper, L. N. & Schrieffer, J. R. Microscopic Theory of Superconductivity. *Phys. Rev.* **106**, 162–164 (1957).

[75] Bardeen, J., Cooper, L. N. & Schrieffer, J. R. Theory of Superconductivity. *Phys. Rev.* **108**, 1175–1204 (1957).

[76] Aebi, P. Fermi surface mapping by Angle-Scanned Photoemission. In Drechsler, S.-L. & Mishonov, T. M. (eds.) *High-T_c Superconductors and Related Materials: Materials Science, Fundamental Properties, and Some Future Electronic Applications*, 39–50 (Kluwer, Dordrecht, 2001).

[77] Campuzano, J. C., Norman, M. R. & Randeria, M. Photoemission in the high-T_c superconductors. In Bennemann, K.-H. & Ketterson, J. P. (eds.) *The Physics of Superconductors* (Springer, Heidelberg, 2004). `cond-mat/0209476`.

[78] Scalapino, D. J. The case for $d_{x^2-y^2}$ pairing in the cuprate superconductors. *Phys. Rep.* **250**, 329–367 (1995).

[79] Poilblanc, D. & Scalapino, D. J. Calculation of $\Delta(k,\omega)$ for a two-dimensional t-J cluster. *Phys. Rev. B* **66**, 052513 (2002).

[80] Schrieffer, J. R. Ward's identity and the suppression of spin fluctuation superconductivity. *J. Low Temp. Phys.* **99**, 397–402 (1995).

[81] Plakida, N. M. *High-Temperature Superconductivity* (Springer, Berlin, 1995).

[82] Annett, J. F., Goldenfeld, N. & Leggett, A. J. Constraints on the pairing state of the cuprate superconductors. *J. Low Temp. Phys.* **105**, 473–482 (1996).

[83] Lee, P. A. Cuprate superconductors from the viewpoint of strong correlation. *J. Low Temp. Phys.* **105**, 581–589 (1996).

[84] Koltenbah, B. E. C. & Joynt, R. Material-specific gap function in the t-J model of high-temperature superconductors. *Rep. Prog. Phys.* **60**, 23–56 (1997).

[85] Li, Q. P., Koltenbah, B. E. C. & Joynt, R. Mixed s-wave and d-wave superconductivity in high-T_c systems. *Phys. Rev. B* **48**, 437–455 (1993).

[86] Ruvalds, J. Theoretical prospects for high-temperature superconductors. *Supercond. Sci. Technol.* **9**, 905–926 (1996).

[87] Brusov, P. *Mechanisms of High Temperature Superconductivity*, vol. 1 and 2 (Rostov State University Publishing, Rostov on Don, 1999). 2815 references, 701 pages.

[88] Wilson, J. A. Again 'why layered, square-planar, mixed-valent cuprates alone?' - further pursuit of the 'chemical' negative-U route to the HTSC mechanism. *J. Phys.: Condens. Matter* **12**, R517–R547 (2000).

[89] Wilson, J. A. Developments in the negative-U modelling of the cuprate HTSC systems. *J. Phys.: Condens. Matter* **13**, R945–R977 (2001).

[90] Wilson, J. A. Bosonic mode interpretation of novel scanning tunnelling microscopy and related experimental results, within boson-fermion modelling of cuprate high-temperature superconductivity. *Phil. Mag.* **84**, 2183–2216 (2003). `cond-mat/0304661`.

[91] Szotek, Z., Gyorffy, B. L., Temmerman, W. M., Andersen, O. K. & Jepsen, O. Quasiparticles in d-wave superconductors within density functional theory. *J. Phys.: Condens. Matter* **13**, 8625–8652 (2001).

[92] Chubukov, A. V., Pines, D. & Schmalian, J. A spin fluctuation model for *d*-wave superconductivity. In Bennemann, K.-H. & Ketterson, J. B. (eds.) *The Physics of Superconductors*, vol. 1 (Springer, Heidelberg, 2002). `cond-mat/0201140`.

[93] Plakida, N. M., Anton, L., Adam, S. & Adam, G. Exchange and spin-fluctuation pairing in the two-band Hubbard model – Application to cuprates. In Annett, J. F. & Kruchinin, S. (eds.) *New Trends in Superconductivity*, vol. 67 of *NATO Science Series II*, 29–38 (Kluwer, Dordrecht, 2002). `cond-mat/0104234`.

[94] Plakida, N. M., Hayn, R. & Richard, J.-L. Two-band singlet-hole model for the copper oxide plane. *Phys. Rev. B* **51**, 16599–16607 (1995).

[95] Friedel, J. & Kohmoto, M. On the nature of antiferromagnetism in the CuO_2 planes of oxide superconductors. *Eur. Phys. J. B* **30**, 427–435 (2002).

[96] Carlson, E. W., Emery, V. J., Kivelson, S. A. & Orgad, D. Concepts in high-temperature superconductivity. In Bennemann, K.-H. & Ketterson, J. B. (eds.) *The Physics of Superconductors*, vol. 1 (Springer, Heidelberg, 2002). `cond-mat/0206217`.

[97] Rigamonti, A., Borsa, F. & Carretta, P. Basic aspects and main results of NMR-NQR spectroscopies in high-temperature superconductors. *Rep. Prog. Phys.* **61**, 1367–1439 (1998).

[98] Mishonov, T. M., Indekeu, J. O. & Penev, E. S. The $3d$-to-$4s$-by-$2p$ highway to superconductivity in cuprates. *Int. J. Mod. Phys. B* **16**, 4577–4585 (2002). `cond-mat/0206350`.

[99] Schubin, S. & Wonsowsky, S. On the electron theory of metals. *Proc. R. Soc. London Ser. A* **145**, 159–180 (1934).

[100] Zener, C. Interaction Between the *d* Shells in the Transition Metals. *Phys. Rev.* **81**, 440–444 (1951).

[101] Zener, C. Interaction between the *d*-Shells in the Transition Metals. II. Ferromagnetic Compounds of Manganese with Perovskite Structure. *Phys. Rev.* **82**, 403–405 (1951).

[102] Zener, C. Interaction between the *d*-Shells in the Transition Metals. III. Calculation of the Weiss Factors in Fe, Co, and Ni. *Phys. Rev.* **83**, 299–301 (1951).

[103] Vonsovskii, S. V. *Magnetism* (Wiley, New York, 1974).

[104] Vonsovskii, S. V., Izyumov, Y. A. & Kurmaev, E. Z. *Superconductivity of Transition Metals, Their Alloys and Compounds* (Springer, Berlin, 1982).

[105] Craik, D. *Magnetism: Principles and Applications* (Wiley, Chichester, 1995).

[106] Yosida, K. *Theory of Magnetism* (Springer, Berlin, 1996). p. 181, p. 319.

[107] Yosida, K. On *s*-*d* and *s*-*f* Interactions. In Rado, G. T. & Suhl, H. (eds.) *Magnetism*, vol. IIB, 215–291 (Academic Press, San Diego, CA, 1966).

[108] Schubin, S. P. In Vonsovskii, S. V. & Katsnelson, M. I. (eds.) *Selected Papers on Theoretical Physics*, 375 (Ural branch RAS, Ekaterinburg, 1991).

[109] Vizgin, V. P. The nuclear shield in the 'thirty-year war' of physicists against ignorant criticism of modern physical theories. *Phys. Usp.* **42**, 1259–1283 (1999).

[110] Andreev, A. F. *et al.* In memory of Sergei Vasil'evich Vonsovskii. *Sov. Phys. Usp.* **42**, 101–102 (1999).

[111] Izumov, Y. A. *Basic Models in Quantum Theory of Magnetism* (Ural branch RAS, Ekaterinburg, 2002).

[112] See also N. N. Bogoliubov *Selected works. Vol 1. Lectures on quantum statistics* (Gordon and Breach Science Publishers, New York, 1967); (Naukova Dumka, Kiev, 1970)(in Russsian); (Ryadanskaya Shkola, Kiev, 1948)(in Russian); It is interesting to note that in the 1948 lectures Bogoliubov has implicitly introduced what is now known as the Hubbard model, as well as referred to Vonsovskii's works.

[113] Heitler, W. & London, F. Interaction of neutral atoms and homopolar binding according to the quantum mechanics. *Z. Phys.* **44**, 455–472 (1927).

[114] Mishonov, T. M., Donkov, A. A., Koleva, R. K. & Penev, E. S. Superconducting gap anisotropy within the framework of a simple exchange model for layered cuprates. The theory of HTSC. *Bulgarian J. Phys.* **24**, 114–125 (1997). `cond-mat/0001033`.

[115] Mishonov, T. M., Groshev, A. V. & Donkov, A. A. Pairing in layered cuprates by two-electron exchange between adjacent oxygen ions. *Bulgarian J. Phys.* **25**, 62–80 (1998).

[116] Mishonov, T. M., Wallington, J. P., Penev, E. S. & Indekeu, J. O. Reduced pairing Hamiltonian for interatomic two-electron exchange in layered cuprates. *Mod. Phys. Lett. B* **16**, 693–699 (2002). `cond-mat/0205616`.

[117] Bloch, F. Über die Quantenmechanik der Elektronen in Kristallgittern. *Z. Phys.* **52**, 555–600 (1928).

[118] Hückel, E. Die Elektronenkonfiguration des Benzols und verwandter Verbindungen. I. *Z. Phys.* **70**, 204–286 (1931).

[119] Hückel, E. Quantentheorie der induzierten Polaritäten. II. *Z. Phys.* **72**, 310 (1931).

[120] Hückel, E. Quantentheoretische Beiträge zum Problem der aromatischen und ungesttigten Verbindungen. III. *Z. Phys.* **76**, 628–648 (1931).

[121] Slater, J. *Quantum Theory of Molecules and Solids* (McGraw-Hill, New York, 1963). Vol. 1, Sec. 2.3, Chap. 3, Chap. 9, and Vol. 2.

[122] Labbe, J. & Bok, J. Superconductivity in alkaline-earth-substituted lanthanum cuprate. *Europhys. Lett.* **3**, 1225–1230 (1987).

[123] Mishonov, T. & Penev, E. Tight-binding modelling of the electronic band structure of layered superconducting perovskites. *J. Phys.: Condens. Matter* **12**, 143–159 (2000). `cond-mat/0001049`.

[124] Heisenberg, W. Mehrkörperproblem und Resonanz in der Quantenmechanik. *Z. Phys.* **38**, 411–426 (1926).

[125] Heisenberg, W. Zur Theorie des Ferromagnetismus. *Z. Phys.* **49**, 619–636 (1928).

[126] Vonsovskii, S. V. & Svirskii, M. S. Superconductivity of a ferromagnetic substance with weak exchange interaction. *Zh. Éksp. Teor. Fiz.* **39**, 384–392 (1960). [Sov. Phys.–JETP **12**, 272–277 (1961)].

[127] Morkovskii, Y. *Fiz. Met. Metalloved.* **13**, 940 (1962).

[128] Sarma, G. On the influence of a uniform exchange field acting on the spins of the conduction electrons in a superconductor. *J. Phys. Chem. Solids* **24**, 1029–1032 (1963).
[129] Vonsovskii, S. V. & Svirsky, M. S. Theory of superconductivity of electron system containing singlet and triplet pairs. *Phys. Status Solidi* **9**, 267–280 (1965).
[130] Dunin, S. Z. *Sov. Phys. Solid State* **14**, 559–560 (1972).
[131] Izyumov, Y. A. & Skryabin, Y. N. Problem of the coexistence of superconductivity and ferromagnetism. *Phys. Status Solidi B* **61**, 9–64 (1974).
[132] Bennemann, K.-H. & Garland, J. W. Occurrence of magnetism in superconductors. *Int. J. Magn.* **1**, 97–122 (1979).
[133] Lee, T.-K., Izymov, Y. A. & Birman, J. L. Theory of the coexistence of superconductivity and magnetism: The superconducting transition temperature. *Phys. Rev. B* **20**, 4494–4505 (1979).
[134] Tonchev, N. S. & Brankov, I. G. On the *s-d* model for coexistence of ferromagnetism and superconductivity. *Phys. Status Solidi B* **102**, 179–187 (1980).
[135] Izyumov, Y. A. Magnetism and superconductivity in strongly correlated systems. *Sov. Phys. Usp.* **34**, 935–957 (1991).
[136] Nielsen, J. R. (ed.) *Niels Bohr, Collected Works* (North-Holland, New York, 1976). In 3 volumes.
[137] Bohr, N. The Constitution of Atoms and Molecules. I and II. *Phil. Mag.* **26**, 1–25 (1913).
[138] Bohr, N. On the Constitution of Atoms and Molecules. Part II. Systems containing only a Single Nucleus. *Phil. Mag.* **26**, 476–502 (1913).
[139] Born, M. *Vorlesungen über Atommechanik* (Springer, Berlin, 1925). P. 331, Fig. 40.
[140] Langmuir, I. The Structure of the Helium Atom. *Phys. Rev.* **17**, 339–353 (1921).
[141] Van Vleck, J. H. The normal helium atom and its relation to the quantum theory. *Phil. Mag.* **44**, 842–869 (1922).
[142] Read, F. H. A new class of atomic states: the 'Wannier-ridge' resonances. *Aust. J. Phys.* **35**, 475–499 (1982).
[143] Read, F. H. A modified Rydberg formula. *J. Phys. B: At. Mol. Phys.* **10**, 449–458 (1977).
[144] Brunt, J. N. H., Read, F. H. & King, G. C. The realisation of high energy resolution using the hemispherical elect rostatic energy selector in electron impact spectrometry. *J. Phys. E: Sci. Instrum.* **10**, 134–139 (1977).
[145] Buckman, S. J., Hammond, P., Read, F. H. & King, G. C. Highly-excited double Rydberg states of He^-. *J. Phys. B: At. Mol. Phys.* **16**, 4039–4047 (1983).
[146] Thomson, J. J. Forces between atoms and chemical affinity. *Phil. Mag.* **27**, 757–789 (1914).
[147] Blackemore, J. *Solid State Physics* (Cambridge University Press, Cambridge, 1985), 2 edn.

[148] Ogg, R. A. Bose-Einstein Condensation of Trapped Electron Pairs. Phase Separation and Superconductivity of Metal-Ammonia Solutions. *Phys. Rev.* **69**, 243–244 (1946).

[149] Schafroth, M. R. Superconductivity of a Charged Ideal Bose Gas. *Phys. Rev.* **100**, 463–475 (1955).

[150] Lewis, G. N. The atom and the molecule. *J. Am. Chem. Soc.* **38**, 762–785 (1916).

[151] Langmuir, I. The arrangement of electrons in atoms and molecules. *J. Am. Chem. Soc.* **41**, 868–934 (1919).

[152] Langmuir, I. Isomorphism, isosterism and covalence. *J. Am. Chem. Soc.* **41**, 1543–1559 (1919).

[153] Parson, A. L. A magneton theory of the structure of the atom. vol. 65 of *Smithsonian Miscellaneous Collection*, 1–80 (Smithsonian, Washington, DC, 1915). Publication No 2371.

[154] Lewis, G. N. The conservation of photons. *Nature* **118**, 874–875 (1926).

[155] Wilson, S. *Electron Correlations in Molecules* (Clarendon, Oxford, 1984). Section 7.3.

[156] Herring, C. & Flicker, M. Asymptotic exchange coupling of two hydrogen atoms. *Phys. Rev.* **134**, A362–A366 (1964).

[157] Herring, C. Critique of the Heitler-London method of calculating spin couplings at large distances. *Rev. Mod. Phys.* **34**, 631–645 (1962).

[158] Herring, C. Direct exchange between well-separated atoms. In Rado, G. T. & Suhl, H. (eds.) *Magnetism*, vol. IIB, 76 (Academic, San Diego, CA, 1966).

[159] Herring, C. Exchange interactions among itinerant electrons. In Rado, G. T. & Suhl, H. (eds.) *Magnetism*, vol. IV (Academic, San Diego, 1966).

[160] Patil, S. H. & Tang, K. T. *Asymptotic Methods in Quantum Mechanics. Application to Atoms, Molecules and Nuclei* (Springer, Berlin, 2000).

[161] Dreizler, R. M. & Gross, E. K. U. *Density Functional Theory* (Springer, Berlin, 1990).

[162] Oliveira, L. N., Gross, E. K. U. & Kohn, W. Density-Functional Theory for Superconductors. *Phys. Rev. Lett.* **60**, 2430–2433 (1988).

[163] Kurth, S., Marques, M., Lüders, M. & Gross, E. K. U. Local Density Approximation for Superconductors. *Phys. Rev. Lett.* **83**, 2628–2631 (1999).

[164] Gor'kov, L. P. & Krotkov, P. L. Spin relaxation and antisymmetric exchange in n-doped III-V semiconductors. *Phys. Rev. B* **67**, 033203 (2003).

[165] Brovetto, P., Maxia, V. & Salis, M. On electron pairing in unconventional superconductors. *Eur. Phys. J. B* 85–94 (2000).

[166] Dirac, P. A. M. Quantum mechanics of many-electron systems. *Proc. R. Soc. A* **123**, 714–733 (1929).

[167] Dalitz, R. H. (ed.) *The Collected Works of P. A. M. Dirac* (Cambridge University Press, Cambridge, 1995). P 389.

[168] Dirac, P. A. M. *The Principles of Quantum Mechanics*, chap. IX. Systems Containing Several Similar Particles (Clarendon, Oxford, 1930).

[169] Feynman, R. P., Leighton, R. B. & Sands, M. *The Feynman Lectures on Physics*, vol. III (Addison-Wesley, Reading, MA, 1966). Eq. (12.16), Fig. 4-15.

[170] Schwinger, J. In Englert, B.-G. (ed.) *Quantum Mechanics: Symbolism of Atomic Measurements*, 160 (Academic Press, San Diego, CA, 2001). Eq. (3.5.14).

[171] Harris, A. B. & Lange, R. V. Single-Particle Excitations in Narrow Energy Bands. *Phys. Rev.* **157**, 295–314 (1967).

[172] Spalek, J. & Honig, J. M. Metal-insulator transitions, exchange interactions and real space pairing. An introduction to the theory of strongly correlated electron systems. In Narlikar, A. (ed.) *Studies of High Temperature Superconductors*, vol. 8, 1–67 (Nova Science Publishers, New York, 1991).

[173] Fulde, P. *Electron Correlations in Molecules and Solids* (Springer, Berlin, 1995), 3rd edn.

[174] Ketterson, J. B. & Song, S. N. *Superconductivity* (Cambridge University Press, Cambridge, 1999).

[175] Mishonov, T. & Groshev, A. Two-electron exchange between adjacent oxygen atoms as a possible origin of the pairing in layered cuprates. *Physica B* **194–196**, 1427–1428 (1994).

[176] Tsuei, C. C. & Kirtley, J. R. Pairing symmetry in cuprate superconductors. *Rev. Mod. Phys.* **72**, 969 (2000).

[177] Markowitz, D. & Kadanoff, L. P. Effect of Impurities upon Critical Temperature of Anisotropic Superconductors. *Phys. Rev.* **131**, 563–575 (1963).

[178] Anderson, P. W. Superconductivity in high T_c Cuprates: The cause is no longer a mystery. *Phys. Scr. T* **102**, 10–12 (2002). `cond-mat/0201429`.

[179] Pokrovskii, V. L. Thermodynamics of anisotropic superconductors. *Zh. Éksp. Teor. Fiz.* **40**, 641 (1961). [Sov. Phys.–JETP **13**, 447–450 (1961), Eq. (14)].

[180] Mesot, J., Norman, M. R., Ding, H., Randeria, M., Campuzano, J. C., Paramekanti, A., Fretwell, H. M., Kaminski, A., Takeuchi, T., Yokoya, T., Sato, T., Takahashi, T., Mochiku, T. & Kadowaki, K. Superconducting Gap Anisotropy and Quasiparticle Interactions: A Doping Dependent Photoemission Study. *Phys. Rev. Lett.* **83**, 840–843 (1999).

[181] Ding, H., Norman, M. R., Campuzano, J. C., Randeria, M., Bellman, A. F., Yokoya, T., Takahashi, T., Mochiku, T. & Kadowaki, K. Angle-resolved photoemission spectroscopy study of the superconducting gap anisotropy in $Bi_2Sr_2CaCu_2O_{8+x}$. *Phys. Rev. B* **54**, R9678–R9681 (1996).

[182] Ding, H., Yokoya, T., Campuzano, J. C., Takahashi, T., Randeria, M., Norman, M. R., Mochik, T., Kadowaki, K. & Giapintzakis, J. Spectroscopic evidence for a pseudogap in the normal state of underdoped high-T_c superconductors. *Nature* **382**, 51–54 (1996).

[183] Randeria, M. & Campuzano, J. C. High-T_c superconductors: New insights from Angle-Resolved Photoemission (1997). (Varenna Lectures), Fig. 10, `cond-mat/9709107`.

[184] Ding, H., Campuzano, J. C., Norman, M. R., Randeria, M., Yokoya, T., Takahashi, T., Takeuchi, T., Mochiku, T., Kadowaki, K., Guptasarma, P. & Hinks, D. G. ARPES study of the superconducting gap and pseudogap in $Bi_2Sr_2CaCu_2O_{8+x}$. *J. Phys. Chem. Solids* **59**, 1888–1891 (1998). Fig. 1.

[185] Zeiger, H. J. & Pratt, G. W. *Magnetic Interactions in Solids* (Clarendon, Oxford, 1973). P. 53, p. 324.

[186] Kondo, J. Resistance Minimum in Dilute Magnetic Alloys. *Prog. Theor. Phys.* **32**, 37–49 (1964).

[187] Kondo, J. Anomalous Scattering Due to *s-d* Interaction. *J. Appl. Phys.* **37**, 1177–1180 (1966).

[188] Ruderman, M. A. & Kittel, C. Indirect Exchange Coupling of Nuclear Magnetic Moments by Conduction Electrons. *Phys. Rev.* **96**, 99 (1954).

[189] Kasuya, T. A Theory of Metallic Ferro- and Antiferromagnetism on Zener's Model. *Prog. Theor. Phys.* **16**, 45–57 (1956).

[190] Yosida, K. Magnetic Properties of Cu-Mn Alloys. *Phys. Rev.* **106**, 893–898 (1957).

[191] de Gennes, P. G. Sur les proprietes des metaux des terres rares. *C. R. Acad. Sci., Paris* **274**, 1836–1838 (1958).

[192] de Gennes, P. G. Polarisation de charge (ou de spin) au voisinage d'une impureté dans un alliage. *J. Physique Radium* **23**, 630–636 (1962).

[193] Elliott, R. J. In Rado, G. T. & Suhl, H. (eds.) *Magnetism*, vol. IIA (Academic Press, San Diego, CA, 1965).

[194] Abrikosov, A. A. & Falkovsky, L. A. The nature of the ferromagnetism due to the introduction of an impurity. *Zh. Éksp. Teor. Fiz.* **43**, 2230–2233 (1962).

[195] Tsvelick, A. M. & Wiegmann, P. B. Exact results in the theory of magnetic alloys. *Adv. Phys.* **32**, 453–713 (1983).

[196] Kramers, H. A. The interaction of magnetogenic atoms in paramagnetic crystals. *Physica* **1**, 182–192 (1934).

[197] Anderson, P. W. Exchange in insulators: superexchange, direct exchange and double exchange. In Rado, G. T. & Suhl, H. (eds.) *Magnetism*, vol. I (Academic Press, San Diego, CA, 1963).

[198] Kondo, J. *g*-shift and anomalous Hall effect in gadolinium metals. *Prog. Theor. Phys.* **28**, 846–856 (1962).

[199] Anderson, P. W. Localized Magnetic States in Metals. *Phys. Rev.* **124**, 41–53 (1961).

[200] Friedel, J. Some electrical and magnetic properties of metallic solid solutions. *Can. J. Phys.* 1190–1210 (1956).

[201] de Faget de Casteljau, P. & Friedel, J. Étude de la résistivité et du pouvoir thermoélectrique des impuretés dissoutes dans les métaux nobles. *J. Physique Radium* **17**, 27–32 (1956).

[202] Friedel, J. Electronic structure of transition metals and alloys and of heavy metals. *J. Physique Radium* **19**, 573–581 (1958).

[203] Muto, T. & Kobayasi, S. On indirect Knight shift and NMR in ferromagnetic metals: I. General formulation and quantitative discussion. *J. Phys. Soc. Japan* **19**, 1837–1849 (1964).

[204] Muto, T., Kobayasi, S. & Hayakawa, H. On indirect Knight shift and NMR in ferromagnetic metals: II. Numerical calculation of NMR frequency and hyperfine field in Fe. *J. Phys. Soc. Japan* **20**, 389–395 (1965).

[205] Muto, T., Kobayasi, S. & Hayakawa, H. On indirect Knight shift and NMR in ferromagnetic metals: III. Numerical calculation of NMR frequency and hyperfine field in Ni, Co and Fe. *J. Phys. Soc. Japan* **20**, 1167–1173 (1965).

[206] Freeman, A. J. & Watson, R. E. Hyperfine interactions in magnetic materials. In Rado, G. T. & Suhl, H. (eds.) *Magnetism*, vol. IIA, 167–305 (Academic Press, San Diego, CA, 1966).

[207] Ganguly, B. N. Nuclear-Spin Relaxation and Knight Shift in Transition Metals. *Phys. Rev. B* **8**, 1055–1060 (1973).

[208] Anderson, P. W. In praise of unstable fixed points: the way things actually work. *Physica B* **318**, 28–32 (2002).

[209] Vollhardt, D., Blümer, N., Held, K. & Kollar, M. Metallic ferromagnetism - an electronic correlation phenomenon. In Baberschke, K., Donath, M. & Nolting, W. (eds.) *Band Ferromagnetism: Ground-State and Finite-Temperature Phenomena*, vol. 580 of *Springer Lecture Notes in Physics*, 191–207 (Springer, Heidelberg, 2001). cond-mat/0012203.

[210] Sekitani, T., Naito, M., Sato, H. & Miura, N. Interplay of Cooper and Kondo singlet formations in high-T_c cuprates (2001). cond-mat/0111422.

[211] Park, K. Kondo effect of nonmagnetic impurities and coexisting charge order in cuprate superconductors. *Phys. Rev. B* **67**, 094513 (2003). 0203142.

[212] Kadowaki, K. & Woods, S. B. Universal relationship of the resistivity and specific heat in heavy-Fermion compounds. *Solid State Commun.* **58**, 507–509 (1986).

[213] Nakamae, S., Behnia, K., Mangkorntong, N., Nohara, M., Takagi, H., Yates, S. J. C. & Hussey, N. E. Electronic ground state of heavily overdoped nonsuperconducting $La_{2-x}Sr_xCuO_4$. *Phys. Rev. B* **68**, 100502 (2003).

[214] de Nobel, J. The discovery of superconductivity. *Phys. Today* **49**, 40–42 (1996).

[215] Ginzburg, V. L. The problem of high-temperature superconductivity. *Contemp. Phys.* **9**, 355–374 (1968).

[216] Ginzburg, V. L. The problem of high-temperature superconductivity. II. *Usp. Fiz. Nauk* **101**, 185–215 (1970). [Sov. Phys. Usp. **13**, 335–351 (1970)].

[217] Ginzburg, V. L. High-temperature superconductivity. *J. Polym. Sci.: C* **29**, 3–26 (1970).

[218] Ginzburg, V. L. High-temperature superconductivity—dream or reality? *Usp. Fiz. Nauk* **118**, 315–324 (1976). [Sov. Phys. Usp. **19**, 174–179 (1976)].

[219] Ginzburg, V. L. In Ginzburg, V. L. & Kirznits, D. A. (eds.) *High-Temperature Superconductivity* (Consultants Bureau, New York, 1982). Translated from Russian.

[220] Ginzburg, V. L. & Kirzhnits, D. A. *Zh. Éksp. Teor. Fiz.* **46**, 397 (1964). [Sov. Phys.–JETP **20**, 1549 (1965)].

[221] Ginzburg, V. L. On surface superconductivity. *Phys. Lett.* **13**, 101–102 (1964).

[222] Akhiezer, A. I. & Pomeranchuk, I. Y. Interaction between conduction electrons in ferromagnets. *Zh. Éksp. Teor. Fiz.* **36**, 859 (1959). [Sov. Phys.–JETP **36**, 605–607 (1959)].

[223] Paretti, J. Superconductivity of transition elements. *Phys. Lett.* **2**, 275–276 (1962).

[224] Kondo, J. Superconductivity in transition metals. *Prog. Theor. Phys.* **29**, 1–9 (1963).

[225] Privorotskii, I. A. The problem of pairings with nonozero angular momentum in a Fermi system. *Zh. Éksp. Teor. Fiz.* **44**, 1401 (1963). [Sov. Phys.–JETP **17**, 942–947 (1963)].

[226] Garland, J. W. Mechanisms for Superconductivity in the Transition Metals. *Phys. Rev. Lett.* **11**, 111–114 (1963).

[227] Akhiezer, A. I. & Akhiezer, I. A. On problem of coexistence of superconductivity and ferromagnetism. *Zh. Éksp. Teor. Fiz.* **43**, 2208–2216 (1962). [Sov. Phys.–JETP. **16** (1962)].

[228] Vonsovskiĭ, S. V. & Svirskii, M. S. Effect of multiplicity of $d(f)$ shells on electron interaction in crystals. *Zh. Éksp. Teor. Fiz.* **47**, 1354 (1964). [Sov. Phys.–JETP **20**, 914 (1965)].

[229] Geilikman, B. T. The electron mechanism of superconductivity. *Usp. Fiz. Nauk* **88**, 327 (1966). [Sov. Phys. Usp. **9**, 142–152 (1966)].

[230] Jensen, M. A. Magnetism and superconductivity. In Rado, G. T. & Suhl, H. (eds.) *Magnetism*, vol. I, 215–291 (Academic Press, San Diego, CA, 1966).

[231] Matthias, B. T. Superconductivity; II. Th eFacts. *Science* **144**, 378–381 (1964).

[232] Ginzburg, V. L. High-temperature superconductivity (history and general review). *Usp. Fiz. Nauk* **161**, 1–11 (1991). [Sov. Phys. Usp. **34**, 283–288 (1991)].

[233] Ginzburg, V. L. High-temperature superconductivity: Some remarks. *Prog. Low. Temp. Phys.* **12**, 1–44 (1989).

[234] Anderson, P. W. *A Career in Theoretical Physics* (World Scientific, Singapore, 1994). pp. 637–656, p. 570, p. 584.

[235] Lee, Y. S., Birgeneau, R. J., Kastner, M. A., Endoh, Y., Wakimoto, S., Yamada, K., Erwin, R. W., Lee, S.-H. & Shirane, G. Neutron-scattering study of spin-density wave order in the superconducting state of excess-oxygen-doped La_2CuO_{4+y}. *Phys. Rev. B* **60**, 3643–3654 (1999). `cond-mat/9902157`.

[236] Savici, A. T., Fudamoto, Y., Gat, I. M., Ito, T., Larkin, M. I., Uemura, Y. J., Luke, G. M., Kojima, K. M., Lee, Y. S., Kastner, M. A., Birgeneau, R. J. & Yamada, K. Muon spin relaxation studies of incommensurate magnetism and superconductivity in stage-4 $La_2CuO_{4.11}$ and $La_{1.88}Sr_{0.12}CuO_4$. *Phys. Rev. B* **66**, 0202037 (2002).

[237] Ramirez, A. P. Superconductivity: Cohabitation in the cuprates. *Nature* **399**, 527–528 (1999).

[238] Bozovic, I., Logvenov, G., Verhoeven, M. A. J., Caputo, P., Goldobin, E. & Geballe, T. H. No mixing of superconductivity and antiferromagnetism in a high-temperature superconductor. *Nature* **422**, 873–875 (2003).

[239] Kirkpatrick, T. R. & Belitz, D. Coexistence of ferromagnetism and superconductivity. *Phys. Rev. B* **67**, 024515 (2003).

[240] Powell, B. J., Annett, J. F. & Györffy, B. L. Competition between disorder and exchange splitting in superconducting $ZrZn_2$. *J. Phys.: Condens. Matter* **15**, L235–L241 (2003).

[241] Overhauser, A. W. Spin density waves in an electron gas. *Phys. Rev.* **128**, 1437–1452 (1962).

[242] Proust, C., Boaknin, E., Hill, R. W., Taillefer, L. & Mackenzie, A. P. Heat Transport in a Strongly Overdoped Cuprate: Fermi Liquid and a Pure *d*-Wave BCS Superconductor. *Phys. Rev. Lett.* **89**, 147003 (2002). `cond-mat/0202101`.

[243] Landau, L. D. Theory of the Fermi liquid. *Zh. Éksp. Teor. Fiz.* **30**, 1058–1064 (1956). [*Sov. Phys.–JETP* **3**, 920–925 (1957)].

[244] Landau, L. D. Theory of the Fermi liquid. *Zh. Éksp. Teor. Fiz.* **32**, 59 (1957). [*Sov. Phys.–JETP* **5**, 101 (1957)].

[245] Presland, M. R., Tallon, J. L., Buckley, R. G., Liu, R. S. & Flower, N. E. General trends in oxygen stoichiometry effects on T_c in Bi and Tl superconductors. *Physica C* **176**, 95–105 (1991).

[246] Anderson, P. W. Spin-charge separation is the key to the high T_c cuprates. *Physica C* **341–348**, 9–10 (2000).

[247] Pines, D. & Nozières, P. *The Theory of Quantum Liquids*, vol. 1. Normal Fermi Liquids (Benjamin, San Francisco, CA, 1966).

[248] Kadanoff, L. P. *Statistical Physics. Statics, Dynamics and Renormalization* (World Scientific, Singapore, 2000). P. 128.

[249] Landau, L. D. The theory of superfluidity of helium II. *Zh. Éksp. Teor. Fiz.* **11**, 592 (1941). [J. Phys. USSR **5**, 71 (1941)].

[250] Anderson, P. W. *Concepts in Solids* (Benjamin, San Francisco, CA, 1963).

[251] Feng, D. L., Armitage, N. P., Lu, D. H., Damascelli, A., Hu, J. P., Bogdanov, P., Lanzara, A., Ronning, F., Shen, K. M., Eisaki, H., Kim, C., Shen, Z.-X., Shimoyama, J.-i. & Kishio, K. Bilayer Splitting in the Electronic Structure of Heavily Overdoped $Bi_2Sr_2CaCu_2O_{8+\delta}$. *Phys. Rev. Lett.* **86**, 5550–5553 (2001).

[252] Anderson, P. W. A battery of smoking guns (for the NFL-interlayer theory of high T_c cuprates). *J. Phys. Chem. Solids* **56**, 1593–1596 (1995).

[253] Hussey, E., Abdel-Jawad, M., Carrington, A., Mackenzie, A. P. & Balicas, L. A coherent three-dimensional Fermi surface in a high-transition-temperature superconductor. *Nature* **425**, 814–817 (2003).

[254] Anderson, P. W. *c*-Axis Electrodynamics as Evidence for the Interlayer Theory of High-Temperature Superconductivity. *Science* **279**, 1196–1198 (1998).

[255] Tamasaku, K., Nakamura, Y. & Uchida, S. Charge dynamics across the CuO_2 planes in $La_{2-x}Sr_xCuO_4$. *Phys. Rev. Lett.* **69**, 1455–1458 (1992).

[256] Buisson, O. & Doria, M. M. Plasma modes in HTS superconductors within the anisotropic London model. In Drechsler, S.-L. & Mishonov, T. (eds.) *High-T_c Superconductors and Related Materials: Materials Science, Fundamental Properties, and Some Future Electronic Applications*, 215 (Kluwer, Dordrecht, 2001).

[257] Buisson, O., Xavier, P. & Richard, J. Observation of propagating plasma modes in a thin superconducting film. *Phys. Rev. Lett.* **73**, 3153–3156 (1994).
[258] Buisson, O. Plasma modes in thin superconducting films. In Drechsler, S.-L. & Mishonov, T. (eds.) *High-T_c Superconductors and Related Materials: Materials Science, Fundamental Properties, and Some Future Electronic Applications*, 229 (Kluwer, Dordrecht, 2001).
[259] Pauling, L. *Nature of the Chemical Bond* (Cornell University Press, Ithaca, NY, 1940).
[260] Pauling, L. Influence of valence, electronegativity, atomic radii, and crest-trough interaction with phonons on the high-temperature copper oxide superconductors. *Phys. Rev. Lett.* **59**, 225–227 (1987).
[261] Eyring, H., Walter, J. & Kimball, G. E. *Quantum Chemistry* (Wiley, New York, 1944). Ch. 12, p. 212.
[262] Goodenough, J. B. *Magnetism and the Chemical Bond* (Wiley, New York, 1963).
[263] de Gennes, P. G. *Simple Views on Condensed Matter* (World Scientific, Singapore, 1998).
[264] de Gennes, P. G. Role of double exchange in copper oxides of mixed valency. *C. R. Acad. Sci., Paris* **305**, 345–348 (1987).
[265] Bogoliubov, N. N., Tolmachev, V. V. & Shirkov, D. V. *A New Method in the Theory of Superconductivity* (Consultants Bureau, New York, 1959).
[266] Mishonov, T. M. & Mishonov, M. T. Simple model for the linear temperature dependence of the electrical resistivity of layered cuprates. *Physica A* **278**, 553–562 (2000). `cond-mat/0001031`.
[267] Hussey, N. E. The normal state scattering rate in high-T_c cuprates. *Eur. Phys. J. B* **31**, 495–507 (2003).
[268] Sandeman, K. G. & Schofield, A. J. Model of anisotropic scattering in a quasi-two-dimensional metal. *Phys. Rev. B* **63**, 094510 (2001). `cond-mat/0007299`.
[269] Kaminski, A., Fretwell, H. M., Norman, M. R., Randeria, M., Rosenkranz, S., Chatterjee, U., Campuzano, J. C., Mesot, J., Sato, T., Takahashi, T., Terashima, T., Takano, M., Kadowaki, K., Li, Z. Z. & Raffy, H. Momentum anisotropy of the scattering rate in cuprate superconductors. *Phys. Rev. B* **71**, 014517 (2005).
[270] Yusof, Z. M., Wells, B. O., Valla, T., Fedorov, A. V., Johnson, P. D., Li, Q., Kendziora, C., Jian, S. & Hinks, D. G. Quasiparticle Liquid in the Highly Overdoped $Bi_2Sr_2CaCu_2O_{8+\delta}$. *Phys. Rev. Lett.* **88**, 167006 (2002).
[271] Ziman, J. M. *Principles of the Theory of Solids* (Cambridge University Press, Cambridge, 1972), 2nd edn.
[272] Abrikosov, A. A. *Fundamentals of the Theory of Metals* (North-Holland, Amsterdam, 1988).
[273] Paglione, J. & Greene, R. L. High-temperature superconductivity in iron-based materials. *Nature Phys.* **6**, 645–658 (2010).
[274] Zaanen, J. High-temperature superconductivity: Quantum salad dressing. *Nature* **415**, 379–381 (2002).

[275] Zaanen, J. High-temperature superconductivity: Stripes defeat the Fermi liquid. *Nature* **404**, 714–715 (2000).
[276] Coleman, P. Superconductors: Opening the gap. *Nature* **392**, 134–135 (1998).
[277] Hudson, E. W., Lang, K. M., Madhavan, V., Pan, S. H., Eisaki, H., Uchida, S. & Davis, J. C. Interplay of magnetism and high-Tc superconductivity at individual Ni impurity atoms in $Bi_2Sr_2CaCu_2O_{8+\delta}$. *Nature* **411**, 920–924 (2001).
[278] Gonzalez, J. C. G., Aguiar, J. A., Quezado, S., Dezanetie, L. M., Chu, C. W. & Dominguez, A. B. Synthesis, structure and superconductivity of $[Y_{0.75}Ca_{0.25}]SrBaCu_{3-x}(PO_4)_xO_y$ with $0.0 \leq x \leq 0.30$. *Physica C* **354**, 441–443 (2001).
[279] Carbotte, J. P., Schachinger, E. & Basov, D. N. Coupling strength of charge carriers to spin fluctuations in high-temperature superconductors. *Nature* **401**, 354–356 (1999).
[280] Powell, B. J., Annett, J. F. & Györffy, B. L. The Behavior of a Triplet Superconductor in a Spin Only Magnetic Field (2002). `cond-mat/0203087`.
[281] Mathur, N. D., Grosche, F. M., Julian, S. R., Walker, I. R., Freye, D. M., Haselwimmer, R. K. W. & Lonzarich, G. G. Magnetically mediated superconductivity in heavy fermion compounds. *Nature* **394**, 39–43 (1998).
[282] Jourdan, M., Huth, M. & Adrian, H. Superconductivity mediated by spin fluctuations in the heavy-fermion compound UPd_2Al_3. *Nature* **398**, 47–49 (1999).
[283] Saxena, S. S., Agarwal, P., Ahilan, K., Grosche, F. M., Haselwimmer, R. K. W., Steiner, M. J., Pugh, E., Walker, I. R., Julian, S. R., Monthoux, P., Lonzarich, G. G., Huxley, A., Sheikin, I., Braithwaite, D. & J., F. Superconductivity on the border of itinerant-electron ferromagnetism in UGe_2. *Nature* **406**, 587–592 (2000).
[284] Pavarini, E., Dasgupta, I., Saha-Dasgupta, T., Jepsen, O. & Andersen, O. K. Band-Structure Trend in Hole-Doped Cuprates and Correlation with $T_{c\ \mathrm{max}}$. *Phys. Rev. Lett.* **87**, 047003 (2001).
[285] Mishonov, T. M. Comment on "Band-Structure Trend in Hole-Doped Cuprates and Correlation with $T_{c\ \mathrm{max}}$". `cond-mat/0406357`.
[286] Stoll, E. P., Meier, P. F. & Claxton, T. A. On the distribution of intrinsic holes in cuprates. *J. Phys.: Condens. Matter* **15**, 7881–7889 (2003).
[287] Mishonov, T. M. & Stoev, M. LCAO model for 3D Fermi surface of high-T_c cuprate $Tl_2Ba_2CuO_{6+\delta}$. `cond-mat/0504290`.
[288] Farkas, C., Brothwell, S., Barnwell, J. & Fleishman, S. American Superconductor (2000). In-Depth Report RC#60133602 (Merrill Lynch).
[289] Bozovic, I. Atomic-layer engineering of superconducting oxides: yesterday, today, tomorrow. *IEEE Trans. Appl. Supercond.* **11**, 2686–2695 (2001).
[290] Anderson, P. W. Brainwashed by Feynman? *Phys. Today* **53**, 11–12 (2000).
[291] Rosner, H. *Electronic structure and exchange integrals of low-dimensional cuprates*. Ph.D. thesis, Technische Universitt Dresden (1999).
[292] Pajda, M., Kudrnovský, J., Turek, I., Drchal, V. & Bruno, P. Ab initio calculations of exchange interactions, spin-wave stiffness constants, and Curie temperatures of Fe, Co, and Ni. *Phys. Rev. B* **64**, 174402 (2001).

[293] Pokrovskii, V. L. & Ryvkin, M. S. *Zh. Éksp. Teor. Fiz.* **43**, 92 (1962). [Sov. Phys.–JETP **16**, 67 (1963)].

[294] Gor'kov, L. P. & Melik-Barkhudarov, T. K. Microscopic Derivation of the Ginzburg-Landau Equations for an Anisotropic Superconductor. *Zh. Éksp. Teor. Fiz.* **45**, 1493 (1963). [Sov. Phys.–JETP **18**, 1031–1034 (1964)].

[295] Kogan, V. G. Macroscopic anisotropy in superconductors with anisotropic gaps. *Phys. Rev. B* **66**, 020509 (2002). cond-mat/0204038; cond-mat/0208362; Phys. Rev. Lett. **89**, 237005 (2002); V. G. Kogan and S. L. Bud'ko, Physica C **385**, 131 (2003); V. G. Kogan and N. V. Zhelezina, cond-mat/0311331.

[296] Markowitz, D. & Kadanoff, L. P. Effect of Impurities upon Critical Temperature of Anisotropic Superconductors. *Phys. Rev.* **131**, 563–575 (1963).

[297] Mishonov, T. & Penev, E. Thermodynamics of anisotropic-gap and multiband clean BCS superconductors. *Int. J. Mod. Phys. B* **16**, 3573–3586 (2002). cond-mat/0206118.

[298] Mishonov, T. M., Penev, E. S. & Indekeu, J. O. Comment on "Anisotropic *s*-wave superconductivity in MgB_2". *Phys. Rev. B* **66**, 066501 (2002). cond-mat/0204545.

[299] Mishonov, T. M., Penev, E. S. & Indekeu, J. O. Comment on "Anisotropic *s*-wave superconductivity: Comparison with experiments on MgB_2" by A. I. Posazhennikova et al. *Europhys. Lett.* **61**, 577–578 (2003). cond-mat/0205104.

[300] Mishonov, T., Drechsler, S.-L. & Penev, E. Influence of the Van Hove singularity on the specific heat jump in BCS superconductors. *Mod. Phys. Lett. B* **17**, 755–762 (2003). cond-mat/0209192.

[301] Gorter, C. J. & Casimir, H. B. G. *Phys. Z.* **35**, 963 (1934).

[302] Ginzburg, V. L. & Landau, L. D. On the theory of superconductivity. *Zh. Éksp. Teor. Fiz.* **20**, 1064 (1950). English translation in *Men of Physics: L. D. Landau*, edited by D. ter Haar (Pergamon, New York, 1965), Vol. 1, pp. 138–167; L.D. Landau, Phys. Z. der Sowjet Union **11**, 26 (1937); **11**, 129 (1937).

[303] Mishonov, T. M., Penev, E. S., Indekeu, J. O. & Pokrovsky, V. L. Specific-heat discontinuity in impure two-band superconductors. *Phys. Rev. B* **68**, 104517 (2003). cond-mat/0209342.

[304] Bouquet, F., Wang, Y., Sheikin, I., Toulemonde, P., Eisterer, M., Weber, H. W., Lee, S., Tajima, S. & Junod, A. Unusual effects of anisotropy on the specific heat of ceramic and single crystal MgB_2. *Physica C* **385**, 192–204 (2003). Cond-mat/0210706, Figs. 2, 3 and 7; F. Bouquet, Y. Wang, R.A. Fisher, D.G. Hinks, J.D. Jorgensen, A. Junod, and N.E. Phillips, Europhys. Lett. **56**, 856 (2001).

[305] Gonnelli, R. S., Daghero, D., Ummarino, G. A., Stepanov, V. A., Jun, J., Kazakov, S. M. & Karpinski, J. Independent determination of the two gaps by directional point-contact spectroscopy in MgB_2 single crystals. *Phys. Rev. Lett.* **89**, 247004 (2002). cond-mat/0209472.

[306] Moskalenko, V. A. Superconductivity of metals taking into account overlapping of the energy bands. *Fiz. Met. Metalloved.* **8**, 503–513 (1959). [Phys. Met. Metallogr. (USSR) **8**, 25 (1959)].

[307] Mishonov, T. M., Pokrovsky, V. L. & Wei, H. Thermodynamics of MgB_2 described by the weak-coupling two-band BCS model. *Phys. Rev. B* **71**, 012514 (2005). `cond-mat/0312210`.

[308] Golubov, A. A., Kortus, J., Dolgov, O. V., Jepsen, O., Kong, Y., Andersen, O. K., Gibson, B. J., Ann, K. & Kremer, R. K. *J. Phys.: Condens. Matter* **14**, 1353–1360 (2002).

[309] Lifshitz, E. M. & Pitaevskii, L. *Statistical Physics, Part 2*, vol. 9 of *Course on Theoretical Physics* (Pergamon Press, Oxford, 1981).

[310] Abrikosov, A. A., Gor'kov, L. P. & Dzyaloshinskii, I. E. *Methods of Quantum Field Theory in Statistical Mechanics* (Dover, New York, 1975).

[311] Hussey, N. E. *Adv. Phys.* **51**, 1685–1771 (2002). Fig. 25; D.A. Bonn, S. Kamal, K. Zhang, R. Liang, D.J. Baar, E. Klein, and W.N. Hardy, Phys. Rev. B **50**, 4051 (1994), Fig. 3.

[312] Migdal, A. B. *Qualitative Methods in Quantum Theory* (Westview Press, Boulder, 2000).

[313] Fisher, R. A., Li, G., Lashley, J. C., Bouquet, F., Phillips, N. E., Hinks, D. G., Jorgensen, J. D. & Crabtree, G. W. *Physica C* **385**, 180–191 (2003). Fig. 10.

[314] Choi, H. J., Cohen, M. L. & Louie, S. G. *Physica C* **385**, 66–74 (2003). Figs. 2, 6-10; H.J. Choi, D. Roundy, H. Sun, M.L. Cohen, and S.G. Louie, Phys. Rev. B **66**, 020513(R) (2002).

[315] Mishonov, T. M. On the theory of type-I superconductor surface-tension and twinning-plane-superconductivity. *J. Phys. (Paris)* **51**, 447–457 (1990).

[316] Rumer, Y. B. & Ryvkin, M. S. *Thermodynamics, Statistical Physics, and Kinetics* (Mir, Moskow, 1980). Eq. (62. 34-39); A.A. Abrikosov and I.K. Khalatnikov, Usp. Fiz. Nauk **65**, 551 (1958); Adv. Phys. **8**, 45 (1959).

[317] Mishonov, T. M., Klenov, S. I. & Penev, E. S. Temperature dependence of specific heat and penetration depth of anisotropic-gap Bardeen-Cooper-Schrieffer superconductors for a factorizable pairing potential. *Phys. Rev. B* **71**, 024520 (2005). `cond-mat/0212491`.

[318] Abrikosov, A. A. Metal-insulator transition in layered cuprates (SDW model). *Physica C* **391**, 147–159 (2003).

[319] Daghero, D., Gonnelli, R. S., Ummarino, G. A., Stepanov, V. A., Jun, J. & Karpinski, S. M. *Physica C* **385**, 255–263 (2003).

[320] Clem, J. R. Effects of Nonmagnetic Impurities upon Anisotropy of the Superconducting Energy Gap. *Phys. Rev.* **148**, 392–401 (1966). **153**, 449 (1967); Ann. Phys. (N.Y.) **40**, 268 (1966).

[321] Mishonov, T. & Penev, E. Bernoulli potential, Hall constant and Cooper pairs effective masses in disordered BCS superconductors. *Int. J. Mod. Phys. B* **17**, 2883–2895 (2003). `cond-mat/0302168`.

[322] Golubov, A. A., Brinkman, A., Dolgov, O. V., Kortus, J. & Jepsen, O. *Phys. Rev. B* **66**, 05424 (2002).

[323] Zehetmayer, M., Weber, H. W. & Schachinger, E. *J. Low Temp. Phys.* **133**, 407–420 (2003).

[324] Wälte, A., Fuchs, G., Müller, K.-H., Handstein, A., Nenkov, K., Narozhnyi, V. N., Drechsler, S.-L., Shulga, S. & Schultz, L. Evidence for strong electron-phonon coupling in $MgCNi_3$ (2004). (unpublished), `cond-mat/0402421`.

[325] Lifshitz, E. M. & Pitaevskii, L. *Physical Kinetics*, vol. 10 of *Course on Theoretical Physics* (Nauka, Moskow, 1979). (in Russian), Eqs. (85.2) and (85.5); L.D. Landau and E.M. Lifshitz, *Electrodynamics of Continuous Media*, Vol. 8 of *Course on Theoretical Physics* (Moskow, Nauka, 1992), (in Russian), Eq. (22.13).

[326] Lifshitz, I. M., Azbel, M. Y. & Kaganov, M. I. *Elektronnaya Teoriya Metallov* (Nauka, Moskow, 1971). Eq. (27.28), English translation: *Electron Theory of Metals* (Consultants Bureau, New York, 1973).

[327] Carrington, A. & Manzano, F. *Physica C* **385**, 205–214 (2003). F. Manzano, A. Carrington, N.E. Hussey, S. Lee, A. Yamamoto, and S. Tajima, Phys. Rev. Lett. **88**, 047002 (2002).

[328] Samuely, P., Szabo, P., Kačmarčik, J., Klein, T. & Jansen, A. G. M. *Physica C* **385**, 244–254 (2003).

[329] Mackenzie, A. P. & Maeno, Y. *Rev. Mod. Phys.* **75**, 657–712 (2003). Fig. 48, Eqs. (D6) and (D7).

[330] Zhitomirsky, M. E. & Rice, T. M. Interband Proximity Effect and Nodes of Superconducting Gap in Sr_2RuO_4. *Phys. Rev. Lett.* **87**, 057001 (2001).

[331] NishiZaki, S., Maeno, Y. & Mao, Z. Effect of impurities on the Specific Heat of the spin-triplet superconductor Sr_2RuO_4. *J. Low Temp. Phys.* **117**, 1581–1585 (1999).

[332] Litak, G., Annett, J. F., Györffy, B. L. & Wysokiński, K. I. Horizontal line nodes in superconducting Sr_2RuO_4. *phys. stat. sol. (b)* **241**, 983–989 (2004). J. F. Annett, B.L. Györffy, G. Litak, and K.I. Wysokiński, Eur. Phys. J. B **36**, 301 (2003).

[333] Bonalde, I., Yanoff, B. D., Salamon, M. B., Harlingen, D. J. V., Chia, E. M. E., Mao, Z. Q. & Maeno, Y. (2000). Preprint, cited in Ref. [334]; I. Bonalde, B.D. Yanoff, D.J. Van Harlingen, M.B. Salamon, and Y. Maeno, Phys. Rev. Lett. **85**, 4775 (2000).

[334] Dahm, T., Won, H. & Maki, K. (2000). (unpublished), `cond-mat/0006301`.

[335] Deguchi, K., Mao, Z. Q., Yaguchi, H. & Maeno, Y. *Phys. Rev. Lett.* **92**, 047002 (2004).

[336] Farrell, D. E., Bonham, S., Foster, J., Chang, Y. C., Jiang, P. Z., Vandervoort, K. G., Lam, D. J. & Kogan, V. G. Giant superconducting anisotropy in $Bi_2Sr_2Ca_1Cu_2O_{8+\delta}$. *Phys. Rev. Lett.* **63**, 782–785 (1989).

[337] Martin, S., Fiory, A. T., Fleming, R. M., Espinosa, G. P. & Cooper, A. S. Anisotropic critical current density in superconducting $Bi_2Sr_2CaCu_2O_8$ crystals. *Appl. Phys. Lett.* **54**, 72–74 (1989).

[338] Nücker, N., Romberg, H., Nakai, S., Scheerer, B., Fink, J., Yan, Y. F. & Zhao, Z. X. Plasmons and interband transitions in $Bi_2Sr_2CaCu_2O_8$. *Phys. Rev. B* **39**, 12379–12382 (1989).

[339] Gurvitch, M., Valles, J. M., Cucolo, A. M., Dynes, R. C., Garno, J. P., Schneemeyer, L. F. & Waszczak, J. V. Reproducible tunneling data on chemically etched single crystals of $YBa_2Cu_3O_7$. *Phys. Rev. Lett.* **63**, 1008–1011 (1989).

[340] Mooij, J. E. & Schön, G. Propagating plasma mode in thin superconducting filaments. *Phys. Rev. Lett.* **55**, 114–117 (1985).

[341] Noh, T. W., Kaplan, S. G. & Sievers, A. J. Far-infrared sphere resonance in isolated superconducting particles. *Phys. Rev. B* **41**, 307–326 (1990).

[342] Kneschaurek, P., Kamgar, A. & Koch, J. F. Electronic levels in surface space charge layers on Si(100). *Phys. Rev. B* **14**, 1610–1622 (1976).

[343] Ando, T., Fowler, A. B. & Stern, F. Electronic properties of two-dimensional systems. *Rev. Mod. Phys.* **54**, 437–672 (1982).

[344] Helm, M., Colas, E., England, P., DeRosa, F. & S. J. Allen, J. Observation of grating-induced intersubband emission from GaAs/AlGaAs superlattices. *Appl. Phys. Lett.* **53**, 1714–1716 (1988).

[345] Nagaoka, T., Matsuda, Y., Obara, H., Sawa, A., Terashima, T., Chong, I., Takano, M. & Suzuki, M. Hall Anomaly in the Superconducting State of High-T_c Cuprates: Universality in Doping Dependence. *Phys. Rev. Lett.* **80**, 3594–3597 (1998).

[346] D'Anna, G., Berseth, V., Forró, L., Erb, A. & Walker, E. Hall Anomaly and Vortex-Lattice Melting in Superconducting Single Crystal $YBa_2Cu_3O_{7-\delta}$. *Phys. Rev. Lett.* **81**, 2530–2533 (1998).

[347] Khomskii, D. I. & Freimuth, A. Charged Vortices in High Temperature Superconductors. *Phys. Rev. Lett.* **75**, 1384–1386 (1995).

[348] Blatter, G., Feigel'man, M., Geshkenbein, V., Larkin, A. & van Otterlo, A. Electrostatics of Vortices in Type-II Superconductors. *Phys. Rev. Lett.* **77**, 566–569 (1996).

[349] Hayashi, N., Ichioka, M. & Machida, K. Relation between Vortex Core Charge and Vortex Bound States. *J. Phys. Soc. Jpn.* **67**, 3368–3371 (1998).

[350] Kato, Y. Charging Effect on Hall Conductivity of single Vortex in Type II Superconductors. *J. Phys. Soc. Jpn.* **68**, 3798–3801 (1998).

[351] Bozovic, I. & Eckstein, J. N. Atomic-Level Engineering of Cuprates and Manganites. *Appl. Surf. Sci.* **113–114**, 189–197 (1997).

[352] Eckstein, I. B. J. N. & Virshup, G. F. Superconducting Oxide Multilayers and Superlattices: Physics, Chemistry, and Nano-engineering. *Physica C* **235–240**, 178–181 (1994).

[353] Bozovic, I. & Eckstein, J. N. Superconducting Superlattices. In Ginsberg, D. M. (ed.) *Physical Properties of Superconductors*, vol. 5, 99–207 (World Scientific, Singapore, 1996).

[354] Corson, J., Mallozzi, R., Orenstein, J., Eckstein, J. N. & Bozovic, I. Vanishing of phase coherence in underdoped $Bi_2Sr_2CaCu_2O_{8+x}$. *Nature* **398**, 221–223 (1999).

[355] Millis, A. J. High-temperature superconductivity: Research enters a new phase. *Nature* **398**, 193–194 (1999).

[356] Mallozzi, R., Orenstein, J., Eckstein, J. N. & Bozovic, I. High-Frequency Electrodynamics of $Bi_2Sr_2CaCu_2O_{8+\delta}$: Nonlinear Response in the Vortex State. *Phys. Rev. Lett.* **81**, 1485–1488 (1998).

[357] Ronning, F., Kim, C., Feng, D. L., Marshall, D. S., Loeser, A. G., Miller, L. L., Eckstein, J. N., Bozovic, I. & Shen, Z.-X. Photoemission Evidence for a Remnant Fermi Surface and a *d*-Wave-Like Dispersion in Insulating $Ca_2CuO_2Cl_2$. *Science* **282**, 2067–2072 (1998).

[358] Mishonov, T. M. Theory of Cooper-pair mass spectroscopy by the current-induced contact-potential difference. *Phys. Rev. B* **50**, 4009–4014 (1994).

[359] Mishonov, T. M. & Zahariev, N. I. Predicted new electrodynamic effect for thin films of layered cuprates - surface Hall currents induced by electric induction $D_{\perp \mathrm{film}}$. *Superlatt. Microstruct.* **26**, 57–60 (1999).

[360] McKee, R. A., Walker, F. J. & Chisholm, M. F. Crystalline Oxides on Silicon: The First Five Monolayers. *Phys. Rev. Lett.* **81**, 3014–3017 (1998).

[361] Fiory, A. T., Hebard, A. F., Mankiewich, P. M. & Howard, R. E. Renormalization of the Mean-Field Superconducting Penetration Depth in Epitaxial $YBa_2Cu_3O_7$ Films. *Phys. Rev. Lett.* **61**, 1419–1422 (1988).

[362] Omel'yanchuk, A. N. & Beloborod'ko, S. I. Bernoulli effect and contact potential difference in superconductors. *Sov. J. Low Temp. Phys.* **9**, 572–574 (1983). [Fiz. Nizk. Temp. **9**, 1105–1108 (1983)].

[363] Rogers, C. T., Myers, K. E., Eckstein, J. N. & Bozovic, I. Brownian motion of vortex-antivortex excitations in very thin films of $Bi_2Sr_2CaCu_2O_8$. *Phys. Rev. Lett.* **69**, 160–163 (1992).

[364] Greiter, M., Wilczek, F. & Witten, E. *Mod. Phys. Lett. B* **3**, 903 (1989).

[365] Simkin, M. V. Bernoulli effect in superconductors and Cooper-pair mass spectroscopy (1996). `cond-mat/9605036`.

[366] Lewis, H. W. Search for the Hall Effect in a Superconductor. I. Experiment. *Phys. Rev.* **92**, 1149–1151 (1953).

[367] Lewis, H. W. Search for the Hall Effect in a Superconductor. II. Theory. *Phys. Rev.* **100**, 641–645 (1955).

[368] London, F. *Suprefluids* (Wiley, New York, 1950). P. 70.

[369] Fiory, A. T., Hebard, A. F., Eick, R. H., Mankiewich, P. M., Howard, R. E. & O'Malley, M. L. Fiory *et al.* reply. *Phys. Rev. Lett.* **67**, 3196 (1991).

[370] Fiory, A. T., Hebard, A. F., Eick, R. H., Mankiewich, P. M., Howard, R. E. & O'Malley, M. L. Metallic and superconducting surfaces of $YBa_2Cu_3O_7$ probed by electrostatic charge modulation of epitaxial films. *Phys. Rev. Lett.* **65**, 3441–3444 (1990). Erratum: *ibid.* **66**, 845 (1991).

[371] Mott, S. N. F. (1982). Private communication.

[372] Ausloos, M. & Varlamov, A. A. (eds.). *Fluctuation Phenomena in High Temperature Superconductors*, vol. 32 of *NATO Science Partnership Sub-Series: 3* (Kluwer, Dordrecht, 1997).

[373] Landau, L. D. & Lifshitz, E. M. *Statistical Physics Part II*, vol. 9 of *Course on Theoretical Physics* (Pergamon Press, New York, 1980). Section 49.

[374] Tinkham, M. *Introduction to Superconductivity* (McGraw-Hill, New York, 1996), 2nd edn.

[375] Cyrot, M. *Rep. Prog. Phys.* **36**, 103 (1973).

[376] Bulaevskii, L. N. *Usp. Fiz. Nauk* **111**, 553 (1973). [Sov. Phys. Uspekhi **111**, 939 (1974)].

[377] Skocpol, W. J. & Tinkham, M. Fluctuations near superconducting phase transitions. *Rep. Prog. Phys.* **38**, 1094–1097 (1975).

[378] Bulaevskii, L. N., Ledvij, M. & Kogan, V. G. *Phys. Rev. Lett.* **68**, 3773 (1992).

[379] Bulaevskii, L. N., Ledvij, M. & Kogan, V. G. Distorted vortex in Josephson-coupled layered superconductors. *Phys. Rev. B* **46**, 11807–11812 (1992).

[380] Song, Q., Halperin, W. P., Tonge, L., Marks, T. J., Ledvij, M., Kogan, V. G. & Bulaevskii, L. N. Low temperature fluctuations of vortices in layered superconductors. *Phys. Rev. Lett* **70**, 3127–3130 (1993).

[381] Li, Q., Suenaga, M., Bulaevskii, L. N., Hikata, T. & Sato, K. Thermodynamics of superconducting $Bi_2Sr_2Ca_2Cu_3O_{10}$ in the critical fluctuation region near the $H_{c2}(T)$ line. *Phys. Rev. B* **48**, 13865–13870 (1993).

[382] Bulaevskii, L., Kolesnikov, N. N., Schegolev, I. F. & Vyasilev, O. M. Effect of vortex fluctuations on ^{205}Tl spin-lattice relaxation in the mixed state of $Tl_2Ba_2CuO_6$. *Phys. Rev. Lett.* **71**, 1891–1894 (1993).

[383] Bulaevskii, L. N. & Maley, M. P. Low temperature specific heat of the vortex lattice in layered superconductors. *Phys. Rev. Lett.* **71**, 3541–3544 (1993).

[384] Bulaevskii, L. N., Cho, J. H., Maley, M. P., Kes, P. H., Li, Q., Suenaga, M. & Ledvij, M. Effect of quantum fluctuations of vortices on reversible magnetization in high-T_c superconductors. *Phys. Rev. B* **50**, 3507–3510 (1994).

[385] Tesanovic, Z. Extreme type-II superconductors in a magnetic field: A theory of critical fluctuations. *Phys. Rev. B* **59**, 6449–6474 (1999).

[386] Pierson, S. W., Katona, T. M., Tesanovic, Z. & Valls, O. T. *Phys. Rev. B* **53**, 8638 (1996).

[387] Tesanovic, Z., Xing, L., Bulaevskii, L. N., Li, Q. & Suenaga, M. *Phys. Rev. Lett.* **69**, 3563 (1992).

[388] Lawrence, W. E. & Doniach, S. In Kanda, E. (ed.) *Proceedings of the Twelfth International Conference on Low-Temperature Physics, Kyoto, Japan, 1970*, 361 (Keigatu, Tokyo, 1971).

[389] Maki, K. & Thompson, R. S. Fluctuation conductivity of high-T_c superconductors. *Phys. Rev. B* **39**, 2767–2770 (1989).

[390] Hikami, S. & Larkin, A. I. Magnetoresistance of high temperature superconductors. *Mod. Phys. Lett. B* **2**, 693–698 (1988).

[391] Landau, L. D. & Lifshitz, E. M. *Quantum Mechanics*, vol. 3 of *Course on Theoretical Physics* (Pergamon Press, Oxford, 1978). Secs. 23 and 112.

[392] Schmidt, H. Onset of superconductivity in time dependent Ginzburg-Landau theory. *Z. Phys.* **216**, 336 (1968).

[393] Schmid, A. Diamagnetic Susceptibility at the Transition to the Superconducting State. *Phys. Rev.* **180**, 527–529 (1969).

[394] Schmidt, V. V. In *Proceedings of the 10th International Conference on Low Temperature Physics*, vol. 2C, 205 (VINITI, Moskow, 1967).

[395] Ramallo, M. V. *Multilayered Gaussian-Ginzburg-Landau Theory for Thermal Fluctuations in High Temperature Superconductors under Weak Magnetic Fields.* Ph.D. thesis, University of Santiago de Compostela (1998).

[396] Balestrino, A. A. V. G., Milani, E. & Livanov, D. V. *Adv. Phys.* **48**, 655 (1999).
[397] Tsuzuki, T. *Phys. Lett. A* **37**, 159 (1971).
[398] Yamayi, K. *Phys. Lett. A* **38**, 43 (1972).
[399] Ramallo, M. V., Pomar, A. & Vidal, F. *Phys. Rev. B* **54**, 4341 (1996).
[400] Díaz, C. T. A., Jegoudez, J., Pomar, A., Ramallo, M. V., Revcolevschi, A., Veira, J. A. & Vidal, F. *Physica C* (1993).
[401] Pomar, A., Díaz, A., Ramallo, M. V., Torrón, C., Veira, J. A. & Vidal, F. *Physica C* **218**, 257 (1993).
[402] Pomar, A., Ramallo, M. V., Mosqueira, J., Torrón, C. & Vidal, F. Fluctuation-induced in-plane conductivity, magnetoconductivity, and diamagnetism of $Bi_2Sr_2CaCu_2O_8$ single crystals in weak magnetic fields. *Phys. Rev. B* **54**, 7470–7480 (1996).
[403] Torrón, C., Díaz, A., Pomar, A., Veira, J. A. & Vidal, F. *Phys. Rev. B* **49**, 13143 (1994).
[404] Pomar, A. *Fluctuations-induced conductivity, magnetoconductivity and diamagnetism in single crystals of multilayered copper oxide superconductors.* Ph.D. thesis, Universidade de Santiago de Compostela (1995). In Spanish.
[405] Aslamazov, L. G. & Larkin, A. I. Effect of Fluctuations on the Properties of a Superconductor Above the Critical Temperature. *Fiz. Tverdogo Tela* **10**, 1104–1111 (1968). [Sov. Phys.-Solid State **10**, 875–880 (1968)].
[406] Ferrell, R. A. *J. Low Temp. Phys.* **1**, 241 (1969).
[407] Thouless, D. J. *Ann. Phys. (N.Y.)* **10**, 553 (1960).
[408] Klemm, R. A. *Phys. Rev. B* **41**, 2073 (1990).
[409] Ramallo, M. V., Torrón, C. & Vidal, F. *Physica C* **230**, 97 (1994).
[410] Abrahams, E. & Woo, J. W. F. Phenomenological theory of rounding of resistive transition of superconductors. *Phys. Lett. A* **27**, 117 (1968).
[411] Abrahams, E. & Tsuneto, T. Time Variation of the Ginzburg-Landau Order Parameter. *Phys. Rev.* **152**, 416–432 (1966).
[412] Schmid, A. *Phys. Kondens. Materie* **5** (1966).
[413] Caroli, C. & Maki, K. Fluctuations of the Order Parameter in Type-II Superconductors. I. Dirty Limit. *Phys. Rev.* **159**, 306–315 (1967).
[414] Fulde, P. & Maki, K. *Phys. Kondens. Materie* **8** (1989).
[415] Yip, S.-K. *J. Low. Temp. Phys.* **81**, 129 (1990).
[416] Yip, S.-K. *Phys. Rev. B* **41**, 2612 (1990).
[417] Woo, J. W. F. & Abrahams, E. *Phys. Rev.* **169**, 407 (1968).
[418] Ramallo, M. V., Carballeira, C., na, J. V., Veira, J. A., Mishonov, T., Pavuna, D. & Vidal, F. Relaxation time of the Cooper pairs near T_c in cuprate superconductors. *Europhys. Lett.* **48**, 79–85 (1999).
[419] Redi, M. Ph.D. thesis, Rutgers University (1969).
[420] Abrahams, E., Prange, R. E. & Stephen, M. J. Effect of a magnetic field on fluctuations above T_c. *Physica* **55**, 230–233 (1971).
[421] Dorin, V. V., Klemm, R. A., Varlamov, A. A., Buzdin, A. I. & Livanov, D. V. *Phys. Rev. B* **48**, 12951 (1993).
[422] Rebolledo, J. M. V. *What can and what cannot be said about paraconductivity in high temperature superconductors.* Ph.D. thesis, University of

Santiago de Compostela (1999). In Spanish; To our knowledge, this is the first (dated Feb. 10th, 1999) determination of the energy cutoff parameter c.

[423] Dorsey, A. T. Linear and nonlinear conductivity of a superconductor near T_c. *Phys. Rev. B* **43**, 7575–7585 (1991).

[424] Aslamazov, L. G. & Varlamov, A. A. *J. Low Temp. Phys.* **38**, 223 (1980).

[425] Tsuzuki, T. & Koyanagi, M. In *Proceedings of the 12th International Conference on Low Temp. Physics, Kyoto, 1970*, 249 (Academic Press of Japan, Tokyo, 1971).

[426] Wynn, P. On a device for computing the $e_m(S_n)$ tranformation. *Math. Comp.* **10**, 91–96 (1956).

[427] Shanks, D. Non-linear transformations of divergent and slowly convergent sequences. *J. Math. Phys.* **34**, 1–42 (1955).

[428] Baker, G., Jr. & Graves-Morris, P. Encyclopedia of Mathematics and its Applications. vol. 13, chap. Padé Approximants (Addison-Wesley, 1981). Table 3, p. 78.

[429] Mishonov, T. M. & Penev, E. S. Thermodynamics of Gaussian fuctuations and paraconductivity in layered superconductors. *Int. J. Mod. Phys. B* **14**, 3831–3879 (2000). `cond-mat/0004023`.

[430] Mehra, J. *The Beat of a Different Drum: The Life and Science of Richard Feynman* (Oxford University Press, Oxford, 1994).

[431] Hutchings, E., Leighton, R., Feynman, R. P. & Hibbs, A. *'Surely You're Joking, Mr. Feynman!': Adventures of a Curious Character* (W. W. Norton & Company, New Yorkxsxs, 1997).

[432] Baraduc, C., Henry, J.-Y., Buzdin, A., Brison, J.-P. & Puech, L. Fluctuations in quasi-2D superconductors under magnetic-field - the case of $YBa_2Cu_3O_{7-\delta}$. *Physica C* **248**, 138–146 (1995).

[433] Bronshtein, I. & Semendiaev, K. *Manual of Mathematics* (Mir, Moscow, 1977). P. 185.

[434] Ramond, P. *Field Theory: A Modern Primer* (Addison-Wesley, Madrid, 1988), 2nd edn. Secs. 3.4–3.7.

[435] Ray, D. B. & Singer, I. M. R-torsion and the Laplacian on Riemannian manifolds. *Adv. Math.* **7**, 145–210 (1971).

[436] Dowker, J. S. & Critchley, R. Effective Lagrangian and energy-momentum tensor in de Sitter space. *Phys. Rev. D* **13**, 3224–3232 (1976).

[437] Corrigan, E., Goddard, P., Osborn, H. & Templeton, S. Zeta-function regularization and multi-instanton determinants. *Nucl. Phys. B* **159**, 469–496 (1979).

[438] Ivich, A. *The Riemann zeta-function* (Wiley, New York, 1985).

[439] Gollub, J. P., Beasley, M. R., Callarotti, R. & Tinkham, M. Fluctuation-Induced Diamagnetism above T_c in Superconductors. *Phys. Rev. B* **7**, 3039–3058 (1973).

[440] Patton, B. R., Ambegaokar, V. & Wilkins, J. W. On diamagnetic susceptibility of a superconductor above transition temperature. *Solid State Commun.* **7**, 1287 (1969).

[441] Lee, P. A. & Payne, M. G. Theory of Fluctuation-Induced Diamagnetism in Superconductors. *Phys. Rev. Lett.* **26**, 1537–1541 (1971).
[442] Lee, P. A. & Payne, M. G. Pair Propagator Approach to Fluctuation-Induced Diamagnetism in Superconductors-Effects of Impurities. *Phys. Rev. B* **5**, 923–931 (1972).
[443] Kurkijärvi, J., Ambegaokar, V. & Eilenberger, G. Fluctuation-Enhanced Diamagnetism in Superconductors above the Transition Temperature. *Phys. Rev. B* **5**, 868–879 (1972).
[444] Maki, H., K.and Takayama. Dynamical fluctuation of the order parameter and diamagnetism of superconductors. I. Dirty limit. *J. Low. Temp. Phys* **5**, 313–323 (1971).
[445] Maki, K. Fluctuation-Induced Diamagnetism in Dirty Superconductors. *Phys. Rev. Lett.* **30**, 648–652 (1973).
[446] Carretta, P., Lascialfari, A., Rigamonti, A., Rosso, A. & Varlamov, A. Superconducting fluctuations and anomalous diamagnetism in underdoped $YBa_2Cu_3O_{6+x}$ from magnetization and ^{63}Cu NMR-NQR relaxation measurements. *Phys. Rev. B* **61**, 12420–12426 (2000).
[447] Carballeira, C., Mosqueira, J., Revcolevschi, A. & Vidal, F. *Phys. Rev. Lett.* **84**, 3157 (2000).
[448] Mishonov, T. Comment on "First observation for a cuprate superconductor of fluctuation-induced diamagnetism well inside the finite-magnetic-field range". Submitted to *Phys. Rev. Lett.* (unpublished), `cond-matt/0007178`.
[449] Mishonov, T. In Kossowsky, R., Methfessel, S. & Wohlleben, D. (eds.) *Physics and Materials Science of High Temperature Superconductors*, vol. 181 of *NATO Science Series: E* (Kluwer, Dordrecht, 1990).
[450] Mishonov, T. M., Penev, E. S. & Ivanov, I. P. Fluctuation torque for high-temperature superconductors with strong anisotropy (1993). Department of Theoretical Physics, Faculty of Physics, University of Sofia (unpublished).
[451] Ullah, S. & Dorsey, A. T. Effect of fluctuations on the transport properties of type-II superconductors in a magnetic field. *Phys. Rev. B* **44**, 262–273 (1991).
[452] Klemm, R. A., Beasley, M. R. & Luther, A. Fluctuation-Induced Diamagnetism in Dirty Three-Dimensional, Two-Dimensional, and Layered Superconductors. *Phys. Rev. B* **8**, 5072–5081 (1973).
[453] Koshelev, A. E. Ginzburg-Landau theory of fluctuation magnetization of two-dimensional superconductors. *Phys. Rev. B* **50**, 506–516 (1994).
[454] Patashinskii, A. Z. & Pokrovsky, V. L. *Fluctuation Theory of Phase Transitions* (Pergamon, Oxford, 1979). Pp. viii, 30, 194.
[455] Livanov, D. V., Varlamov, A. A., Putti, M., Cimberle, M. R., & Ferdeghini, C. Evidence of critical fluctuations from the magnetoconductivity data in $Bi_2Sr_2Ca_2Cu_3O_{10+x}$ phase. *Eur. Phys. J. B* **18**, 401–404 (2000).
[456] Prange, R. E. Diamagnetic Susceptibility at the Transition to the Superconducting State. *Phys. Rev. B* **1**, 2349–2350 (1970).
[457] Wolfram, S. *The Mathematica Book* (Wolfram Media/Cambridge University Press, 1999), 4th edn. P. 1469.

[458] Booth, J. C., Wu, D. H., Qadri, S. B., Skelton, E. F., Osofsky, M. S., Piqué, A. & Anlage, S. M. *Phys. Rev. Lett.* **77**, 4438 (1996).
[459] Anlage, S. M., Mao, J., Booth, J. C., Wu, D. H. & Peng, J. L. Fluctuations in the microwave conductivity of $YBa_2Cu_3O_{7-\delta}$ single crystals in zero dc magnetic field. *Phys. Rev. B* **53**, 2792–2796 (1996).
[460] Larkin, A. & Varlamov, A. Fluctuation Phenomena in Superconductors. In Bennemann, K.-H. & Ketterson, J. B. (eds.) *Conventional and High-T_c Superconductors*, vol. I of *The Physics of Superconductors*, 95 (Springer, Berlin, 2003). `cond-mat/0109177`.
[461] Glatz, A., Vinokur, V. M. & Varlamov, A. A. Fluctuation Conductivity of Two-Dimensional Superconductor in Magnetic field: a Complete Picture. Unpublished.
[462] Konsin, P., Sorkin, B. & Ausloos, M. Electric Field Effects in High-T_c Superconductors. In Ausloos, M. & Varlamov, A. A. (eds.) *Fluctuation Phenomena in High Temperature Superconductors*, vol. 32 of *NATO Science Partnership Sub-Series: 3*, 91–100 (Kluwer, Dordrecht, 1997).
[463] Konsin, P. & Sorkin, B. Electric field effects in high-T_c cuprates. *Phys. Rev. B* **58**, 5795–5802 (1998).
[464] Konsin, P., Sorkin, B. & Ausloos, M. *Supercond. Sci. Technol.* **11**, 1 (1998).
[465] Varlamov, A. A. & Ausloos, M. Fluctuation Phenomena in Superconductors. In Ausloos, M. & Varlamov, A. A. (eds.) *Fluctuation Phenomena in High Temperature Superconductors*, vol. 32 of *NATO Science Partnership Sub-Series: 3*, 3–41 (Kluwer, Dordrecht, 1997).
[466] Ausloos, M., Clippe, P. & Laurent, C. *Solid State Commun.* **73**, 137 (1990).
[467] Konsin, P., Sorkin, B. & Ausloos, M. Electric Field Effects in High-T_c Cuprates with Different Bulk and Surface Conductivities. In Ausloos, M. & Kruchinin, S. (eds.) *Symmetry and Pairing in Superconductors*, vol. 63 of *NATO Science Partnership Sub-Series: 3*, 151–160 (Kluwer, Dordrecht, 1999).
[468] Ausloos, M., Gillet, F., Laurent, C. & Clippe, P. *Z. Phys. B* **84**, 13 (1991).
[469] Ausloos, M., Patapis, S. K. & Clippe, P. In Kossowsky, R., Raveau, B., Wohlleben, D. & Patapis, S. K. (eds.) *Physics and Materials Science of High Temperature Superconductors II*, 755–785 (Kluwer, Dordrecht, 1992).
[470] Sergeenkov, S. A., Gridin, V. V., de Villiers, P. & Ausloos, M. *Phys. Scr.* **49**, 637 (1994).
[471] Vos, K., Dixon, J. M. & Tuszyński, J. A. Magnetic-field penetration in superconductors studied using the Landau-Ginzburg model and nonlinear techniques. *Phys. Rev. B* **44**, 11933–11950 (1991).
[472] Tuszyński, J. A. & Wierzbicki, A. Non-Gaussian approach to critical fluctuations in the Landau-Ginzburg model and finite-size scaling. *Phys. Rev. B* **43**, 8472–8481 (1991).
[473] Mishonov, T. M. & Damianov, D. C. Kinetic equation for fluctuation Cooper pairs applied to fluctuation Hall effect in thin superconducting films. *Czech. J. Phys.* **46**, (Suppl. S2) 631 (1996).
[474] Damianov, D. C. & Mishonov, T. M. Fluctuation paraconductivity within the framework of time-dependent Ginzburg-Landau theory. *Superlat. Microstruct.* **21**, 467 (1997).

[475] Abrahams, E., Redi, M. & Woo, J. W. F. Effect of Fluctuations on Electronic Properties above the Superconducting Transition. *Phys. Rev. B* **1**, 208–213 (1970).

[476] Kemoklidze, M. P. & Pitaevskiĭ, L. P. On the Dynamics of a Superfluid Fermi Gas. *Zh. Éksp. Teor. Fiz.* **50**, 243–250 (1966). [Sov. Phys.–JETP **23**, 160 (1966)].

[477] Boltzmann, L. *Vorlesungen über Gastheorie* (Amnrosius Barth, Leipzig, 1912).

[478] Flamm, D. Ludwig Boltzmann – A Pioneer of Modern Physics (1997). `physics/9710007`.

[479] Lifshitz, E. M. & Pitaevskii, L. *Physical Kinetics*, vol. 10 of *Course on Theoretical Physics* (Pergamon, Oxford, 1981).

[480] Fiory, A. T., Hebard, A. F. & Glaberson, W. I. Superconducting phase transitions in indium/indium-oxide thin-film composites. *Phys. Rev. B* **28**, 5075–5087 (1983).

[481] Mishonov, T., Posazhennikova, A. & Indekeu, J. Fluctuation conductivity in superconductors in strong electric fields. *Phys. Rev. B* **65**, 064519 (2002). `cond-mat/0106168`.

[482] Gor'kov, L. P. *Zh. Éksp. Teor. Fiz. Pis. Red.* **11**, 52–56 (1970). [Sov. Phys.–JETP Lett. **11**, 32–35 (1970)].

[483] Michotte, S., Mátéfi-Tempfli, S. & Piraux, L. Investigation of superconducting properties of nanowires prepared by template synthesis. *Supercond. Sci. Technol.* **16**, 557–561 (2003).

[484] Michotte, S., Piraux, L., Dubois, S., Pailloux, F., Stenuit, G., & Govaerts, J. *Physica C* **377**, 267 (2002).

[485] Sharifi, F., Herzog, A. V. & Dynes, R. C. Crossover from two to one dimension in in situ grown wires of Pb. *Phys. Rev. Lett.* **71**, 428–431 (1993).

[486] Bezryadin, A., Lau, C. N. & Tinkham, M. *Nature* **404**, 971 (2000).

[487] Prober, D. E., Feuer, M. D. & Giordano, N. Fabrication of 300-Å metal lines with substrate-step techniques. *Appl. Phys. Lett.* **37**, 94–96 (1980).

[488] Martin, C. R. Nanomaterials - a membrane-based synthetic approach. *Science* **266**, 1961–1966 (1994).

[489] Yi, G. & Schwarzacher, W. Single crystal superconductor nanowires by electrodeposition. *Appl. Phys. Lett.* **74**, 1746–1748 (1999).

[490] Dubois, S., Michel, A., Eymery, J. P., Duvail, J. L. & Piraux, L. Fabrication and properties of arrays of superconducting nanowires. *J. Mater. Res.* **14**, 665–671 (1999).

[491] Webb, W. W. & Warburton, R. J. Intrinsic Quantum Fluctuations in Uniform Filamentary Superconductors. *Phys. Rev. Lett.* **20**, 461–465 (1968).

[492] Meyer, J. & von Minnigerode, G. *Phys. Lett A* **38**, 529 (1972).

[493] Skocpol, W. J., Beasley, M. R. & Tinkham, M. Phase-slip centers and nonequilibrium processes in superconducting tin microbridges. *J. Low. Temp. Phys.* **16**, 145–167 (1974).

[494] Kramer, L. & Baratoff, A. Lossless and Dissipative Current-Carrying States in Quasi-One-Dimensional Superconductors. *Phys. Rev. Lett.* **38**, 518–521 (1977).

[495] Kramer, L. & Watts-Tobin, R. J. Theory of dissipative current-carrying states in superconducting filaments. *Phys. Rev. Lett.* **40**, 1041–1044 (1978).

[496] Watts-Tobin, R. J., Krähenbühl, Y. & Kramer, L. Nonequilibrium theory of dirty, current-carrying superconductors: phase-slip oscillators in narrow filaments near T_c. *J. Low Temp. Phys.* **42**, 459–501 (1981).

[497] Maneval, J. P., Boyer, F., Harrabi, K., & Ladan, F. R. *J. Supercond.* **14**, 347 (2001).

[498] Kramer, L. & Rangel, R. Structure and properties of the dissipative phase-slip state in narrow superconducting filaments with and without inhomogeneities. *J. Low Temp. Phys.* **57**, 391–414 (1984).

[499] Silva, E., Marcon, R., Fastampa, R., Giura, M. & Sarti, S. High-Frequency Fluctuational Conductivity in $YBa_2Cu_3O_{7-\delta}$. *J. Low Temp. Phys.* **131**, 831–835 (2003).

[500] Silva, E., Marcon, R., Sarti, S., Fastampa, R., M. Giura, M. B. & Cucolo, A. M. Microwave fluctuational conductivity in $YBa_2Cu_3O_{7-\delta}$. *Eur. Phys. J. B* **37**, 277–284 (2004).

[501] Puica, I. & Lang, W. Non-Ohmic critical fluctuation Hall conductivity of layered superconduc tors in strong electric fields. *Phys. Rev. B* **70**, 092507 (2004).

[502] Mishonov, T. M. Magnetoplasma waves in thin high-temperature-superconducting layers. *Phys. Rev. B* **42**, 6715–6716 (1990).

[503] Karrai, K., Choi, E., Dunmore, F., Liu, S., Ying, X., Li, Q., Venkatesan, T., Drew, H. D., Li, Q. & Fenner, D. B. Far-infrared magneto-optical activity in type II superconductors. *Phys. Rev. Lett.* **69**, 355–358 (1992).

[504] Mishonov, T. M. Technique for measuring the Cooper-pair velocity, density, and mass using Doppler splitting of the plasmon resonance in low-dimensional superconductor microstructures. *Phys. Rev. B* **50**, 4004–4008 (1994).

[505] Ziman, J. M. *Electrons and Phonons* (Oxford University Press, Oxford, 1960).

[506] Lifshitz, I. M. & Kaganov, M. I. *Sov. Phys. Usp.* **2**, 831 (1960). *ibid* **5**, 878 (1963), *ibid* **8**, 805 (1966) [Usp. Fiz. Nauk **69**, 419 (1959), *ibid* **78**, 411 (1962), *ibid* **87**, 389 (1965)].

[507] Ashcroft, N. W. & Mermin, N. D. *Solid State Physics* (Holt, Rinehart and Winston, New York, 1976). Ch. 16.

[508] Varlamov, A. A. & Reggiani, L. Nonlinear fluctuation conductivity of a layered superconductor: Crossover in strong electric fields. *Phys. Rev. B* **45**, 1060–1063 (1992). Eqs. (1), (6).

[509] Hurault, J. P. Nonlinear Effects on the Conductivity of a Superconductor above Its Transition Temperature. *Phys. Rev.* **179**, 494–496 (1969).

[510] Varlamov, A. A. & Yu, L. Double-crossover behavior of fluctuation conductivity for multilayered superconducting films. *Phys. Rev. B* **44**, 7078–7080 (1991).

[511] Puica, I., Lang, W., Peruzzi, W., v Lemmermann, K., Pedaring, J. D. & Bäuerle, D. High electric field study of the superconducting transition of $YBa_2Cu_3O_{7-x}$. *Supercond. Sci. Technol.* **17**, S543–S547 (2004).

[512] Gurvitch, M. & Fiory, A. T. Resistivity of $La_{1.825}Sr_{0.175}CuO_4$ and $YBa_2Cu_3O_7$ to 1100 K: Absence of saturation and its implications. *Phys. Rev. Lett.* **59**, 1337–1340 (1987).
[513] Martin, S., Fiory, A. T., Fleming, R. M., Schneemeyer, L. F. & Waszczak, J. V. Temperature Dependence of the Resistivity Tensor in Superconducting $Bi_2Sr_{2.2}Ca_{0.8}Cu_2O_8$ Crystals. *Phys. Rev. Lett.* **60**, 2194–2197 (1988).
[514] Ong, N. P. Cuprates Fall into a Gap. *Science* **273**, 321–322 (1996).
[515] Allen, P. B., Fisk, Z. & Migliori, A. Normal State Transport and Elastic Properties of High-T_c Materials and Related Compounds. In Ginsberg, D. M. (ed.) *Physical Properties of High-Temperature Superconductors*, vol. I, 213 (World Scientific, Singapore, 1989).
[516] Ong, N. P. The Hall Effect and its Relation to other Transport Phenomena in the Normal State of the High-Temperature Superconductors. In Ginsberg, D. M. (ed.) *Physical Properties of High-Temperature Superconductors*, vol. II, 459 (World Scientific, Singapore, 1990).
[517] Batlogg, B. In Bedell, K. S., Coffey, D., Meltzer, D. E., Pines, D. & Schrieffer, J. R. (eds.) *High Temperature Superconductivity; Proc. Los Alamos Symp. 1989*, 37 (Addison-Wesley, Redwood City, CA, 1990).
[518] Maki, K. & Won, H. Fermi liquid theory with antiparamagnon. *Physica B* **206–207**, 650–653 (1995).
[519] Virosztek, A. & Ruvalds, J. Nested-Fermi-liquid theory. *Phys. Rev. B* **42**, 4064–4072 (1990).
[520] Rieck, C. T., Little, W. A., Ruvalds, J. & Virosztek, A. *Solid State Commun.* **88**, 325 (1993).
[521] Varma, C. M., Littlewood, P. B., Schmitt-Rink, S., Abrahams, E. & Ruckenstein, A. E. Phenomenology of the normal state of Cu-O high-temperature superconductors. *Phys. Rev. Lett.* **63**, 1996–1999 (1989).
[522] Moshchalkov, V. V. *Physica B* **163**, 59 (1990). Physica C **156**, 473 (1988).
[523] V. V. Mochchalkov, L. T. & Vanacken, J. *Europhys. Lett* **46**, 75 (1999).
[524] Abrikosov, A. A. Resonant tunneling in high-T_C superconductors. *Usp. Fiz. Nauk* **168**, 683–695 (1998). [Sov. Phys. Usp. **41**, 605 (1998)].
[525] Orear, J. *Fundamental Physics* (Willey, New York, 1967). Chap. 15, Sec. 11, Appendix 2, Fig. 322.
[526] Carlson, E. W., Kivelson, S. A., Emery, V. J. & Manousakis, E. Classical Phase Fluctuations in High Temperature Superconductors. *Phys. Rev. Lett.* **83**, 612–615 (1999).
[527] Aslamazov, L. G. & Varlamov, A. A. *Astonishing Physics*, vol. 63 of *Library Quant* (Nauka, Moscow, 1988). In Russian, pp. 27–31.
[528] Landau, L. D. & Lifshitz, E. M. *Electrodynamics of Continuous Media*, vol. 8 of *Course on Theoretical Physics* (Pergamon, New York, 1974).
[529] Berger, H., Forró, L. & Pavuna, D. On linear resistivity from ~ 1 to 10^3 K in $Sr_2RuO_{4\delta}$single crystals grown by flux technique. *Europhys. Lett.* **41**, 531–534 (1998).
[530] Pavuna, D., Berger, H. & Forró, L. Linear resistivity from ~ 1 to 1050 K in $Sr2RuO_4$ single crystals grown by the flux technique. *J. Eur. Ceram. Soc.* **19**, 1515–1518 (1999).

[531] Pavuna, D., Forró, L. & Berger, H. Anomalous Linear Resistivity from 1 to 10^3 K in $Sr_2RuO_{4-\delta}$. In *et al.*, S. E. B. (ed.) *AIP Conference Proceedings of HTS99 Miami Conference*, 483 (Woodbury, 1999).

[532] Analytis, J. G., Abdel-Jawad, M., Balicas, L., French, M. M. J. & Hussey, N. E. Angle-dependent magnetoresistance measurements in $Tl_2Ba_2CuO_{6+\delta}$ and the need for anisotropic scattering. *Phys. Rev. B* **76**, 104523 (2007).

[533] Mishonov, T. M., Pachov, G. V., Genchev, I. N., Atanasova, L. A. & Damianov, D. C. Kinetics and Boltzmann kinetic equation for fluctuation Cooper pairs. *Phys. Rev. B* **68**, 054525 (2003).

[534] Mishonov, T. M. & Mishonov, M. T. Generation of Electric Oscillations by Continuous, Supercooled Superconductrors with an Applied Voltage (2003). W02004079893 (unpublished).

[535] Tucker, J. R. & Halperin, B. I. Onset of Superconductivity in One-Dimensional Systems. *Phys. Rev. B* **3**, 3768–3782 (1971).

[536] Chow, W. F. *Principles of Tunnel Diode Circuits* (Wiley, New York, 1964).

[537] Enns, R. H. & McGuire, G. C. *Computer Algebra Recipes. A Gourmet's Guide to the Mathematical Models of Science* (Springer, Berlin, 2001). Sec. 7.4.1, p. 445, Fig. 7.22, "The Bizarre World of the Tunnel Diode Oscillator".

[538] Press, W. H., Teukolsky, S. A., Vetterling, W. T. & Flannery, B. P. *Numerical Recipes in Fortran 77: the Art of Scientific Computing*, vol. 1 of *Fortran Numerical Recipes* (Cambridge University Presss, Cambridge, 2001), 2nd edn. Sec. 16.1, p. 704, "Runge-Kutta Method".

[539] Ferguson, B. & Zhang, X.-C. Materials for teraherz science and technology. *Nature Materials* **1**, 26 (2002).

[540] Mueller, E. R. Terahertz radiation: applications and sources. *The Industrial Physicist* **9**, 27–29 (2003).

[541] Camera 'looks' through clothing. BBC News Technology (2008). URL `http://news.bbc.co.uk/2/hi/technology/7287135.stm`.

[542] Mishonov, T. M. Acoustic-phonon emission by two-dimensional plasmons. *Phys. Rev. B* **43**, 7787–7791 (1991).

[543] Tachiki, M., Iizuka, M., Minami, K., Tejima, S. & Nakamura, H. Emission of continuous coherent terahertz waves with tunable frequency by intrinsic Josephson junctions. *Phys. Rev. B* **71**, 134515 (2005).

Index